Rethinking the Republic of Letters

# Rethinking the Republic of Letters

*Memory and Identity in*
*Early Modern Learned Communities*

*Koen Scholten*

Routledge
Taylor & Francis Group
LONDON AND NEW YORK

The research and this resulting publication have received funding from the European Research Council (ERC) under the European Union's Horizon 2020 research and innovation programme (grant agreement no. 724972).

First published in 2025 by Amsterdam University Press Ltd.

Published 2025 by Routledge
4 Park Square, Milton Park, Abingdon, Oxon OX14 4RN
605 Third Avenue, New York, NY 10158

*Routledge is an imprint of the Taylor & Francis Group, an informa business*



ISBN: 9789048559855 (hbk)
ISBN: 9781041185499 (pbk)
ISBN: 9781003702993 (ebk)
NUR 784

Cover illustration: Jan Luyken (print maker), *Bibliotheek vanden Groot Hartoogh van Toscanen*. Amsterdam: Jan Claesz ten Hoorn, 1698. Etching, 169 × 284mm. Rijksmuseum: RP-P-1896-A-19368-287.

Cover design: Coördesign, Leiden

DOI: 10.5117/9789048559855

*Voor mijn ouders*

# Table of Contents

# List of Figures

# List of Tables

# Introduction: The Republic of Letters as an Imagined Community

**Abstract:** The introduction redefines the early modern Republic of Letters as an imagined community, inspired by Benedict Anderson's concept. It moves away from essentialist definitions and explores how early modern scholars viewed themselves within a transnational commonwealth of learning. Through media like letters, biographies, and travel writings, scholars imagined a Europe-wide intellectual society. The introduction outlines historiographical debates around the term and provides a framework for studying scholarly identity and memory, positioning these as crucial to understanding the Republic of Letters. It argues that media representations were key in shaping this imagined community, setting the stage for the book's exploration of scholarly virtues and the role of commemoration in learned communities.

**Keywords:** imagined community; Republic of Letters; scholarly identity; memory

The 'Republic of Letters' is not a self-evident term. It implies two associations in the common interpretations of *republic* and *letters* that do not correspond to their early modern meanings. First, the Republic of Letters was not a republic in the political sense, that is, a state in which power is held by the people rather than a monarch. Second, the Republic of Letters did not solely revolve around letters. The Latin term *respublica literaria* that was used in the early modern period consists of two parts: *respublica*, which can meaningfully be translated as 'the common cause' or the 'commonwealth'; and *literaria*, which designates learning. The *respublica literaria* is then the 'learned commonwealth'. From its apparent first use in 1417, the term implied a European commonwealth that was dedicated to the pursuit of knowledge and learning.[1]

1 The first occurrence of the term *respublica literaria* appears to be in a letter from the Venetian humanist Francesco Barbaro to Poggio Bracciolini from 1417. The letter mainly consists

Scholten, Koen. *Rethinking the Republic of Letters: Memory and Identity in Early Modern Learned Communities.* Taylor & Francis Group, 2025
DOI 10.5117/9789048559855_INTRO

This book aims to offer an understanding of the Republic of Letters as an imagined community by looking at its media representations in early modern Europe. Since the book builds on a broad array of historical studies of the Republic of Letters, the first part of this introduction will present an overview of the historiography of the Republic of Letters and the diverse ways historians have used the term, as well as the challenges that arise from the current historiography. Even when we agree on the meaning of the Republic of Letters as a commonwealth of learning, it remains a term that is hard to define due to the many interpretations of the term by both historical actors and historians. Such an imprecise definition might not be a problem in general speech and writing; however, it poses challenges for a sustained theoretical study in which the Republic of Letters is the subject, such as this one. Below, I outline how we can meaningfully apply Benedict Anderson's concept of an 'imagined community' to the Republic of Letters with the aim of providing the present study of the Republic of Letters with a theoretical approach that has largely been absent from historiography.

This introduction addresses four main points. The first section presents an overview of the historiography of the Republic of Letters and the multitude of approaches to it in secondary literature. The second section presents examples of the usage of the term in historical literature and the challenges its imprecise use poses for a theoretical approach. The third section will introduce a theoretical approach to the Republic of Letters that serves as a backbone for the present study. Theorising the Republic of Letters as an imagined community has direct consequences for method and source selection: I argue that media representations of the Republic of Letters are key for studying the Republic of Letters as an imagined community. Consequently,

of praise for Bracciolini's revival of old classic texts. The Republic of Letters was invoked to refer to the pursuit of resuscitating texts from antiquity, which Barbaro thought benefited the public good. This letter was first noted in Elizabeth L. Eisenstein, *The Printing Press as an Agent of Change: Communications and Cultural Transformations in Early-Modern Europe*, vol. 1 (New York: Cambridge University Press, 1980), p. 137, n. 287. The letter appears in a 1743 printed edition of letters; see Francesco Barbaro et al., *Francisci Barbari et aliorum ad ipsum Epistolae* [...] (Brescia: Joannes-Maria Rizzardi, 1743), ep. I, pp. 1–8. On p. 5, Barbaro is in the midst of a highly rhetorical panegyric on Poggio: 'An quisquam ita invidus erit, ut vos nimium exornari a me censeat? Quos autem orno? Eos nempe, qui huic litterariae Reip. plurima adjumenta, atque ornamenta contulerunt' (Will anyone be jealous, because I [Barbaro] bestow too much honour on you [Bracciolini]? But on whom do I bestow honour? On those, who contributed the most resources and ornaments to this commonwealth of learning). My translation; cf. the English translation in *Two Renaissance Book Hunters: The Letters of Poggius Bracciolini to Nicolaus de Niccolis*, trans. and annot. Phylis Walter Goodhart Gordan (New York: Columbia University Press, 1974), appendix, letter IV, pp. 196–203. Subsequently cited in Marc Fumaroli, 'The Republic of Letters', *Diogenes* 36 (1988): 129–152 (p. 136).

in this study I draw attention to two aspects of the early modern learned world that are central to understanding these media representations: scholarly identity and scholarly memory. The fourth and fifth sections of this introduction will relate the concepts of scholarly identity and scholarly memory to the study of the Republic of Letters as an imagined community. In summary, this introduction rethinks and redefines the Republic of Letters and proposes an alternative approach to the investigation of the historical formation of learned communities.

## Historiography

The historiography of the Republic of Letters is vast and highlights a wide variety of aspects of the learned world. Despite an expansive corpus of literature, the simple question 'What was the Republic of Letters?' remains elusive, since the term 'Republic of Letters' appears to be a moving target. This fluidity has rarely been acknowledged, which has led to essentialist definitions that seek to identify some time-transcending essence or ideal.

Scholars hold varying ideas of the Republic of Letters depending on the period they study. When focusing on the fifteenth and sixteenth centuries, scholars generally consider the Republic of Letters to be a humanist community contingent with the revival of the Latin and Greek languages, especially in Italy. The main actors in this narrative are Italian humanists such as Petrarch, Dante, and Lorenzo Valla; as well as Desiderius Erasmus and Thomas More; and finally, the German, French, and Swiss educational reformers north of the Alps. This idea of the *respublica lit(t)eraria* is founded on the currency of the Ciceronian concept of friendship (*amicitia*) in letters. In practice, humanist scholars maintained intimate relationships with fellow scholars through letters.[2] This Republic of Letters was both the network and community through which ideas spread and the backbone of the Renaissance as well as the Reformation, both of which relied on intense

2 Fritz Schalk, 'Von Erasmus' Res publica literaria zur Gelehrtenrepublik der Aufklärung', in *Studien zur französischen Aufklärung*, ed. Schalk, 143–163 (Frankfurt a.M.: Vittorio Klostermann, 1977); Hanan Yoran, *Between Utopia and Dystopia: Erasmus, Thomas More, and the Humanist Republic of Letters* (Lanham, MD: Lexington Books, 2010); Marc Fumaroli, *The Republic of Letters*, trans. Lara Vergnaud (New Haven, CT: Yale University Press, 2018); Constance M. Fury, *Erasmus, Contarini, and the Religious Republic of Letters* (Cambridge: Cambridge University Press, 2005). The term 'Republic of Letters' has also been applied to the Islamic learned world; see Muhsin J. al-Musawi, *The Medieval Islamic Republic of Letters: Arabic Knowledge Construction* (Notre Dame, IN: University of Notre Dame Press, 2015).

communication between scholars and theologians. According to some historians who study these fifteenth- and sixteenth-century scholars, the Reformation constituted a loss of the Republic of Letters and of the open exchange of knowledge.[3]

However, scholars who study the seventeenth century present different conceptions of the Republic of Letters. Significantly, the term gained an increasing popularity outside humanist circles in this period. Especially in the second half of the seventeenth century, the term was used as a metaphor to denote a larger community of scholars and scientists. The rise of literary and learned journals at the end of the seventeenth century undoubtedly helped to create a platform for the idea of a Republic of Letters. Pierre Bayle's short-lived but hugely popular journal *Nouvelles de la République des Lettres*, printed between 1684 and 1687, and Samuel Masson's *l'Histoire critique de la République des Lettres*, printed between 1712 and 1718, are prime examples of this development. Historians often consider Bayle's *République des Lettres* to be a realisation of the Enlightenment ideal of the Republic of Letters in print. During this period, the Latin term *respublica literaria* also found its way into vernacular languages such as Italian, French, English, and Dutch.[4] The eighteenth century gave rise to the popular view of the learned world as an independent entity with its own enlightened citizens. The term became commonplace when referring to the learned world at large. In much of the nineteenth century, the Republic of Letters became an increasingly powerful trope. 'Letters' came to refer to literature. Amidst the increasing separation of the humanities from the sciences, the Republic of Letters became associated with literature and literary figures, and an elite literary class.

These general historical differences in the use of the term often greatly affect the approach taken by historians who study different centuries. Modern historians' usages of the term *Republic of Letters*, as well as their approach to the subject, can rarely be positioned in clear-cut historiographical traditions. Historians often use the term in an imprecise manner. To offer some insight into the various usages of the term, I have identified three main approaches to and interpretations of the Republic of Letters:

3 For a general overview of historical conceptions of the Republic of Letters, see Dirk van Miert, 'What Was the Republic of Letters? A Brief Introduction to a Long History (1417–2008)', *Groniek* 204–205 (2010): 269–287.

4 Italian: *Repubblica delle lettere*; French: *République des Lettres*; English: *Republic of Letters*; Dutch: *Republiek der letteren*.

1. Social approach: The Republic of Letters is an unspecific social category for the learned world, including scholars, theologians, universities, booksellers, monasteries, and anything related to learning and knowledge.
2. Moral ideal approach: The Republic of Letters is a normative ideal of the learned world, mostly defined by historical actors themselves, that fostered the production and exchange of knowledge. Historians who adopt this interpretation usually focus on conduct, civility, internal frictions and struggles, and the so-called moral economy of knowledge.
3. Network approach: The Republic of Letters is the exchange between learned men and women, especially in letters, but also through journals and books. This approach often takes the term quite literally as the corpus of letters the scholars produced and is also prevalent in recent scholarship that approaches the Republic of Letters as an epistolary network.

In a sense, all these approaches of the Republic of Letters have their own, albeit overlapping, historiographies. Historians sometimes use these interpretations of the Republic of Letters interchangeably, but most studies do have a specific focus that fits into one of these three approaches. The division serves to untangle and evaluate the different approaches to the Republic of Letters by historians. Let me briefly explain these three historiographies further.

First, the scholarship that considers the Republic of Letters as the 'entire learned world' can be roughly divided into two categories: studies that use the term in an offhand way as a generic social category and studies that present the Republic of Letters as the subject of study. Studies in the latter category often present an overarching narrative of the history of the (European) learned world, such as Françoise Waquet and Hans Bots's *La République des Lettres*.[5] When a book carries the term in its title it often foreshadows a study of the social and cultural aspects of science and learning. *La République des Lettres* presents a narrative with a wide scope and focuses on 'time', 'spaces', 'citizens', and 'dynamics' within the Republic of Letters. Waquet and Bots thereby present it as a transhistorical entity and reality. In a more recent monograph, Bots casts the net somewhat wider and places scholars, universities, epistolary exchange, books and booksellers, and the periodical press all within the Republic of Letters.[6]

5 Hans Bots and Françoise Waquet, *La République des Lettres* (Paris: Belin, 1997).

6 Hans Bots, *Republiek der Letteren: De Europese intellectuele wereld, 1500–1760* (Nijmegen: Uitgeverij Vantilt, 2018). For a consideration of the role of literary and learned societies in the European learned world at large, see Arjan van Dixhoorn and Susie Speakman Sutch, eds., *The Reach of the Republic of Letters: Literary and Learned Societies in Late Medieval and Early Modern Europe*, 2 vols. (Leiden; Boston: Brill, 2008).

Another example of such a wide interpretation of the term can be found in the volume titled *Res Publica Litteraria*, edited by Sebastian Neumeister and Conrad Wiedemann, which contains chapters on topics ranging from schools and universities, interpretations of Aristotle, confessional disputes, as well as the social position of learned women.[7] The editors explain that the common denominator in the volume is *Gelehrtenkultur*, or learned culture, and thus interpret the Republic of Letters as a broad term for the learned world and its culture.[8] In all of these examples, the Republic of Letters refers to the 'world of learning', including everyone and everything involved in the production of knowledge. Most of the scholarship on the Republic of Letters has followed this broad definition and focuses on the cultural history of learning and science in general.

A second sizable share of scholarship focuses on the Republic of Letters as a normative ideal, closely related to Enlightenment ideals such as equality and reason. This second interpretation also studies the Republic of Letters as a learned culture, but focuses more on the moral aspects of knowledge. Studies that take this approach often argue that the Republic of Letters encompassed the ideal that scholars reinforced and called upon to smooth the way for a transconfessional and transnational exchange of knowledge. Within this dynamic, members shared a sense of civil behaviour and morality that functioned as a catalyst for the exchange of knowledge. However, scholars have not taken this ideal at face value and argue that it had its practical limits. For example, Anne Goldgar's *Impolite Learning* studied the behaviour of scholars in letters and the ways in which this ideal worked in practice, showing that conduct and moral codes were embedded in parts of scholarly culture and stressing the limits of the rosy eighteenth-century ideal of open science.[9]

In other studies that take such a moral and behavioural approach, the Republic of Letters is closely tied to the Enlightenment and the study thereof. Since the Republic of Letters gained high cultural visibility from the end of the seventeenth century onwards, propelled by French-speaking scholars such as Pierre Bayle, it became an essential part of the cultural history of the Enlightenment.[10] For Enlightenment historians, the Republic of Letters

7 Sebastian Neumeister and Conrad Wiedemann, eds., *Res Publica Litteraria: Die Institutionen der Gelehrsamkeit in der frühen Neuzeit*, 2 vols. (Wiesbaden: Harrassowitz, 1987).

8 Ibid., vol. I, p. xi.

9 Anne Goldgar, *Impolite Learning: Conduct and Community in the Republic of Letters, 1680–1750* (New Haven, CT; London: Yale University Press, 1995).

10 For a foundational work of this historiography, see Annie Barnes, *Jean le Clerc (1657–1736) et la République des Lettres* (Paris: E. Droz, 1938).

is thus understood in the context of the appearance of strict social and moral norms, usually referred to as civility and polite sociability. In this context, the Republic of Letters is often taken as the moral and epistemic framework that dictated proper behaviour in the learned world, which was then perpetuated in journals, most notably Bayle's *République des Lettres*. Thus, the Republic of Letters became a conceptual mixture of the learned world and a Habermasian 'public sphere'.[11] Dena Goodman, for example, considered the Republic of Letters to have been an important prerequisite for the so-called Enlightenment: 'In the eighteenth century, the Republic of Letters was composed of French men and women, philosophes and salonnières, who worked together to attain the ends of philosophy, broadly conceived as the project of Enlightenment.'[12] In studies of the European Enlightenment such as Goodman's, the Republic of Letters is thus both a broad term for learned culture including journals, salons, and polite conversation, but also for the epistemic and moral norms that prescribed how knowledge could and should be exchanged.[13] Other studies sometimes refer to it as the moral economy of knowledge.[14] Goodman's study, however, also takes the Republic of Letters to be the entire cultural and social context of the Enlightenment. Historical studies do not always stipulate their approach clearly, and often move between these first two approaches and usages of the Republic of Letters.

A third historiographical strand has focused on the communication, especially epistolary exchanges, between scholars and called this the Republic of Letters. Eighteenth-century scholars referred to their epistolary output as their *litterarum commercium* ('exchange of letters') or *epistolicum*

11 In referring to the 'public sphere' I am citing Jürgen Habermas, *The Structural Transformation of the Public Sphere: An Inquiry into a category of Bourgeois Society*, trans. Thomas Burger (Cambridge: Polity Press, 1989).

12 Dena Goodman, *The Republic of Letters: A Cultural History of the French Enlightenment* (Ithaca, NY; London: Cornell University Press, 1994), p. 9.

13 See also Dena Goodman, 'Governing the Republic of Letters: The Politics of Culture in the French Enlightenment', *History of European Ideas* 13, no. 3 (1991): 183–199; Martin Mulsow, *Die unanständige Gelehrtenrepublik: Wissen, Libertinage und Kommunikation in der Frühen Neuzeit* (Stuttgart; Weimar: J. B. Metzler, 2007).

14 See, e.g., Ulrich Johannes Schneider, ed., *Kulturen des Wissens im 18. Jahrhundert* (Berlin; New York: Walter de Gruyter, 2008), part VI, 'Die moralische Ökonomie des Wissens' (pp. 259–296); Lorraine Daston, 'The Moral Economy of Science', in *Constructing Knowledge in the History of Science*, ed. Arnold Thackray, 3–24 (Chicago: University of Chicago Press, 1995); Marian Füssel, '"On the Means of Becoming Famous in the Learned World": Practices in Scholarly Constitution of Status and the Emergence of a Moral Economy of Knowledge in the Eighteenth Century', in *Scholars in Action: The Practice of Knowledge and the Figure of the Savant in the 18th Century*, vol. 1, ed. André Holenstein et al., 123–144 (Leiden; Boston: Brill, 2013).

*commercium* ('epistolary exchange').[15] Historians have taken up this term to refer to modes of communication within the learned world.[16] The Republic of Letters is then represented as an epistolary 'web' or 'network' that tied early modern scholars together throughout time.[17] This type of scholarship includes the study of epistolary contacts as social networks which are increasingly mapped through the use of digital tools. The Republic of Letters is presented as the epistolary network that can be 'reassembled' and visualised with digital network analytical tools.[18]

This short survey of the historiography of the Republic of Letters shows the broad array of approaches to, and interpretations of, the 'Republic of Letters'. This multiplicity of approaches and meanings runs the risk of equivocation, ambiguity, and incompatibility between historical studies. Historians generally do not acknowledge this disparity of approaches, nor do they clearly define their own approach to the Republic of Letters. While this disparity is often of little consequence in studies that use the phrase 'Republic of Letters' in a general meaning, for the present study it is important to gain a clear vision of all these meanings. In this book, the theoretical conceptualisation of the Republic of Letters has direct consequences for the sources and the methods. The next section will address two challenges in the historiography of the Republic of Letters: uncertainty about whether the Republic of Letters is an actor's category or a historiographical category,

15 See, e.g., the following collections of letters: Joseph Stepling et al., *Litterarum Commercium* (Bratislava: Guil. Theoph. Kornius, 1782); John Collins et al., *Commercium epistolicum D. Johannis Collins, et aliorum, de analysi promota* (London: J. Johnson & J. Watts, 1722).

16 See, e.g., Hans Bots and Françoise Waquet, eds., *Commercium Litterarium: La Communication dans la République des Lettres / Forms of Communications in the Republic of Letters, 1600–1750* (Amsterdam; Maarssen: APA-Holland University Press, 1994); Ulrich Johannes Schneider, ed., *Kultur der Kommunikation: Die europäische Gelehrtenrepublik im Zeitalter von Leibniz und Lessing* (Wiesbaden: Harrassowitz, 2005). For a consideration of the role of letters in the exchange of knowledge in the early modern period, see Dirk van Miert, ed., *Communicating Observations in Early Modern Letters (1500–1675): Epistolography and Epistemology in the Age of the Scientific Revolution* (London: Warburg Institute; Turin: Nino Aragno, 2013).

17 David S. Lux and Harold J. Cook, 'Closed Circles or Open Networks? Communicating at a Distance During the Scientific Revolution', *History of Science* 36 (1998): 179–211; Laurence Brockliss, *Calvet's Web: Enlightenment and the Republic of Letters in Eighteenth-Century France* (Oxford: Oxford University Press, 2002).

18 See, e.g., Howard Hotson and Thomas Wallnig, eds., *Reassembling the Republic of Letters in the Digital Age: Standards, Systems, Scholarship* (Göttingen: Göttingen University Press, 2019); Ingeborg van Vugt, 'Using Multi-Layered Networks to Disclose Books in the Republic of Letters', *Journal of Historical Network Research* 1, no. 1 (2017): 25–51; Javier Ureña-Carrion et al., 'Communication Now and Then: Analyzing the Republic of Letters as a Communication Network', *Applied Network Science* 7 (2022).

and the emphasis on the dichotomy between the ideal and reality of the Republic of Letters in historical studies.

## Whose Republic of Letters?

Much discussion about the nature of the Republic of Letters has been caused by differences in its usage as a historiographical category by modern historians and its disparate use by early modern actors. From the second half of the seventeenth century onwards the term 'Republic of Letters' enjoyed ample cultural visibility in books, journals, disputations, and correspondence. Consequently, much of the scholarship has focused on this period. In those cases, the Republic of Letters is often conceptualised broadly as the learned and intellectual world, including universities, booksellers, and everyone who might identify as learned in one way or another (Approach 1: the general social group). To equate the Republic of Letters with the entire intellectual world and its practices creates a tension between what the term meant for early modern actors and what it means for us, historians, today. As a result, the sixteenth-century Republic of Letters is often understood through a conceptualisation of the eighteenth-century Republic of Letters. This eighteenth-century lens is then often used to look at the 'roots' of the Republic of Letters, which are then found in the sixteenth century.

Since most historical studies do not define the Republic of Letters precisely, it is hard to clearly position historical studies and condense the field into a categorisation of the historiography. In 1989, Françoise Waquet rightfully asked 'what is the Republic of Letters?', and discerned the main characteristics of the Republic of Letters expressed by eighteenth-century scholars: it is a universal, sovereign, and multiconfessional state, made up of equal citizens, and it is also equated by some with the corpus of scholarly letters or an intellectual community.[19] The eighteenth-century use of the term seems to have as much variance as the usages by modern historians. Given the historical change of the meaning of the term itself, and the eighteenth-century historical specificity of these conceptions, we need to distinguish two questions: what *is* the Republic of Letters and what *was* the Republic of Letters?

19 Françoise Waquet, 'Qu'est-ce que la République des Lettres? Essai de sémantique historique', *Bibliothèque de l'école des chartes* 147, no. 1 (1989): 473–502; a similar essay is also printed in Bots and Waquet, *République des Lettres*, pp. 11–27.

This conceptual challenge has been addressed and lamented over before. Herbert Jaumann shared his frustration with the cursory use of the term, and particularly its use as a catchphrase in titles and subtitles of books. Jaumann bemoaned the fact that 'the would-be subject of analysis, namely *Republic of Letters* proper, was never sufficiently discriminated from other items of investigation like "Scholarly Culture" in general.'[20] Two decades later, little appears to have changed. The Republic of Letters scholarship has been clouded by this conceptual unclarity. I have selected two articles, one from *Renaissance Quarterly* and the other from a recent issue of the *Journal of the History of Ideas*, with which to illustrate the common use of the term as a catchphrase and the uncertainty about whether the term is a historiographical or an actors' category.[21] The point is not to discredit the scholarship presented in these publications, but rather, to illustrate the variety of the use of the term. In particular, I want to point out how easily authors move between the actors' category and the historiographical category denoting 'the learned world' at large.

In his article in *Renaissance Quarterly*, Timothy Kircher characterised the correspondence of Lapo da Castiglionchio the Younger as Lapo being 'At Play in the Republic of Letters'.[22] Here we get the impression that the term is used to denote the learned world, or more specifically, the literary exchange between learned men. The only English instance of the term 'Republic of Letters' in the article refers to a way of life and what goes on between learned men. Lapo's correspondence can be read, according to Kircher, as having a playful purpose of 'exposing the uncertainties of life in the Republic of Letters.'[23] While in this instance the meaning of the term is not made explicitly clear, we can infer it denotes the 'learned world'. The difference between historiographical and actors' category becomes fuzzy, however, when Kircher introduces a Latin variation in italics: 'Lapo's words reveal the psychological dimension of the tension between private

20 Herbert Jaumann, '*Respublica litteraria / Republic of letters*. Concept and Perspectives of Research', in *Die europäische Gelehrtenrepublik*, ed. Jaumann, 11–19. Wolfenbütteler Forschungen 96 (Wiesbaden: Harrassowitz, 2001), p. 12.

21 Another example is Bots, *De Republiek der Letteren*, treated extensively in my review: 'De Republiek der Letteren als de totale intellectuele wereld?', *Tijdschrift voor Geschiedenis* 131, no. 4 (2018): 713–714. See also Thijs Weststeijn, *The Visible World: Samuel van Hoogstraten's Art Theory and the Legitimation of Painting in the Dutch Golden Age* (Amsterdam: Amsterdam University Press, 2008), esp. ch. 1, 'Samuel van Hoogstraten in the Republic of Letters', pp. 25–79.

22 Timothy Kircher, 'At Play in the Republic of Letters: The Correspondence of Lapo da Castiglionchio the Younger', *Renaissance Quarterly* 71, no. 3 (2018): 841–867.

23 Ibid., p. 853.

*otium* and public *amicitia* in a life within the *respublica litterarum*.'[24] At this point, as a reader, it becomes harder to infer whether these terms in italics come directly from Lapo's correspondence, and are thereby actors' categories, or whether Kircher prefers to present these concepts in Latin for stylistic reasons.[25] These nuances do not change the conclusions of Kircher's article, but indicate the general use of the term in critical and high-quality historical literature. For a study that takes the Republic of Letters as its subject, however, such instances in which the author seemingly moves between historiographical and actors' category and does not comment upon the meaning of the term feeds the desire for a comprehensive theoretical approach.

The second example, from the *Journal of the History of Ideas*, shows the uncertainty as to whether the term is an actors' category or historiographical category. Richard Oosterhoff used the term in the subtitle of his article on the reception of Pythagoras and the cultural status of mathematics: 'Pythagoras in the *Res publica litterarum* of French Renaissance Mathematics'.[26] Oosterhoff's aim is to analyse the reception of Pythagoras's texts in a series of overlapping learned communities: 'Following Pythagoras through epistolary cultures in monasteries, universities, and finally the new literary academies, this narrative is structured around three overlapping communities of scholars important to the history of mathematics and philosophy in sixteenth-century France.'[27] The term 'Republic of Letters' never appears in the main text, neither in English nor in Latin. Given the scope and the content of the article, one is led to believe that the *res publica litterarum* mentioned in the subtitle refers to the disparate learned communities Oosterhoff describes: monasteries, universities, and literary academies. While the usage of the term in the subtitle is not problematic for the argument in the paper, one could argue with Jaumann that it is used rather imprecisely. Such usage leads one to ponder whether this wide interpretation of the *res publica litterarum* comes from the primary sources or the secondary literature.

24 Ibid., pp. 847–848.

25 Additionally, if the term *respublica litterarum* is an actors' category, it begs the question of whether there is a difference between the common terms *respublica litteraria* and *respublica litterarum*. These nuances would be important for a *Begriffsgeschichte* of the Republic of Letters. Since little research has been carried out on the use of the term by fifteenth-century Italian humanists, it would be significant if Lapo were using this term in his correspondence.

26 Richard J. Oosterhoff, 'From Pious to Polite: Pythagoras in the *Res publica litterarum* of French Renaissance Mathematics', *Journal of the History of Ideas* 74, no. 4 (2013): 531–552.

27 Ibid., pp. 532–533.

Together, Kircher's and Oosterhoff's articles demonstrate the usage of the term as a generic social category without a clear indication if the actors in their case studies used the term. Historians of the early modern period subscribe to the idea of a Republic of Letters as an abstract notion of the learned world and use it widely. This is the first conceptual difficulty: the phrase is often used as a catchphrase instead of a clear descriptor, since its use moves between an actors' category and a historiographical category.

The second difficulty comes from a frequently established dichotomy between the *ideal* and the *reality* of the Republic of Letters. Historians have often used this dichotomy to describe the dynamics within the Republic of Letters. In their seminal monograph on the topic, Hans Bots and Françoise Waquet reach this conclusion numerous times: 'the Republic of Letters appears, upon final analysis, as the place of inevitable tensions between the ideas and facts, utopia and reality'.[28] This trope of a dichotomy between ideal and facts, utopia and reality, became a common way of thinking about the Republic of Letters in the 1990s.[29] This demarcation has been prevalent in historical studies ever since and continues to be repeated and reinvented. In 2005, Robert Mayhew lamented the state of the historiography, and reintroduced the reader to the dichotomy: 'What they [the studies on the Republic of Letters] do not address is the tension identified at the outset between the *ideal* and the *reality* of the Republic of Letters.'[30] Unsurprisingly, Mayhew does not cite any of the literature from the 1980s and 1990s that regularly stressed this dichotomy.[31] Mayhew's proposal is an example of the never-ending discussion about the difference between the ideal and reality of the Republic of Letters and the lack of a clear historiography of the Republic of Letters. The dichotomy between ideal and reality implicitly

28 '[L]a République des Lettres apparaît, en dernière analyse, comme le lieu de tensions inévitables entre les idées et les faits, l'utopie et la réalité': Bots and Waquet, *République des Lettres*, p. 27; similar instances can be found on pp. 90, 111, 114–115, 120–125, 141, 159–160. This dichotomy was also the point of departure for Bots's inaugural address; see Hans Bots, *Republiek der Letteren: Ideaal en Werkelijkheid* (Amsterdam: APA-Holland Universiteits Pers, 1977).

29 See, e.g., Lorraine Daston, 'The Ideal and Reality of the Republic of Letters in the Enlightenment', *Science in Context* 4, no. 2 (1991): 367–386; Goldgar, *Impolite Learning*. While Goldgar does not explicitly comment on this dichotomy, she stresses how ideals of conduct triumphed over quarrels of religion or political philosophy, see esp. ch. 4, 'Of Two Minds: Religion, Politics, and the Competition of Values', pp. 174–218.

30 Robert Mayhew, 'Mapping Science's Imagined Community: Geography as a Republic of Letters, 1600–1800', *The British Journal for the History of Science* 38, no. 1 (2005): 73–92 (p. 76). My italics.

31 There is no reference to, or citation of, Bots and Waquet. There is a citation of Daston and a mention of Goldgar.

defines the Republic of Letters as both the ideal of open science and the grubby reality of polemical scholars hampering the free flow of knowledge.

The difficulty with this demarcation between the ideal and reality of the Republic of Letters is that it is often not made explicit *whose* ideal or reality are at stake. The introduction of the term '*the* ideal of the Republic of Letters' presumes the existence of a universal ideal. It avoids the question: did the ideal of openly sharing knowledge exist, and if so, to what extent was it universal, transnational, and transconfessional? Or is this supposed ideal rather a modern ideal of the Republic of Letters? There is always, and by definition,[32] an opposition between the reality and ideal of concepts, including the Republic of Letters if one interprets it as an ideal of openly sharing knowledge. And it begs the question, which actors create such ideals and realities? In other words, whose ideal is at stake here? Sixteenth-century humanists, clergymen, scholars who are well-connected or peripheral, modern historians, or, as often seems to be the case, eighteenth-century Enlightenment thinkers?

This kind of contrast between the ideal and reality may reinforce the view that the Republic of Letters was a failure, since, so the argument goes, *the* ideal was never achieved because *the* reality was in fact rather messy. In short, the dichotomy equates the reality with a failed ideal, because the early modern discourse was different from praxis. Talking about *the* ideal of Republic of Letters, as if it were free from temporal and spatial differences, poses a challenge to the project of historicising it as a commonwealth of learning. Modern historians should avoid the example of eighteenth-century learned men, who projected the ideals and moral system of their learned world onto preceding decades, and even centuries, and created a narrative about the origin and perpetuation of the Republic of Letters as an ideal. Given the conceptual challenges that reappear in the historiography, the next section will propose a comprehensive theoretical approach to the Republic of Letters that negates the problems of the historiographical/actors' category confusion and the ideal/reality dichotomy for the present study.

## The Republic of Letters as an Imagined Community

This section proposes the study of the Republic of Letters as an imagined community. Historians have linked the Republic of Letters to the concept

32 According to the Oxford English Dictionary, *ideal* means 'existing only in idea'; *reality* means 'what is real rather than imagined or desired'.

of an imagined community before, albeit in a cursory fashion.[33] Anthony Grafton, for example, argued that '[t]he Republic of Letters *imagined itself* as Europe's first egalitarian society, even if it did not always enact these high ideals in the grubby reality of its intellectual and professional practices.'[34] Elsewhere, Robert Mayhew used the concept to explain disparate views of the Republic of Letters. Mayhew looked at the geographical spread of citations in the works of two early modern scholars: Peter Heylyn's *Microcosmus* (1621) and Pinkerton's *Mercurio Peruviano* (1802).[35] Mayhew took the practice of citation as a reflection of an imagination of the Republic of Letters and concluded: 'In sum, the relationship between the age, language, and authority accorded to an author was profoundly different in Heylyn and Pinkerton and led them to have different historical geographies of the republic of letters.'[36] While citation patterns do offer us some insight into Heylyn and Pinkerton's views of the learned world and authority, in this introduction I argue that there is more to be gained from conceptualising the Republic of Letters as an imagined community. But before we delve deeper into the implications of such a conceptualisation of the Republic of Letters, let us first reconsider Benedict Anderson's original conceptualisation of the imagined community.

Benedict Anderson introduced the concept of an imagined community as a new way of thinking about the emergence of the nation state.[37] He proposed that the nation is an imagined community, because 'the nation is always conceived as a deep, horizontal comradeship.'[38] Through a national language, Anderson contends, we imagine a community to which we belong, which gives us a sense of belonging and connection with people we will never meet, but whom we know are there. This national community consists of peers, with whom we share a country, history, and customs, as well as a language. It is important to note that *imagined* in this sense does not mean 'made up'.

33 See, e.g., Ludmilla Jordanova, 'Science and Nationhood: Cultures of Imagined Communities', in *Imagining Nations*, ed. Geoffrey Cubitt, 192–211 (Manchester: Manchester University Press, 1998); Peter Burke, 'Erasmus and the Republic of Letters', *European Review* 7, no. 1 (1999): 5–17 (p. 8): 'The republic [of letters] was what we might call an invisible or "imagined community".'

34 Anthony Grafton, 'A Sketch Map of a Lost Continent: The Republic of Letters', *Republic of Letters: A Journal for the Study of Knowledge, Politics, and the Arts*, vol. 1:1 (2009), pp. 1–2. My italics.

35 Robert Mayhew, 'British Geography's Republic of Letters: Mapping an Imagined Community, 1600–1800', *Journal of the History of Ideas* 65, no. 2 (2004): 251–276.

36 Ibid., p. 271.

37 Benedict Anderson, *Imagined Communities: Reflections on the Origin and Spread of Nationalism*, rev. ed. (London: Verso, 2006; originally published in 1983).

38 Ibid., p. 7.

Rather, Anderson explained, '[i]t is *imagined* because the members of even the smallest nation will never know most of their fellow-members, meet them, or even hear of them, yet in the minds of each lives the image of their communion.'[39] Anderson presented a sweeping history in which religious communities crumbled and made way for the nation as an imagined community in the eighteenth century. Central to such an imagined community are media representations that provide the community with imaginations that are able to substitute the distance between members with a feeling of togetherness. For an imagined community to exist and change it must be *imagined* and subsequently *re-imagined.*

The concept of an imagined community is thus deeply entrenched in the study of nationalism. While nationalism is hard to define, many historians have instead diverted to the question of when people developed a collective sense of a 'homeland' or 'nation'. What the nation *was* has turned out to be an unanswerable question and historians have instead focused on the representations and ideas of a homeland, as well as the need by certain individuals to create such a narrative. The prime focus of these studies is the sense of community—the imagined community—which tied them together.[40] The 'homeland' or 'nation' was a perceived imagined community, changing form and shape in ways that were highly dependent on time, place, and individuals. These disparities not only give us an insight into the development of the *idea* of a homeland, but also offer a mirror of culture in general.

This monograph will not address what the Republic of Letters *was*, since it is as unfathomable as the nation state. Rather, this monograph will address what early modern actors *imagined* the Republic of Letters to be. In essence, it argues that the Republic of Letters is the sum of the imaginations of itself. If we are to study the Republic of Letters as an imagined community, we must ask ourselves how imaginations of the learned community were constructed and how they spread. Anderson stressed that 'communities are to be distinguished, not by their falsity/genuineness, but by the style in which they are imagined.'[41] Therefore, if we study the Republic of Letters as an imagined community, the Republic of Letters should not be thought of as being *real* or *ideal*. Rather, we need to focus on the media—that is,

39 Ibid., p. 6.

40 See, e.g., Cubitt, *Imagining Nations*; N. C. F. van Sas, ed., *Vaderland: Een geschiedenis van de vijftiende eeuw tot 1940* (Amsterdam: Amsterdam University Press, 1999); Lotte Jensen, ed., *The Roots of Nationalism: National Identity Formation in Early Modern Europe, 1600–1815* (Amsterdam: Amsterdam University Press, 2016).

41 Anderson, *Imagined Communities*, p. 6.

the means of collective communication—through which these *imaginings* were constructed and shared. From the sixteenth century onwards, the advent of print helped to make the learned communities visible through compendia with icons, prints of tomb memorials, as well as collections of lives of scholars. This widespread cultural visibility helped create a sense of a commonwealth of learning. Therefore, this monograph turns to media representations of the Republic of Letters in sources such as letters, collections of lives, collections of portraits, travelogues, tombs, and memorials. These sources reflect the formation and reformation of imagined learned communities, meant to embody and perpetuate distinct scholarly identities and memory. The next two sections will introduce the concepts of scholarly identity and collective scholarly memory in the context of the Republic of Letters.

## Scholarly Identity

Within the discipline of sociology, the general definition of 'community' is an entanglement of relationships between a group of people, in a certain geographical locale or imagined space, that goes beyond casual acknowledgement. Sociologist John G. Bruhn explains:

> These relationships are closer than casual relationships because the group shares some common goals, values, and, perhaps a way of life that reinforce[s] each other, creates positive feelings, and results in a degree of mutual commitment and responsibility.[42]

Bruhn's observation that a community shares common goals, values, and a way of life, as well as a sense of mutual commitment and responsibility, is applicable to the Republic of Letters. The Republic of Letters then becomes a pan-European imagined community, with the common goal of the advancement of learning, based on distinct scholarly virtues and values. Its members shared a collective duty to ensure and secure the pursuit of knowledge. These goals and values were therefore part of the scholars' identities.

The 1990s saw a surge in works on scholarly identity and the self-fashioning of scholars. Inspired by Stephen Greenblatt's *Renaissance Self-Fashioning*,

42 John G. Bruhn, *The Sociology of Community Connections*, 2nd ed. (Dordrecht: Springer, 2011), p. 12. There is no agreed-upon definition of 'community' in sociology. For a historical overview of the concept of 'community', see ibid., pp. 29–46.

the scholarly identity of early modern scholars was increasingly seen as self-constructed.[43] For example, Lisa Jardine argued 'that Erasmus's European prominence was something in which Erasmus himself made a considerable investment, in terms of effort and imagination.'[44] Jardine emphasised that Erasmus fashioned himself as a man of letters and scholar-saint, placing himself at the centre of the northern humanist world of learning and collective history. Mario Biagioli's bold statement about Galileo Galilei's self-fashioning and patronage at the Florentine court is in line with Jardine's way of thinking. Biagioli suggested 'that patronage is the key to understanding processes of identity and status formation that are the keys to understanding *both* the scientists' cognitive attitudes *and* career strategies.'[45] Biagioli argued that Galileo fashioned his ideas and career to suit contemporary courtly ideals of identity and status. Both Jardine and Biagioli emphasised the *individual* construction of an identity and offered this as a way of explaining the scholarly and scientific successes of Erasmus and Galileo, respectively.

In the wake of these seminal publications, we have seen work on the self-fashioning and self-presentation of scholars in epistolary exchanges,[46] university settings,[47] but also in their pictorial representations.[48] The activity of the formation of a scholarly identity is hereby attributed to the scholar him- or herself. However, rather than putting the perspective and activity of the author first, this monograph will take the learned community as the starting point of the analysis. When one sets out to explain how figures such as Erasmus and Galileo became the subject of cults and are still remembered as pinnacles of learning and science, it is necessary to move away from their individual construction of identity and look at how exemplary figures were collectively constructed, mediated, and remembered by a group, a collective,

43 Stephen Greenblatt, *Renaissance Self-Fashioning: From More to Shakespeare* (Chicago; London: University of Chicago Press, 1980).

44 Lisa Jardine, *Erasmus, Man of Letters: The Construction of Charisma in Print* (Princeton, NJ: Princeton University Press, 1993), p. 5.

45 Mario Biagioli, *Galileo, Courtier: The Practice of Science in the Culture of Absolutism* (Chicago; London: University of Chicago Press, 1993), p. 14. *Galileo, Courtier* was the centre of an open epistolary exchange between Biagioli and Michael H. Shank: see Mario Biagioli, '"Playing with Evidence"', and Michael H. Shank, 'How Shall We Practice History? The Case of Mario Biagioli's *Galileo, Courtier*', both in *Early Science and Medicine* 1, no. 1 (1996): 70–105 and 106–150, respectively.

46 Toon Van Houdt et al., eds., *Self-Presentation and Social Identification: The Rhetoric and Pragmatics of Letter-Writing in Early Modern Times* (Leuven: Leuven University Press, 2002).

47 Richard Kirwan, ed., *Scholarly Self-Fashioning and Community in the Early Modern University* (Farnham; Burlington, VT: Ashgate, 2013).

48 Hole Rößler, 'Character Masks of Scholarship: Self-Representation and Self-Experiment as Practices of Knowledge Around 1770', in *Scholars in Action*, vol. 1, ed. Holenstein et al., pp. 459–480.

a community. While it is true that individuals helped to construct their own identity, this monograph focuses on the collective construction and reconstruction of exemplary figures. The latter helped to establish a sense of collective identity and community for centuries. This monograph assumes that scholarly identity consists of collectively shared, archetypical properties.

In the early modern discussions of the Republic of Letters we find cursory references to *exempla* for the entire learned world. For the seventeenth-century French minister Samuel Masson, it was that clear Gisbert Cuper was a 'sovereign of the powerful and glorious Republic [of Letters]'. Other heroes, princes, and stars of the Republic of Letters were subject to cults, most notably Petrarch, Dante, Desiderius Erasmus, Thomas More, Hugo Grotius, Joseph Scaliger, and Isaac Newton. They were imagined as exemplary figures and ideal members of a scholarly community.[49] Just as saints provided Christian believers with examples of virtuous behaviour, scholarly exempla provided the learned community with models of the intellectual virtues to be imitated and emulated. This study will look closer at the scholarly virtues these *exempla* were meant to represent and ask how they were constructed and hailed in learned communities.[50]

In line with such a collective view, scholarly identity has been the focus of philosophical inquisitions into so-called epistemic virtues.[51] Herman Paul has proposed the concept of scholarly personae, which he defines as 'ideal-typical models of scholarly selfhood', which helped shape the scholarly self.[52] These personae, or exemplary figures, as I will call them,

49 Throughout this work I use the terms 'scholars', 'learned men/women', and 'men and women of letters' interchangeably, in place of the unwieldy but more precise definition given by Constance Fury: 'elite, highly educated [...] men and women who, despite other differences in class and public and private duties, shared the sense that reading, writing, and studying were uniquely valuable activities and who were, in turn, viewed by their peers as admirable because of their dedication to, and skill in, these activities.' Fury, *Erasmus, Contarini and the Religious Republic of Letters*, p. 171, n. 3.

50 This concept has also been introduced in Koen Scholten, 'Introduction: Memory and Identity in Learned Communities', in *Memory and Identity in the Learned World: Community Formation in the Early Modern World of Learning and Science*, ed. Koen Scholten et al., 1–27 (Leiden; Boston: Brill, 2022).

51 Jeroen van Dongen and Herman Paul, eds., *Epistemic Virtues in the Sciences and the Humanities* (Cham: Springer International Publishing, 2017). The study of epistemic virtues is also closely related to virtue ethics; see, e.g., Andreas Hellerstedt, ed., *Virtue Ethics and Education from Late Antiquity to the Eighteenth Century* (Amsterdam: Amsterdam University Press, 2018). For a historical overview of the virtues of historians specifically, see Herman Paul, *Historians' Virtues: From Antiquity to the Twenty-First Century* (Cambridge: Cambridge University Press, 2022).

52 Herman Paul, 'What is a Scholarly Persona? Ten Theses on Virtues, Skills, and Desires', *History and Theory* 53, no. 3 (2014): 348–371.

shape scholarship itself.[53] The scholarly identities attributed to exemplary figures and role models are key to understanding implicit views of virtues, proper conduct, collective goals, and the mnemonic mechanisms through which the Republic of Letters was imagined as a community. The presumption in this monograph is that when someone portrayed Erasmus as a pious Catholic, as an independent scholar, or as a reviver of critical scholarship preceding the Reformation, this reflected the scholarly identity and values of a community.[54] The portrayer derived a sense of identity from the narrative in which he or she placed Erasmus. For the modern historian, exemplary models, and the narrative in which they play a central role, provide a unique glimpse of the sense of community and identity such portrayers wanted to convey and reinforce. The collective construction of an exemplary figure in media representations is part of what historians call cultural memory and memorialisation. In the next section, I will elaborate on early modern scholarly memory and the sources that allow its study.

## Scholarly Memory

To study the remembrance and collective construction of exemplary figures, and thereby a sense of community, it is helpful to turn to the field of memory studies.[55] To avoid a lengthy elaboration on the ambiguity and diversity of the meaning of 'memory', this monograph will rely on Geoffrey Cubitt's comprehensive definition of the study of memory as 'the study of the means by which a conscious sense of the past, as something meaningfully connected to the present, is sustained and developed within human individuals and human cultures.'[56] Memory is taken to be a specific historical consciousness in a social or cultural group. This conceptualisation is in line with cultural historians who have used the term *collective memory* to describe the ways

53 Herman Paul, 'Performing History: How Historical Scholarship is Shaped by Epistemic Virtues', *History and Theory* 50, no. 1 (2011): 1–19.

54 For a reflection on the disparate identities Erasmus represented, see Scholten, 'Introduction', pp. 1–8.

55 For a multidisciplinary and multitemporal overview of the field, see Susannah Radstone and Bill Schwarz, eds., *Memory: Histories, Theories, Debates* (New York: Farnham University Press, 2010).

56 Geoffrey Cubitt, *History and Memory* (Manchester; New York: Manchester University Press, 2007), p. 9. For an overview of the uses and abuses of collective memory in cultural history, see Alon Confino, 'Collective Memory and Cultural History: Problems of Method', *The American Historical Review* 102, no. 5 (1997): 1386–1403.

in which societies remember their past.[57] Maurice Halbwachs argued that all memory is collective memory, since even individual memory is shaped by its cultural context. Collective memory is, therefore, not the sum of individual memories, but rather a social and cultural environment that shapes individual memory into a coherent collective memory.[58] The advantage of this conception is that it leads away from individual memory and the idiosyncrasies of remembering. The study of collective memory focuses on the ways memory is employed in relation to community formation. Collective memory is thus the way a community remembers its own past.[59]

Memory has most often been studied in a national, modern context. Pierre Nora proposed the term *lieux de mémoire*, or 'sites of memory', to denote memorial heritage essential for a community's identity. In the seven-volume study *Les lieux de mémoire*, Nora and many other authors abundantly showed how monuments, rituals, and symbols helped form a myriad of French national identities.[60] Building on this scholarship, Jay Winter focused solely on war memorials as conveyors of memory, and defined sites of memory simply as 'physical sites where commemoration acts take place'.[61] Together, Nora's and Winter's works exemplify the introduction of the idea that places and spaces such as churches, graveyards, statues, and public squares can convey an imagination of a past, and, more importantly, sustain cultural identities and a social group's sense of unity and individuality.

These ideas of memory and commemoration have successfully been applied beyond the modern and national frame. On the one hand, historians have called for so-called transnational memory studies to overcome borders and national frameworks.[62] On the other hand, historians of early modern

57 For a concise overview of the study and formulation of collective memory in cultural history, see Anna Green, *Cultural History* (Basingstoke; New York: Palgrave Macmillan, 2008), pp. 99–116.
58 Maurice Halbwachs, *On Collective Memory*, ed., trans., and with an introduction by Lewis A. Coser (Chicago; London: University of Chicago Press, 1992). For a consideration of Halbwachs's thought in light of memory studies, see Erika Apfelbaum, 'Halbwachs and the Social Properties of Memory', in Radstone and Schwarz, *Memory*, pp. 77–92.
59 Jeffrey Andrew Barash, *Collective Memory and the Historical Past* (Chicago: University of Chicago Press, 2016).
60 Pierre Nora, ed., *Les lieux de mémoire*, 7 vols. (Paris: Gallimard, 1984–1992). For an introduction to *lieux des mémoire*, see Nora, 'Between Memory and History: Les Lieux de Mémoire', *Representations* 26 (1989): 7–24.
61 Jay Winter, 'Sites of Memory', in Radstone and Schwarz, *Memory*, pp. 312–324 (p. 312). For Jay Winter's seminal work on sites of memory, see Jay Winter, *Sites of Memory, Sites of Mourning: The Great War in European Cultural History* (Cambridge: Cambridge University Press, 2014; originally published in 1995).
62 Chiara De Cesari and Ann Rigney, eds., *Transnational Memory: Circulation, Articulation, Scales* (Berlin: De Gruyter, 2014).

Europe, most notably Judith Pollmann, have opened the field of early modern memory.[63] Pollmann has not only showed that disparate senses of the past co-existed, but that they played a significant role in negotiations of peace and confessional disputes. In the wake of Winter's study of remembrance and war, the field of early modern memory now largely consists of studies that pertain to the interactions between memory and politics and war.[64] Other studies of early modern memory have followed Nora's concept of sites of memory and have focused on public memorials and pantheons—public sites of remembrance to celebrate heroes of statesmanship, military, and the sciences.[65]

Despite the influence of memory studies in historical studies, only a few studies have focused on memory and remembrance in scholarship and science. Pnina Abir-Am and Clark Elliot edited a volume on commemorations of science and scientists, focusing on modern centenaries and festivities that aim to remember and hail scientists and scholars, such as the Copernican Quincentennial or the 1959 Darwin centennial celebration.[66] As many contributors note, these commemorations often had significant political and nationalistic overtones. The process of communication through forms of memory and commemoration in the early modern period, however, has been largely overlooked. Because the common mould of nationalism often makes for an awkward fit, insights from memory studies and the early modern history of science have found little overlap. This is a missed opportunity, since learned men throughout the early modern period were preoccupied with memory, eternal fame, and posterity, which resulted in many memorials, both in print and stone.[67] To study the *imagining* of a learned community,

63 Judith Pollmann, *Memory in Early Modern Europe, 1500–1800* (Oxford: Oxford University Press, 2017); Erika Kuijpers et al., eds., *Memory before Modernity: Practices of Memory in Early Modern Europe* (Leiden: Brill, 2013); Raingard Esser, *The Politics of Memory: The Writing of Partition in the Seventeenth-Century Low Countries* (Leiden; Boston: Brill, 2012); David P. LaGuardia and Cathy Yandell, eds., *Memory and Community in Sixteenth-Century France* (Farnham; Burlington, VT: Ashgate, 2015).

64 Winter, 'Sites of Memory'; Jasper van der Steen, *Memory Wars in the Low Countries, 1566–1700* (Leiden; Boston: Brill, 2015); Andrea Frisch, *Forgetting Differences, Tragedy, Historiography and the French Wars of Religion* (Edinburgh: Edinburgh University Press, 2015).

65 Richard Wrigley and Matthew Craske, eds., *Pantheons: Transformations of a Monumental Idea* (Aldershot; Burlington, VT: Ashgate, 2004); Eveline G. Bouwers, *Public Pantheons in Revolutionary Europe: Comparing Cultures of Remembrance, c. 1790–1840* (Basingstoke; New York: Palgrave Macmillan, 2012).

66 Pnina G. Abir-Am and Clark A. Elliot, eds., *Commemorative Practices in Science: Historical Perspectives on the Politics of Collective Memory* (Chicago: University of Chicago Press, 1999).

67 Peter Sherlock, 'The Reformation of Memory in Early Modern Europe', in Radstone and Schwarz, *Memory*, pp. 30–40.

these memorials offer a representation of the community to which a learned man was thought to belong. Each representation of the learned community acts as a mirror of ideals, aspirations, and collective identity.

Memory is a conscious sense of the past and a memorial is an object—for instance, a text, statue, volume, or a portrait—which conveys such a memory. Scholarly memory, then, is a conscious sense of a scholarly past. When this memory relates to universities, learned societies or a small academy it tells us something about the idea of a learned community of the producer of its memorial. For example, if a collection of lives of scholars—the memorial—contains only Germanic legal scholars, this reveals something about the sense of community of its compiler. The memory—the sense of the past—that this memorial conveys is both local and disciplinary. By comparing different scholarly memories and its corresponding memorials, we can start to understand how the Republic of Letters was imagined.

The sources consulted in this study are mostly memorials conveying scholarly memory. Examples of scholarly memory and memorials are plentiful: memories and citations of bygone scholars in extensive epistolary exchanges; heaps of printed collections of lives of exceptional scholars, or *viri eruditione illustres*, men illustrious on account of their erudition, published throughout Europe; eulogies and funerary orations of pre-eminent scholars and professors, both in print and manuscript form; funerary slabs, tombs, memorials, statues and epitaphs in churches throughout Europe; written histories of scholarship and institutions; or recorded memories of travels undertaken to libraries, collections of curiosities, universities, academies, and learned circles in the form of travelogues. All these sources present their own scholarly memory—that is, a conscious sense of the past and present—which is often part of a shared scholarly memory. These sources point to the fundamental question of this monograph: how did early modern actors *imagine* the learned community?

## Structure, Sources, and Outline

In this monograph, the Republic of Letters will thus be approached as an imagined community. Since a fifteenth-century legal scholar in Bologna would have had a different conception of the learned world to an eighteenth-century *philosophe* in Paris, I will consider these as two separate, local, and temporal representations—or imaginings of the Republic of Letters. This introduction has put forward the idea that all memorials of the learned world—through eulogies, portraits, commemorative coins and medals, funeral orations, or one of the many other conveyors of memory in the early modern learned world—offer us a way to rethink the Republic of Letters as an imagined community. The different ideals, collective identities, and senses of the past these different forms of scholarly memory conveyed and represented will be the subject of the remaining chapters.

To explore how scholars imagined the Republic of Letters, this study is divided into six chapters that each addresses a different source type. Chapters 1 and 2 focus on collections of *vitae* and *elogia*. Chapter 1 aims to distil a typology of collective virtues and values in the Republic of Letters from these collections. With the help of digital textual analysis, it examines a corpus of early modern collections of biographies to see which virtues and characteristics were commonly associated with learned men and women. Chapter 2 looks at the differences in content and geographical representations in these collections. In other words, it asks: who was being remembered where? Chapter 3 focuses on the digital corpus of the correspondence of Hugo Grotius to analyse which virtues and values were commonly exchanged between scholars. Additionally, this chapter looks at the citation of exemplary figures in the correspondence to analyse what virtues and sense of history they represented. Chapter 4 examines what I will call the *peregrinatio literaria* undertaken by scholars, and their experiences, representations, and formations of the Republic of Letters on their perpetual travels throughout Europe. Chapters 5 and 6 consider the material context of scholarly memory by analysing the remembrance of scholars in churches in Florence and Leiden and by comparing the kinds of communities this type of memory represented.

The analyses in the first three chapters rely on qualitative, textual, and quantitative techniques; the use of such textual analysis using computational methods places this section in the field of digital humanities. In Chapter 1, digital analysis is used to identify common tropes and values associated with learned men and women. Digital analysis is particularly useful, since it allows us to consider a relatively large corpus of collections of *vitae* as opposed to close-reading one or two collections. Additionally, digital analysis is particularly well suited to identify recurring language and common tropes. Such digital analysis acts as a divining rod, pointing to regularities and irregularities in the use of language that are worthy of closer attention. In Chapter 2, the analysis focuses on the composition of the collections of *vitae*. Who is included, and which groups of learned men and women were represented and how? The digital analysis used in Chapter 1 is also employed in Chapter 3, focusing on the correspondence of Hugo Grotius. Digital textual analysis is used to quantify values and virtues that the correspondents used to create a sense of identity. Additionally, Chapter 3 uses digital textual analysis to quantify the citations and authors throughout the digitised correspondence and thereby enable a citation analysis. Overall, digital analysis is a central part of this book and was feasible because of the availability of high-quality, 'clean' digitised texts. The full-text versions of the compendia of *vitae* and *elogia* were available in the *Thesaurus Eruditionis* and the Hugo Grotius correspondence was made digitally available in the *ePistolarium* project.[68]

The selection of cases for the digital analysis in the first three chapters was primarily based on the availability of high-quality textual data. The collections of *vitae* and *elogia* that form the majority of the primary source material for the first two chapters are part of the *Thesaurus Eruditionis* corpus. This digital corpus contains compendia of scholarly lives that were published throughout the early modern period. Although the size of the individual compendia varies, the corpus as a whole offers a balanced representation of the genre. The digital text files contain few digitisation errors and the Latin spelling has been standardised to enable quantification with digital analysis software. The Grotius correspondence was selected as the basis of the analysis in Chapter 3 since it is one of the few digitised, standardised, and freely available collections of correspondence by prominent early modern scholars. Therefore, the corpus offers an opportunity to digitally analyse the use of language throughout the scholarly life of Grotius.

The unpublished journal of Joannes Kool (1672–c. 1708) is central to Chapter 4 and offers a distinctive perspective on travelling scholars at the end of the seventeenth century. This source was unknown to historians before 2017 and helps to expand our view of the identity formation of travelling scholars for four reasons. First, since Kool travelled after obtaining his doctoral degree and did not visit universities, the journey is not a *peregrinatio academica* but a scholarly journey, a *peregrinatio literaria*. Second, the travelogue is an account of almost every single day of the year-long journey, as opposed to many accounts that cover short travel periods. Third, at 970 folios in length Kool's journal is the most elaborate and detailed Dutch

68 The *Thesaurus Eruditionis* was compiled by the Deutsche Forschungsgemeinschaft for the CAMENA and TERMINI projects between 1999 and 2008: see http://mateo.uni-mannheim.de/camenahtdocs/ camenaref_e.html. See also Appendix 1. *The Correspondence of Hugo Grotius* is a digital version of *BW*: see http://grotius.huygens.knaw.nl. The *ePistolarium* was part of the *Circulation of Knowledge and Learned Practices in the 17th-Century Dutch Republic* project: see http://ckcc.huygens.knaw.nl.

travel journal handed down from the seventeenth century.[69] Lastly, Kool's journal is unique since it describes scholarly interactions in meticulous detail not seen in other early modern travel journals. Taken together, Kool's journal offers a unique insight into how scholars presented themselves with a distinct scholarly identity on a scholarly journey.

Chapters 5 and 6 consider the remembrance of learned men and women in two different local contexts: the Basilica di Santa Croce in Florence and the Pieterskerk in Leiden. Both churches house numerous grave monuments of learned men and were known as sites of scholarly memory, but were shaped by two widely different political and cultural contexts. The rationale for comparing two churches in two different historical contexts—a Franciscan church in Catholic Florence and a Protestant church in the university town of Leiden—is to showcase the aspects of the scholarly memory culture that were particular and universal. Taken together, the chapters show the remembrance of learned men was shaped by both the local context and the appeal to universal values of a wider Republic of Letters.

The diverse types of sources in this book have been chosen to show the myriad ways through which scholars conveyed a sense of identity and community, or in Anderson's words, 'the style in which they are imagined'.[70] The case studies in this monograph address how learned communities imagined themselves as part of a larger European scholarly community through different processes and media. The following chapters explore how collective identities and communities were represented and sustained in compendia of scholarly lives, correspondence, travelogues, and on funerary monuments in different parts of Europe. In doing so, I hope to offer a novel perspective on the construction of the Republic of Letters and offer an account of how this imagined community was perpetuated in the early modern period.

69 This is based on a comparison with the travel journals registered in the list in Ruud Lindeman, Yvonne Scherf, and Rudolf Dekker, *Reisverslagen van Noord-Nederlanders van de zestiende tot begin negentiende eeuw: een chronologische lijst* (Haarlem: Stichting Egodocument, 1994).

70 Anderson, *Imagined Communities*, p. 6.

# 1. An Inventory of Scholarly Values and Virtues

**Abstract:** This chapter categorises the virtues associated with early modern scholars, using digital textual analysis on biographical collections (*vitae* and *elogia*). It highlights key virtues such as innate brilliance (*ingenium*), learning (*doctrina*), and cardinal virtues like wisdom and justice, as well as Christian virtues like faith and piety. The chapter also discusses how female scholars were marginalised in these representations. By analysing this inventory of virtues, the chapter illustrates how scholarly values evolved and varied by region and period, contributing to a broader understanding of intellectual identity formation within the Republic of Letters.

**Keywords:** Digital text analysis, biography, life-writing, scholarly virtues, epistemic virtues

> But his fault is onely this, that his mind is somewhat too much taken up with his mind, and his thought not loaden with any carriage besides [...] He has not humbled his Meditations to the industry of Complement. [...] He cannot kisse his hand and cry *Madame*, nor talke idle enough to bear her company. His smacking of a Gentlewoman is somewhat too savoury, and he mistakes her nose for her lip. [...] He names this word Colledge too often, and his discourse bears too much on the University. The perplexity of mannerlinesse will not let him feed, and he is sharp set at an argument when he should cut his meat.[1]

John Earle (1601–1665), Bishop of Salisbury, characterised the 'downright scholar' as a pariah and an eccentric. Earle's *Micro-cosmographie*, a satirical

1 John Earle, *Micro-cosmographie; Or, a Piece of the World Discovered; in essays, and characters*, 8th ed. (London: R.D. for P.C., 1664), pp. 112–114.

Scholten, Koen. *Rethinking the Republic of Letters: Memory and Identity in Early Modern Learned Communities*. Taylor & Francis Group, 2025
DOI 10.5117/9789048559855_CH01

collection of characters, was originally published anonymously in 1628 and was enthusiastically reprinted and translated up until the nineteenth century.[2] Earle's representation of the scholar was hardly the first, and certainly not the last, to offer such a stereotype: that is, male, obsessed with his own thoughts, uncivil. This scholar thrives within the confines of the university, but beyond its walls he is an outsider. In a stereotypical fashion, Earle stated that 'The Hermitage of his study, has made him somewhat uncouth in the world, and men make him worse by staring on him. Thus is he silly and ridiculous, and it continues with him for some quarter of a year out of the Universitie.'[3] The caricature reflects late seventeenth-century concerns about the importance of civility for scholars and their conduct.

This stereotype of the scholar as a man living in a bubble has persisted throughout the centuries. The image of the isolated scholar brings to mind Dr. Frankenstein, the learned man in Mary Shelley's *Frankenstein*.[4] In line with Earle's characterisation of the learned man as a fool, Shelley presents the philosopher as a man driven by passion, but with little more knowledge than the common man. She writes, 'The untaught peasant beheld the elements around him, and was acquainted with their practical uses. The most learned philosopher knew little more. He had partially unveiled the Face of Nature, but her immortal lineaments were still a wonder and a mystery.'[5] Ever since, writers have stereotyped scholars and scientists as evil, noble, foolish, inhuman, and also dangerous,[6] and filmmakers have portrayed them as unethical geniuses.[7] Scientists are also thought of as eccentric, lab-coated, bespectacled, Erlenmeyer flask–holding men, if the results of so-called 'draw-a-scientist' tests from the 1990s are to be believed (these tests contain associative drawings that show general attributes of a stereotype). Drawings of scientists often include symbols of research, such as test tubes, flasks, microscopes, and Bunsen burners, as well as symbols of knowledge such as books or filing cabinets.[8] These stereotypes persist

2 John Spurr, 'John Earle', *ODNB*.

3 Earle, *Micro-cosmographie*, p. 115.

4 Mary Shelley, *Frankenstein; Or, the Modern Prometheus*, ed. Maurice Hindle (London: Penguin Books, 2012; originally published in 3 vols, London: Hughes Harding, Mayor and Jones, 1818).

5 Ibid., p. 31.

6 Roslynn D. Haynes, *From Faust to Strangelove: Representations of the Scientist in Western Literature* (Baltimore; London: Johns Hopkins University Press, 1994).

7 Peter Weingart, Claudia Muhl, and Petra Pansegrau, 'Of Power Maniacs and Unethical Geniuses: Science and Scientists in Fiction Film', *Public Understanding of Science* 12, no. 3 (2003): 279–287.

8 David Wade Chambers, 'Stereotypic Images of the Scientist: The Draw-A-Scientist Test', *Science Education* 67, no. 2 (1983): 255–265; Cheryl L. Mason et al., 'Draw-A-Scientist Test: Future

and are important, because they reveal implicit ideas about science and scholarship and its practitioners.

The pursuit of knowledge has always been closely connected with the 'kind of person' who practiced it. As Michael Shortland and Richard Yeo have pointed out, 'in the period from the Scientific Revolution to the late nineteenth century there was an explicit concern with the character, motivations and values of the persons engaged in the pursuit of natural knowledge.'[9] Calls for the reform of knowledge production and organisation are usually accompanied by a call to change the morality and character of the knowledge practitioner. Robert Boyle (1627–1691) envisioned his empirical science with civil, aristocratic experimenters, whom he considered trustworthy.[10] Both Francis Bacon (1561–1626) and René Descartes (1596–1650) wanted a reform in philosophy and consequently devised a completely new persona to carry out the necessary scholarship.[11] Isaac Newton's (1643–1727) moral and intellectual achievements served as an example of the life of an eminent man of science as well as of the efficacy of the inductive method introduced by Francis Bacon.[12] Galileo Galilei's (1564–1642) life and work served as a model for subsequent Galilean disciples and mathematicians in Florence, such as Evangelista Torricelli (1608–1647) and Vincenzo Viviani (1622–1703).[13] Voltaire's (1694–1778) achievements were put forward by French revolutionary thinkers to exemplify the free and politically engaged intellectual, as well as the idea that genius can serve as the basis of political power.[14] It is fair to say that with supposedly

Implications', *School Science and Mathematics* 91, no. 5 (1991): 193–198; Jrène Rahm and Paul Charbonneau, 'Probing Stereotypes Through Students' Drawings of Scientists', *American Journal of Physics* 65, no. 8 (1997): 774–778.

9 Michael Shortland and Richard Yeo, 'Introduction' in *Telling Lives in Science: Essays on Scientific Biography*, ed. Shortland and Yeo, 1–44 (Cambridge: Cambridge University Press, 1996), p. 37.

10 Steven Shapin, *A Social History of Truth: Civility and Science in Seventeenth-Century England* (Chicago; London: University of Chicago Press, 1994), esp. ch. 4, 'Who Was Robert Boyle? The Creation and Presentation of Experimental Identity', pp. 126–192.

11 Stephen Gaukroger, 'Biography as a Route to Understanding Early Modern Natural Philosophy', in *The History and Poetics of Scientific Biography*, ed. Thomas Söderqvist, 37–49, Science, Technology and Culture, 1700–1945 (Aldershot; Burlington, VT: Ashgate, 2007).

12 Richard Yeo, 'Genius, Method, and Morality: Images of Newton in Britain, 1760–1860', *Science in Context* 2, no. 2 (1988): 257–284.

13 Michael Segre, *In the Wake of Galileo* (New Brunswick, NJ: Rutgers University Press, 1991), esp. ch. 3, 'Galileo's Followers', pp. 50–68; Segre, 'Viviani's Life of Galileo', *Isis* 80, no. 2 (1989): 206–231.

14 Darrin M. McMahon, *Divine Fury: A History of Genius* (New York: Basic Books, 2013), pp. 109–112.

new schools of thought, epistemologies, and traditions, new role models and examples emerged.[15]

Thomas Söderqvist pointed to this link between epistemology, exemplary figures and biography in his edited volume on scientific biography, while expressing his amazement at the lack of attention paid to this phenomenon: 'Historians of science are remarkably silent about biography as a product of and contributor to the social and political formation of modern science and its identities.'[16] While in more recent years, historians of science and scholarship have turned their attention toward the genre of biography and life-writing, the lives of scientists and scholars remain relatively understudied.[17]

Early modern lives of scholars convey a sense of collective scholarly identity and are therefore central to this and the next chapter. The emergence of the widespread publication of compendia of lives attests to the rising cultural visibility of the scholar as a type of person and profession. The image of the scholar emerging from these sources reveals both how scholars see themselves and the role they are supposed to have in society. What is more, these collections present the scholar as somebody with distinct character traits and virtues. In the following two chapters, compendia of lives of scholars, so-called *vitae*, will be analysed as mirrors of the collective identity of scholars as well as the scholarly community they were part of.

All of this does not mean that these compendia, when combined, present a coherent and homogeneous scholarly identity, regardless of temporality or spatiality. Instead, the *vitae* reveal a complicated combination of self-perception and a more general, detached perception of scholars, their conduct, and their community. This chapter focuses on the discourse in compendia of lives of scholars, and on the virtues with which scholars were commonly associated. As such, this chapter presents an overview of the characteristics of the archetypical scholar, the scholarly identity

15 Matthew L. Jones, *The Good Life in the Scientific Revolution: Descartes, Pascal, Leibniz, and the Cultivation of Virtue* (Chicago; London: University of Chicago Press, 2006).

16 Thomas Söderqvist, 'Introduction' to *History and Poetics of Scientific Biography*, ed. Söderqvist, p. 3.

17 See, e.g., Shortland and Yeo, *Telling Lives in Science*; Karl Enenkel and Claus Zittel, eds., *Die* Vita *als Vermittlerin von Wissenschaft und Werk: Form- und Funktionanalytische Untersuchungen zu frühneuzeitlichen Biographien von Gelehrten, Wissenschaftlern, Schriftstellern und Künstlern* (Berlin: LIT Verlag, 2013); L. J. Dorsman and P. J. Knegtmans, eds., *De menselijke maat in de wetenschap: de geleerden(auto)biografie als bron voor wetenschaps- en universiteitsgeschiedenis* (Hilversum: Uitgeverij Verloren, 2013).

he embodied, and the associated epistemic virtues based on the language used in the *vitae* to describe scholars and learned men. Thus, this chapter is concerned with the linguistic analysis of the *vitae*. Chapter 2 also focuses on collections of *vitae*, but looks at the different selection strategies and selection criteria that determined whose *vita* deserved a place in a compendium. It presents a quantitative analysis of the content of the collections, with a particular attention to the similarities and differences with respect to geography and chronology. Taken together, Chapters 1 and 2 will present the complicated image of scholarly communities and collective identities the diverse compendia of *vitae* reveal.

The central research question of this chapter is: which central aspects of a collective scholarly identity do collections of *vitae* convey and amplify? Qualitative computational analyses and close reading will identify the common characteristics and virtues associated with learned men. The first section of this chapter introduces the genre—if one can call it that—of collective life-writing in the early modern period: its history, revival in humanist culture, and its increased popularity throughout this period. Here, I will also reflect on the link between *vita*, virtue, and exemplary behaviour. Finally, I will comment on the methodology and the merit of computational linguistic analyses of collections of *vitae*. Section 1.2 contains an overview of the common characteristics and virtues ascribed to learned men in the collections of *vitae*. Learned women often appear in separate, dedicated tomes and are therefore hardly present in these collections. The third section will reflect on the discourse of alterity of learned women in such volumes. Throughout the chapter I argue that the collections of *vitae* reflect a self-conscious attempt to present examples of scholarly virtue. The collections of *vitae* are evidence of a tradition of representing learned men as having a distinct collective identity. Despite the variations in the representations of learned men, the collective life-writing of scholars turns out to be a pan-European phenomenon that helped form the sense of a pan-European learned community.

## 1.1 *Vitae*, Virtue, and Digital Textual Analysis

There has been a sustained tradition of the use of lives for religious, didactic, moral, political purposes, often to promote an individual or collective identity. Many plays and biographies, such as Shakespeare's *Richard III* (c. 1592–1594) or Francis Bacon's *History of the Reign of King Henry VII* (1622), have enabled historians to study the representation and identity formation of both the subject and the playwright or biographer.[18] This chronological telling of a life—the biography—only came to prominence at the end of the seventeenth century, but was founded on a tradition dating back to at least antiquity. Early modern historians therefore tend to use the more inclusive term 'life-writing' to refer to the practice of writing life stories and the texts it produces; it also avoids the implication of a coherent genre such as biography. As such, life-writing is more than a mere chronological narrative of a life and includes a wide array of texts, such as letters to posterity, hereditary notes, hagiographic life stories, and summations of someone's deeds and works.[19] Life-writing is the practice of writing or collecting someone's life. This cultural product will be referred to as a 'life', after the Latin word *vita* commonly used in early modern texts.

Early modern life-writing has received much attention from historians who have approached the topic from a variety of angles. Generally speaking, one can make a distinction between two different types of early modern life-writing: single lives and collective lives.[20] The first, most common and most studied type is the single *vita*, which comprises one life, either autobiographical or written by a disciple, family member, historian, or contemporary. The historical study of single *vitae* has mostly focused on the lives of reformers, humanists, and artists. The second type of life-writing is collective life-writing. Collective, here, is taken to mean representing or

18 Judith H. Anderson, *Biographical Truth: The Representation of Historical Persons in Tudor-Stuart Writing* (New Haven, CT; London: Yale University Press, 1984).

19 For an overview of the many types of early modern life-writing, see Thomas F. Mayer and Daniel R. Woolf, 'Introduction' in *The Rhetorics of Life-Writing in Early Modern Europe: Forms of Biography from Cassandra Fedele to Louis XIV*, ed. Mayer and Woolf, 1–37 (Ann Arbor: University of Michigan Press, 1995), pp. 10–18.

20 Most notably August Buck, ed., *Biographie und Autobiographie in der Renaissance* (Wiesbaden: Otto Harrassowitz, 1983); Mayer and Woolf, *Rhetorics of Life-Writing*; Karl A. E. Enenkel, Betsy de Jong-Crane, and Peter Liebregts, eds, *Modelling the Individual: Biography and Portrait in the Renaissance. With a Critical Edition of Petrarch's Letter to Posterity* (Amsterdam; Atlanta, GA: Rodopi, 1998); Patrick Coleman et al., eds, *Representations of the Self From the Renaissance to Romanticism* (Cambridge; New York; Melbourne: Cambridge University Press, 2000); Peter Burke, 'Representations of the Self from Petrarch to Descartes', in *Rewriting the Self, Histories from the Renaissance to the Present*, ed. Roy Porter, 17–28 (London; New York: Routledge, 1997); Katherine MacDonald, *Biography in Early Modern France, 1540–1630: Forms and Functions* (London: Maney, 2007); Irena Backus, *Life Writing in Reformation Europe: Lives of Reformers by Friends, Disciples and Foes* (Aldershot; Burlington, VT: Ashgate, 2008); James M. Weiss, *Humanist Biography in Renaissance Italy and Reformation Germany: Friendship and Rhetoric* (Farnham: Ashgate, 2010); Patrick Baker, ed., *Biography, Historiography, and Modes of Philosophizing: The Tradition of Collective Biography in Early Modern Europe* (Leiden; Boston: Brill, 2017). The distinction between single and collective lives is based on Enenkel and Zittel, *Die* Vita *als Vermittlerin von Wissenschaft und Werk*.

entailing multiple individuals. Where singular life-writing presents one lived life, collective life-writing presents a collection of lives and thereby a type of life. Collective life-writing mostly entails compendia of *vitae* or *elogia*, often carrying names such as *De viris illustribus* (*On Illustrious Men*) and *Virorum illustrium vitae* (*Lives of Illustrious Men*).[21] This chapter will exclusively focus on one such form of collective life-writing: compendia of lives of illustrious scholars, humanists, and scientists. Some volumes, such as Jacques Auguste de Thou's *Monumenta litteraria* (*Monuments of Learning*, 1640), contain an extensive list of eulogies of perished *viri docti* (learned men); other tomes aim to convey a clear confessional and political agenda, as Cornelia Niekus Moore has observed for early modern Lutheran funeral biographies.[22]

The early modern compilations of *vitae* are part of a long tradition and based on ancient models of Greek and Roman antiquity. In Greek literature, poets and philosophers such as Homer and Socrates were celebrated as exemplary citizens, socially engaged, and wise.[23] Diogenes Laertius's *Lives of the Eminent Philosophers* is the culmination of this Greek tradition of celebrating intellectuals and philosophers.[24] Diogenes's *Lives*, written in the third century AD, can be considered one of the first compendia of lives of learned men. It is rich in anecdotes as well as unique doctrinal details about the Stoics and Epicureans.[25] The ways in which it was disseminated, preserved, and referred to, attest to the exemplary status of this volume.

In Roman antiquity, intellectual activities were considered mostly a private, leisurely matter for the elite called *otium*, as opposed to public work, *negotium*.[26] Collective lives reflect this in the sense that they generally compile the lives of elite rulers. In the Roman life-writing tradition, two writers of compendia of lives stand out: Suetonius and Plutarch. Suetonius wrote *De viris illustribus* (*On Illustrious Men*) and *De vita caesarum* (*On the*

21 For a general introduction to the *vitae* of illustrious men, see Patricia Eichel-Lojkine, *Le siècle des grand hommes: Les recueils de vies d'hommes illustres avec portraits du XVIème siècle* (Leuven; Paris; Sterling, VA: Éditions Peeters, 2001); Tommaso Casini, *Ritratti parlanti: Collezionismo e biografie illustrate nei secoli XVI e XVII* (Florence: Edifir, 2004).

22 Cornelia Niekus Moore, *Patterned Lives: The Lutheran Funeral Biography in Early Modern Germany* (Wiesbaden: Harrassowitz, 2006).

23 Paul Zanker, *The Masks of Socrates: The Image of the Intellectual in Antiquity*, trans. Alan Shapiro (Berkeley; Los Angeles; London: University of California Press, 1995), pp. 40–89.

24 Patricia Cox, *Biography in Late Antiquity: A Quest for the Holy Man* (Berkeley; Los Angeles; London: University of California Press, 1983), pp. 4–12.

25 Diogenes Laertius, *Lives of the Eminent Philosophers*, trans. Pamela Mensch, ed. James Miller (Oxford; New York: Oxford University Press, 2018), pp. xii–xv.

26 Zanker, *Masks of Socrates*, pp. 203–210.

*Life of the Emperors*). These are not coherent attempts to record history. Suetonius's goal was to portray ideal and quintessential traits of statesmen and emperors that matched their deeds (*praxeis*) to their character (*ēthos*). For him, the added value of collecting lives in a compendium was not to offer a chronological narrative, but rather to invite the reader to contemplate the traits and virtues of illustrious men or emperors in comparison with each other. In *De vita caesarum*, Suetonius therefore organised the lives according to thematic rubrics and deliberately matched the deeds and their outer appearance to a preconceived model of a successful emperor.[27]

In Plutarch's *Parallel Lives*, written at the beginning of the second century AD, this moral component of the life is even more explicit. Through a combination of good and bad examples, Plutarch offered inspiring models of behaviour to imitate and avoid. The compendium consistently pairs two lives to illustrate the tensions of virtuous behaviour, providing matter for contemplation on issues such as the dangers of discontent, the contrast between a ruler's appearance and real character, or when to reinforce or challenge moral norms.[28] This morality was not one-sided and dogmatic. Plutarch invited the reader to challenge and contemplate issues of morality and behaviour exemplified by his carefully selected pairs of examples, most of whom were emperors and military leaders.[29] Both Suetonius's and Plutarch's collection of lives proved to be models for the centuries to come. Thus, the inherent link between *vita* and *virtus* was cemented in the practice of collective life-writing. To contemplate the lives, deeds, and characters of illustrious men enabled one to reflect on one's own moral behaviour.

Historically, this Roman life-writing tradition was adopted by early Christian writers, most notably Eusebius and Jerome. Eusebius, well known for his extensive ecclesiastical history, devoted much of his attention to writing the lives of martyrs and of his theological heroes, for example Origen of Alexandria.[30] In a similar vein, Jerome developed a canon of central Christian figures in *De viris illustribus* (*On Illustrious Men*), the first collective hagiography written in Latin.[31] Building on the classical tradition of life-writing, as well as on Jewish and Christian literary heritage, Jerome presented the lives of Anthony, Paul, and Hilarion as moral examples

27 Cox, *Biography in Late Antiquity*, pp. 12–15.

28 Timothy E. Duff, *Plutarch's* Lives*: Exploring Virtues and Vice* (Oxford: Clarendon Press, 1999), pp. 99–240.

29 Ibid., pp. 52–71.

30 Cox, *Biography in Late Antiquity*, pp. 28–29, 69–101.

31 Saint Jerome, *On Illustrious Men*, trans. Thomas P. Halton (Washington, DC: Catholic University of America Press, 1999).

instead of focusing on their historical deeds. Additionally, Jerome created a sense of lineage and presented himself as a successor to Paul: an ascetic, exemplary Christian in the desert.[32] Jerome's *De viris illustribus* can be seen as a continuation of Plutarch's emphasis on morality and exemplary figures, but departs from the Roman tendency to focus on emperors and military leaders. Jerome wanted his collective life-writing to present a new type of man: the Christian saint and martyr. One could argue that both Plutarch and Jerome were obsessed with the potential of collected *vitae* as a model of exemplary behaviour based on a coherent set of virtues and characters.

This tradition of Christian hagiography continued in the medieval period with the proliferation of saints' lives, moral allegories, and fables. Throughout Europe, local as well as widely popular saints were venerated both privately and collectively. Around the twelfth century, these cults of saints were enriched with new texts and visual environments, such as shrines, relics, and liturgies.[33] Our modern notions of saints and sainthood are generally thought to have their origins in this period, since it saw an expansion of the number of canonised saints, as well as the repertoires of venerations. Towards the end of the thirteenth century, the Holy See placed more restrictions on the canonisation of saints, both in terms of numbers and ways of veneration.[34] The saintly life became an example of a virtuous Christian life, as well as of the omnipotence and wisdom of God. Saints were often related to an aristocratic birth, pilgrimages, martyrdom, and saintly virtues such as humility, patience, piety, the humility of spirit, and the practice of good works.[35] The establishment of a canon of saints reflects the late medieval common practice of delivering a moral and religious message through a strictly controlled collection of lives of saints. It is hard to understate the significance and omnipresence of what André Vauchez called a 'hagiographical mentality': saints became representations of perfect virtue with a special place for *exempla*—lively stories meant for edification—of miracles and deeds.[36] For example, in this tradition

32 Susan Weingarten, *The Saint's Saints: Hagiography and Geography in Jerome* (Leiden; Boston: Brill, 2005).

33 Barbara Abou-El-Haj, *The Medieval Cult of Saints: Formations and Transformation* (Cambridge: Cambridge University Press, 1994), pp. 15–17, 35–37. For the development of sainthood in the medieval period, also see Peter Brown, *The Cult of the Saints: Its Rise and Function in Latin Christianity* (Chicago; London: University of Chicago Press, 1981).

34 André Vauchez, *Sainthood in the Later Middle Ages*, trans. Jean Birrell (Cambridge: Cambridge University Press, 1997), pp. 61–84.

35 Ibid., pp. 513–521.

36 Ibid., pp. 527–534.

Saint Martin became the ideal representation of the missionary bishop and Saint Lawrence the archetypical persevering martyr. Seen in this light, hagiography mainly consisted of curated collections of lives of saints with the goal of edifying readers and controlling doctrine. More than anything, the religious community is both the reader and author of hagiography, in the sense that the lives of saints reflect contemporary ideals of sainthood, holiness, and religious values.

This tradition of hagiography then mixed with a humanistic interest in classical antiquity. From the fourteenth century onwards, the Plutarch-inspired compendia of lives enjoyed a revival. Italian humanists were avid readers of Roman collective life-writing, as a source of both historical chronology and moral guidance. Soon, these humanists started to write and compose their own collections of lives of illustrious men. Petrarch's collection of illustrious lives, *De viris illustribus* (c. 1351–1353), is an influential example of this development. Like the Plutarchian works, as well as the works in the hagiographical tradition, Petrarch's *De viris illustribus* is rich in moral overtones. In the preface, he stresses that learning and teaching virtue through *vitae* is valuable 'for the very reason that through the remembrance of virtue we censure vice.'[37] One exemplary life does not suffice, according to Petrarch, because no example is perfect, since even 'the most illustrious minds are often afflicted by nature with some grave defect'.[38] The purpose of his collection is to compare and contemplate the moral behaviour of illustrious men in relation to each other. Petrarch explicitly excluded the trivialities and simply asks any reader interested in such things to 'consult other historians'. At this point in the preface, Petrarch echoes Livy's defence of history as we catch a glimpse of the goals of the collection.

> In my book, he will look for these things [trivialities] in vain, except inasmuch as they can lead to virtues or to the contraries of virtues. For, unless I am mistaken, this is the profitable goal for the historian: to point up to the readers those things that are to be followed and those to be avoided, with plenty of distinguished examples provided on either side. Whatever author would presume to wander outside this boundary, let him know that he is treading on foreign soil and wandering in foreign territory, and let him be reminded to return to the beaten path[.][39]

37 Benjamin G. Kohl, 'Petrarch's Prefaces to *De Viris Illustribus*', *History and Theory* 13, no. 2 (1974): 132–144 (p. 138).

38 Ibid., p. 141.

39 Ibid. For Livy's defense of the use of history, see Livy, *Ab urbe condita*, praefatio 10.

In short, the goal of history according to Petrarch is to teach virtue, and in this context his *De viris illustribus* also pointed to himself and implicitly to his community. Petrarch cemented the centrality of collective life-writing in the humanist tradition and the scholarly realm in general for centuries to come. His message was simple and clear: to become a virtuous man, one needs to contemplate the lives of similarly virtuous men in the past.[40]

We see the emergence of the first volumes specifically dedicated to learned men at the end of the fifteenth century. Johannes Trithemius's (1462–1516) *Cathalogus illustrium virorum Germaniam suis ingeniis et lucubrationibus omnifariam exornantium* (Catalogue of Illustrious Men Who Have Adorned *Germania* in Every Way with their Innate Brilliance and Tireless Studies) was published in 1495, most likely in Mainz.[41] The title explicitly refers to fame, innate brilliance (*ingenium*) and scholarly works (*lucubrationes*). According to Johannes Helmrath, Trithemius's aim was to contribute to monastic reform by advocating for a humanist education. In the *Cathalogus*, Trithemius presented German humanism and provided this flourishing intellectual movement with a long past.[42] In the following century, the compendia of lives of learned men takes flight, with publications such as Johannes Fichard's (1512–1581) *Virorum qui superiori nostroque seculo eruditione et doctrina illustres atque memorabiles fuerunt, Vitae* (Lives of Men Who Were in Our and Previous Centuries Illustrious and Memorable on Account of their Learning and Erudition) in 1536 and Paolo Giovio's (1483/1486–1552) *Elogia doctorum virorum* (Eulogies of Learned Men) in 1557.[43] From the sixteenth century onwards, collections of lives of learned men enjoyed an increase of popularity throughout Europe, with many reprints and editions. That the publication of such compendia was continuous throughout the early modern period is one of the reasons why *vitae* lend themselves so well to digital discourse analysis.

The second reason *vitae* are ideal sources to analyse quantitatively is that they have a strong moral component, as I indicated above. In a way,

40 For moral education and identity formation in Italian humanism, see Patrick Baker, *Italian Renaissance Humanism in the Mirror* (Cambridge: Cambridge University Press, 2015).

41 Johannes Trithemius, *Cathalogus illustrium virorum Germaniam suis ingeniis et lucubrationibus omnifariam exornantium* [...] ([Mainz: Peter von Friedberg], c. 1495).

42 Johannes Helmrath, 'Perception of the Middle Ages and Self-Perception in German Humanism: Johannes Trithemius and the *Cathalogus illustrium virorum Germaniam ... exornantium*', in Baker, *Biography, Historiography, and Modes of Philosophizing*, pp. 177–208.

43 Johannes Fichard, *Virorum qui* [...] *eruditione et doctrina illustres* [...] *fuerunt, Vitae* (Frankfurt a.M.: Christian Egenolph, 1536); Giovio, *Elogia doctorum virorum* [...] (Antwerp: Ioannes Bellerus, 1557).

*vitae* serve as a mirror of cultural values since compilers were obsessed with virtue and with exemplary men. Many of these collections of lives have a strong moral or political agenda, which is clearly visible in the so-called hagiography of humanism such as Trithemius's *Cathalogus*, Giovanni Boccaccio's (1313–1375) *De claris mulieribus* (On Bright Women), or Petrarch's *De viris illustribus* mentioned above. The same is true for the collections of *vitae* of learned men. The similarities and differences between these attempts to shape a collective identity are at the centre of this chapter.

The analyses in this chapter and the next are based on qualitative, textual, and quantitative analyses. The quantitative analysis presented in Chapter 2 entails the counting of the lives included in the volumes, as well as characteristics of the scholars whose lives are included, such as their name, stated nationality, and the number of pages dedicated to their *vita*.[44] In the remainder of this chapter, the textual analysis of the *vitae* entails the identification of tropes and themes in collective scholarly life-writing. The focus will be on the language used to describe learned men in these collections of *vitae* with the aim of answering the following question: What kind of characteristics and virtues were commonly associated with learned men?

With the textual analysis using computational methods, this research is part of a tradition of what is commonly called 'digital humanities' and 'distant reading'. These terms are becoming a commonplace in research in the humanities, but they still deserve some explanation. The value of computational analysis as well as its epistemological and methodological consequences are tentatively being explored not just in linguistics, but also in the history of science.[45] While neither 'digital humanities' nor 'distant reading' are particularly well defined, they often gain meaning in contrast with non-digital humanities or close reading.

To conduct textual analysis with the help of digital tools does not yield the same results as close reading and, indeed, does not aim to do so. Franco Moretti is known as a staunch proponent of the advent of distant reading in literary history. Moretti's stance on the bright future of distance reading for

44 For a general introduction to quantitative analysis in history, see Claire Lemercier and Claire Zalc, *Quantitative Methods in the Humanities: An Introduction*, trans. Arthur Goldhammer (Charlottesville; London: University of Virginia Press, 2019), esp. ch. 2.

45 See, e.g., Manfred D. Laubichler et al., 'Computational History of Knowledge: Challenges and Opportunities', *Isis* 110, no. 3 (2019): 502–512; Abraham Gibson and Cindy Ermus, 'The History of Science and the Science of History: Computational Methods, Algorithms, and the Future for the Field', *Isis* 110, no. 3 (2019): 555–566.

the study of literature is bold and provocative and his thinking on distant reading in relation to close reading is insightful:

> you invest so much in individual texts *only* if you think that very few of them really matter. [...] And if you want to look beyond the canon, [...] close reading will not do it. [...] Distant reading: where distance, let me repeat it, *is a condition of knowledge*: it allows you to focus on units that are much smaller or much larger than the text: devices, themes, tropes—or genres and systems. [...] If we want to understand the system in its entirety, we must accept losing something.[46]

Since the themes and tropes of the *vitae* of learned men are the prime concern of this research, distant reading makes sense. Yet distant reading as a new way to approach texts should not replace close reading or speed up that process. Rather, it allows for a quantitative perspective on a corpus. The analysis employed in this chapter is a combination of digital distant reading as well as close reading. The quantitative analysis of the corpus will be used to flag relevant passages that will inform the close reading and thus act as a divining rod.[47]

The corpus used for the digital textual analysis contains fifteen compendia of scholarly lives printed between 1536 and 1719.[48] The volumes are dispersed evenly over this period and reflect the continuous publication of collections of *vitae* of scholars throughout the early modern period. The size of the collections varies between the relatively small *Templum honoris reseratum* (Opened Temple of Honour, 1673) by Gottlieb Spitzel (1639–1691) with fifty lives and 67,143 words, to the five-volume *Vitae* by Melchior Adam († 1622) with 543 lives and 680,609 words (these works will be examined extensively in Chapter 2.3). On average, each collection contains roughly 170 *vitae*, and some lives reappear in several volumes, such as the lives of popes and influential reformers.

For the discourse analysis of the corpus of compendia, the software tool AntConc offers a unique opportunity to identify tropes and themes in

46 Franco Moretti, *Distant Reading* (London: Verso, 2013), pp. 48–49; original italics.

47 This combination of close and distant reading is also recommended in Rik Hoekstra and Marijn Koolen, 'Data Scopes for Digital History Research', *Historical Methods* 52, no. 2 (2019): 79–94.

48 The full-text versions of the compendia were available in the *Thesaurus Eruditionis der Projekte CAMENA und TERMINI*, projects carried out by the Deutsche Forschungsgemeinschaft, http://mateo.uni-mannheim.de/camenahtdocs/camenaref.html. The corpus with complete bibliographic information can be found in Appendix 1.

digitised collective scholarly life-writing.[49] The early modern texts were edited to plain-text text files and stripped of any paratextual material such as prefaces or indexes. The analysis thus strictly focused on the words used in the *vitae* themselves and the texts can be considered reasonably 'clean' data.[50] AntConc can search texts for frequently used words, word combinations as well as frequently used expressions and phrases. The corpus was analysed using the word frequency and keyword analysis against a reference corpus containing a great variety of early modern texts.[51] In this way, AntConc was able to present words and phrases that are characteristic and unique to the *vitae* genre or individual collections. This keyword list of words and phrases functioned as a guide in the following step: identifying virtues and references to characteristics. To understand how this works, we can look at the keyword analysis of Melchior Adam's *Vitae Medicorum* in Table 1.1. It shows which words are statistically more likely to appear in the volume compared to the rest of the corpus. The words 'medicina' (medicine or cure), 'medicus' (as a noun: doctor or physician; or as an adjective: healing or medical), and 'Galen' in different declinations and conjugations are

49 AntConc 3.5.7 (Macintosh OS X), developed by Laurence Anthony, Faculty of Science and Engineering, Waseda University, Japan. For a short introduction to AntConc and its use, see Shawn Graham et al., *Exploring Big Historical Data: The Historian's Macroscope* (London: Imperial College Press, 2016), pp. 79–83.

50 Data in the humanities are all texts and objects that humanists study. Clean data in the context of text-mining refers to digitised texts with few Optical Character Recognition (OCR) errors or other faulty artefacts. The texts in the corpus at hand have deviations due to OCR/ transcription mistakes, hyphenation, and other artefacts. Since these errors are randomly distributed throughout the texts, it is assumed that they do not significantly alter the outcome of the analysis. See Christof Schöch, 'Big? Smart? Clean? Messy? Data in the Humanities', *Journal of Digital Humanities* 2, no. 3 (2013) for an introduction to data in the humanities. For an exposition on the research cycle and the contingencies of data collection and analysis in the history of science, see Julia Damerow and Dirk Wintergrün, 'The Hitchhiker's Guide to Data in the History of Science', *Isis* 110, no. 3 (2019): 513–521. For a systematic approach to the use of digitised texts in the humanities, see Geoffrey Rockwell and Stéfan Sinclair, *Hermeneutica: Computer-Assisted Interpretation in the Humanities* (Cambridge, MA; London: The MIT Press, 2016). For examples of studies using digital analysis of texts, see Silke Schwandt, 'Digitale Methoden für die Historische Semantik: Auf den Spuren von Begriffen in digitalen Korpora', *Geschichte und Gesellschaft* 44 (2018): 107–134; Joris van Eijnatten and Ruben Ros, 'The Eurocentric Fallacy: A Digital-Historical Approach to the Concepts of "Modernity", "Civilization" and "Europe" (1840–1990)', *International Journal for History, Culture and Modernity* 7 (2019): 686–736; Christopher N. Warren, 'Historiography's Two Voices: Data Infrastructure and History at Scale in the Oxford Dictionary of National Biography (ODNB)', *Journal of Cultural Analytics* (22 November 2018); Henry Jones, 'Searching for Statesmanship: a Corpus-Based Analysis of a Translated Political Discourse', *Polis* 36, no. 2 (2019): 216–241.

51 For the reference corpus, see Appendix 1.

characteristic of the discourse in *Vitae Medicorum*, while other high-ranking keywords such as 'et' and 'anno' might not distinctively characterise the discourse in this tome. This immediately shows the pitfall of this type of keyword analysis: different conjugations in Latin are counted as different words and the analysis picks up much more than the results we are interested in. Therefore, while keyword analysis can be insightful, it cannot be taken at face value. In the last few rows of Table 1.1, we find words in the text that are characteristically absent and have a negative keyness: here 'literis' and 'virgo' stand out. Both words are twenty to thirty times less likely to appear in the *Vitae Medicorum* compared to the rest of the corpus.[52] To make sense of these results, we will have to dig deeper and read the usage of the keywords in the text.

**Table 1.1. Results of the keyword analysis of Melchior Adam's Vitae Medicorum (1620)**

| Rank | Keyword | Keyness | Frequency | Effect |
|---|---|---|---|---|
| 1 | medicinae | 299 | 190 | 5.9 |
| 2 | et | 238 | 5509 | 1.2 |
| 3 | medicinam | 226 | 124 | 7.2 |
| 4 | medica | 202 | 101 | 8.3 |
| 5 | medicus | 201 | 127 | 6.0 |
| 6 | pub | 157 | 124 | 4.6 |
| 7 | medicae | 145 | 84 | 6.7 |
| 8 | medico | 139 | 90 | 5.8 |
| 9 | medicis | 137 | 84 | 6.2 |
| 10 | medici | 136 | 96 | 5.2 |
| 11 | medicum | 132 | 77 | 6.7 |
| 12 | medicam | 125 | 66 | 7.7 |
| 13 | galeni | 115 | 65 | 6.9 |
| 14 | anno | 114 | 601 | 1.6 |
| 15 | medicina | 110 | 63 | 6.8 |
| -1 | quum | -153 | 1 | 0.011 |
| -2 | literis | -115 | 2 | 0.029 |
| -3 | virgo | -66 | 2 | 0.046 |
| -4 | at | -65 | 20 | 0.24 |
| -5 | quia | -58 | 38 | 0.35 |

52 The 'effect' is the relative frequency compared to the rest of the corpus (relative frequency text divided by relative frequency corpus). For 'literis' and 'virgo' this is 0.029 and 0.046, respectively. From the inverse of these ratios we can infer that 'literis' appears around 34 more times in the rest of the corpus and 'virgo' around 22.

The main goal of the keyword analysis is to obtain a quantitative analysis of the discourse employed in the *vitae*, to assess whether they are any differences between works. These quantitative differences can in turn serve as a divining rod and provide a macroscopic view to guide us to relevant passages, topoi, and tropes. There is an interaction with the data that is not a neutral one and data are not used as inert collections of texts that can be unproblematically 'mined' and used as a source.[53] It is precisely the continuous movement between digital analysis with the tool AntConc and close reading that has yielded the results of this chapter.

The next three sections of this chapter (1.2 to 1.4) will assess different aspects of the discourse in the compendia. Section 1.2 will analyse the presence of descriptive language of scholars in the *vitae*, such as ascribed virtues and character traits. It will present the characteristics of an archetypical scholar, such as innate brilliance, cardinal virtues, piety, and faith. The third section examines the particular and peculiar place of women in the early modern discourse on learnedness and virtue, based on the language employed in Johannes Ravisius's *De memorabilibus et claris mulieribus* (On Memorable and Bright Women, 1521). By comparing Ravisius's collection of learned women with the rest of the corpus, which almost exclusively consists of learned men, we can trace the different social and cultural position of women and the expectations they faced. The conclusions of this chapter will be discussed in Section 1.4. This chapter as a whole will argue for the existence of an archetype of a learned man in early modern Europe. This is not to say that this archetype was uniform and resistant to change. Rather, the confessional fragmentation after the Reformation, as well as divergent gender roles and expectations for learned men and women, gave rise to reinterpretations of the archetypical scholar described here.

## 1.2 Typology of Scholarly Virtue

This section presents the most common virtues that emerge from the *vitae* and provides them with historical context and depth. By 'common virtues' I mean the characteristics the texts use to describe learned men and delineate both collective and individual exemplarity. This includes the overlapping categories of moral virtues (which are a foundation for being a morally *good*, learned man), social virtues (collectively accepted norms of good behaviour),

53 This interaction is also dubbed a 'data scope' in Hoekstra and Koolen, 'Data Scopes for Digital History Research'.

as well as culturally determined characteristics (for example, noble descent or having a beard). In general, I am interested in the characteristics, values, and virtues that made a learned man exemplary, exceptional, or representative. The term *virtues* thus refers to the qualities that render someone an excellent, good, and representative member of its kind, in this case a learned man.[54] I presume that such virtues were culturally construed, an essential part of a collective identity, and promoted both implicitly and explicitly in the collections of *vitae*. The identification and study of these virtues can teach us what early modern compilers of *vitae* considered to be unique to learned men and how these collective identities developed, overlapped, and diverged. While a plethora of such characteristics emerges from the texts, a few immediately stand out. Figure 1.1 shows the relative frequency of identified common virtues and characteristics of scholars in the corpus of fifteen collections of *vitae*.[55]

Figure 1.1 requires some scrutiny and commentary. To start, all the words represent the presence of the word in all its different word types, for example as a noun and adjective. All the possibilities of number (singular and plural), declination (nominative, genitive, accusative, dative, ablative, and vocative cases), and gender (masculine, feminine, and neuter) have thus been clustered.[56] This is an attempt to cast a wide net and remove any bias due to the style of the writer or editor of the compendium. The relative frequency

54 This is based on an Aristotelian interpretation of virtue as the 'excellence of character', outlined in the *Nicomachean Ethics*. For an overview of the nature of virtue, see Kevin Timpe and Craig A. Boyd, *Virtues and Their Vices* (New York: Oxford University Press, 2014), pp. 6–9.

55 The scholarly virtues and characteristics were selected based on a thorough reading of the frequency lists and keyword analysis results of the texts.

56 To be specific, the declinations are the nominative, accusative, genitive, ablative, dative, and vocative cases, all in singular and plural forms, and in all genders (masculine, feminine, and neuter). Since the digital texts are standardised, there is no need to include spelling variations with *v* and *u*, for example in 'ingenium' and 'ingenivm'. The word types include the virtues as an abstract noun (*eruditio*), personal noun/adjective (*eruditus*), comparative (*eruditior*), and the superlative (*eruditissimus*). All these word types are included in all cases, numbers, and genders, as outlined before. Verbs and verb tenses (e.g., *erudire*) are not included, although it must be noted that the past participle is included, due to its similarity to the noun/adjective (e.g., *eruditus*). As an example, a search query for 'eruditio' includes the following 49 search terms: eruditio, eruditionis, eruditioni, eruditionem, eruditione, eruditiones, eruditionum, eruditionibus, eruditus, erudita, eruditum, eruditi, eruditae, erudito, eruditam, erudite, eruditorum, eruditarum, eruditis, eruditos, eruditas, eruditior, eruditius, eruditioris, eruditiori, eruditiorem, eruditiores, eruditiora, eruditiorum, eruditioribus, eruditissimus, eruditissimi, eruditissimo, eruditissimum, eruditissmo, eruditissime, eruditissima, eruditissimae, eruditissimam, eruditissima, eruditissimi, eruditissimorum, eruditissimis, eruditissimos, eruditissimi, eruditissimae, eruditissimarum, and eruditissimas.

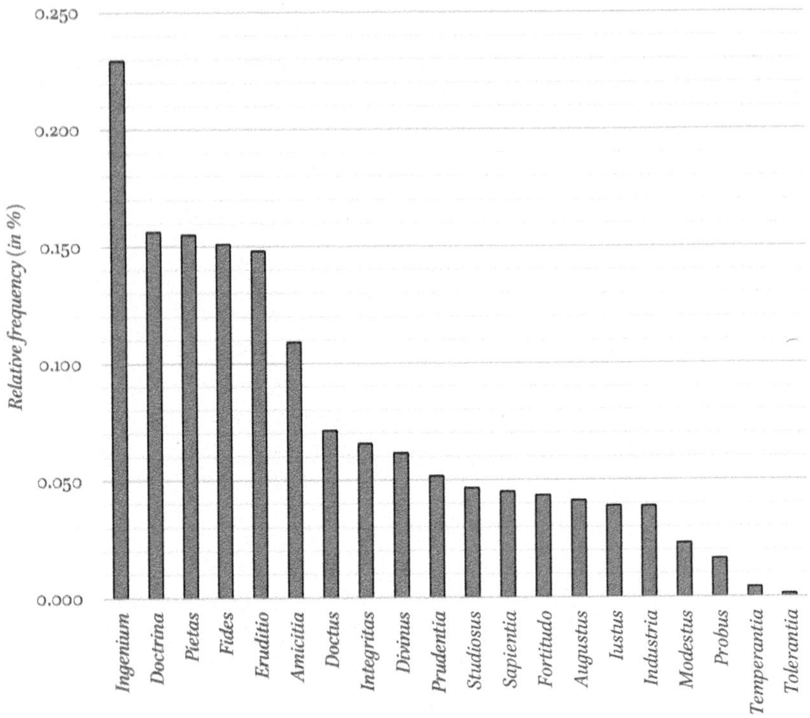


Figure 1.1. The relative word frequency (as a percentage of the total words in the corpus) of virtues on the horizontal axis in the corpus of *vitae*

indicates how often the word occurs as a percentage of all the words in the corpus. All the figures in this chapter that contain relative frequencies of virtues have been compiled in a similar manner.

The most common virtues associated with the learned men in the *vitae* are *ingenium, doctrina, pietas, fides, eruditio,* and *amicitia*. These virtues suggest an early modern philosophy of talent, learning, and the development of an illustrious, learned man. To explore this further, the subsections 1.2.1 through 1.2.3 will look at the uses and employment of these virtues in the *vitae*, all of which are in Figure 1.1. These most common characteristics in the discourse in the *vitae* represent a general discourse of the early modern learned man (*vir doctus*).

### 1.2.1 Innate Brilliance, Learning, and Erudition—*ingenium, doctrina*, and *eruditio*

*Ingenium* is most widely associated with learned men. *Ingenium*, here translated consistently as 'innate brilliance', has its roots in Roman antiquity.

In the minds of illustrious Roman citizens such as Cicero, innate brilliance became a staple of the civic illustrious man in the later Roman Republic. In his *De finibus* (On ends), Cicero explains the distinctive characteristics of *ingenium* compared to other virtues:

> The mind and that dominant part of the mind that is called the intellect, possess many virtues, but these are of two main types: one type consists of those which are implanted by their own nature, and which are called non-volitional; the other of those depend on our volition, and are usually called 'virtues' in the more precise sense. The latter are the preeminent glory and distinction of minds. The former type—the ability to learn and memory—and all the virtues of this type are generally named as 'innate brilliance' [*ingenium*], and they who possess these virtues are called 'geniuses' [*ingeniosi*]. The other class consists of the lofty virtues properly so called, which we speak of as dependent on volition, like prudence [*prudentia*], temperance [*temperantia*], courage [*fortitudo*], justice [*iustitia*], and others of the same kind.

> Animi autem et ius animi partis quae princeps est quaeque mens nominatur plures sunt virtutes, sed duo prima genera, unum earum quae ingenerantur suapte natura appelanturque non voluntariae, alterum earum quae in voluntate positae magis proprio nomine appellari solent, quarum est excellens in animorum laude praestantia. Prioris generis est docilitas, memoria; quae fere omnia appellantur uno ingeni nomine, easque virtutes qui habent ingeniosi vocantur. Alterum autem genus est magnarum verarumque virtutum quas appellamus voluntarias, ut prudentiam, temperantiam, fortitudinem, iustitiam et reliquas generis eiusdem.[57]

The meaningful distinction Cicero draws between innate and volitional virtues continues to be a guiding principle in works from the early modern period. To possess *ingenium* is often described as having a divine gift and, more importantly, one 'implanted by nature'. In contrast, the voluntary, or rather, acquired virtues such as prudence and courage can be learned, acquired, and nourished.

The modern word 'genius' is a conflation of the Latin *ingenium* and *genius*. Where the former referred to an innate disposition of future

57 Cicero, *De finibus bonorum et malorum*, bk. V, ch. XIII, l. 36. Translation based on Cicero, *De finibus bonorum et malorum* [*On Ends*], trans. H. Rackham, 2nd ed., repr. (Cambridge, MA: Harvard University Press; London: Heinemann, 1971), pp. 432–433.

excellence, *genius* referred to a spirit or demon who served as a companion to one's soul. Darrin McMahon observed that in the imagination of Roman writers, 'Men of eminence took their breath from the gods; they were in their nature unnatural; they possessed extraordinary *ingenium*, strange demons, uncommon *genius*.'[58] Throughout the later centuries, *ingenium* and *genius* tended to be conflated. Humanists picked up on this idea of innate talent in *ingenium* and the accompanying spirit or muse of *genius*. In texts such as Juan Huarte's *Examen de Ingenios para las Sciencias* (1575) the excellence of mind constituted of two elements: on the one hand, the angelic muse and the poetic fury as a spin-off of the *genius*, and on the other hand *ingenium*, one's given innate intelligence and brilliance.[59] The characteristic traits of the early modern learned man should be understood in this particular tradition of understanding excellence. I will argue here that the early modern learned man was thought of as a practitioner of his *ingenium*, in both the northern humanistic tradition and the Italian tradition of divine gifts.

*Ingenium* was noted as a distinctive characteristic of learned men in the titles of the collections of *vitae* in the corpus at hand. One can think of the previously mentioned *Cathalogus* by Trithemius, which explicitly characterised learned men as producing *lucubrationes* (literally, works made by candlelight) with their innate brilliance (*suis ingeniis*). Other texts stressed the aspect of learning (*doctrina*), for instance Johannes Fichard's *Vitae*, which speaks of men 'who were illustrious and memorable on account of their erudition and learning'.[60] Another example is Nicolaus Reusner who, according to the full title of his *Icones*, focused on men who were 'illustrious on account of letters', but especially on 'the religious teachings and others on account of good sciences'.[61] The titles of these compendia thus reveal the close association between learned men and the virtue of *ingenium*.

*Ingenium* is generally used to refer to an innate talent in many of the *vitae* of learned men. The virtue of having such brilliance is equalled to having been blessed with a preordained talent. At the same time, *ingenium* is often related to noble or good descent, where this descent is presented as fertile soil for greatness. This imagery of hereditary foundations and learned men

58 McMahon, *Divine Fury*, p. 27. Original italics.

59 Ibid., pp. 56–60.

60 Fichard, *Virorum* [...] *Vitae*.

61 Nicolaus Reusner, *Icones sive Imagines virorum literis illustrium* [...] (Strasbourg: Bernhard Iobin, 1590).

picking its fruits are a commonplace throughout the entire corpus. If we remember Cicero's distinction between non-volatile and volatile virtues, it is insufficient merely to possess predetermined, innate brilliance: it has to be polished as well, that is, a learned man must become *eruditus* (literally, cultivated or polished 'out of what is crude'). To be able to shine, one needs other cultivated minds as companions. Erudite friends should be admired because of their innate brilliance. The theologian Joannes Langus, for example, 'did not only have [Helius] Eobanus [Hessus] as a friend, but he was at the same time an admirer of his learning and innate brilliance'.[62] Similarly, Helius Eobanus Hessus's (1488–1540) relation to other scholars was brought to the fore as a fundamental aspect of his development:

> But out of all these men his closest friend was Martin Hanus. And he remained unfailingly friendly with Peter Eberbach of Erfurt, for whom he had felt particular friendship since his early youth, a young man who excelled in cultivating his innate brilliance and polishing his learning.
>
> Sed ex omnibus familiarissime complectebatur Martinum Hanum: et affectione praecipua in prima adolescentia cognitum, constantissime diligebat Petreium Aperbachum Erphordiensem, iuvenem ingenii cultu et eruditione doctrinae excellentem[.][63]

During childhood and adolescence, young men's minds were cultivated by other men, for instance a teacher or a father. In the case of Johannes Marcellus (1510–1554):

> It appears he made use of his father, a diligent man who was particularly studied in literature. It was he who took care to train his son from the earliest age onwards in the first principles with which a young boy needs to be imbued.
>
> Patre, ut apparet, usus est diligente, inprimisque, studioso litterarum; qui a teneris statim filium iis rudimentis, quibus aetas puerilis imbui debet, erudiendum curavit.[64]

62 '[N]on modo anicum habebat Eobanum; sed doctrinae atque ingenii etiam admiratorem'. Melchior Adam, *Vitae Germanorum Philosophorum* [...], *Vitae* I (Frankfurt a.M.; Heidelberg: Jona Rosa 1615), p. 116.

63 Ibid.

64 Adam, *Vitae* I, 'Ioannes Marcellus', p. 145.

Here, Melchior Adam's account is both descriptive and prescriptive. The passage makes two simple points: first, the point that Marcellus was cultivated from his early years onwards; secondly, and implicitly, that childhood is also the right time to start the cultivation towards erudition ('erudiendum curavit'). The collection of Adam's *Vitae* is a blueprint of how to bring young boys to erudition. Many learned men shared this development, according to the humanistic ideals of education. Throughout the corpus, the agricultural lingo persists: the fertile soil, the crude minds (*rudes mentes*), must be cultivated (*colere*), and polished (*erudire*) to be able to bear fruits of brilliance (*fructus ingenii*), and come to erudition (*eruditio*). This discourse presented the archetypical coming of age of the learned man—from the first gift of *ingenium* until their development through *eruditio*—that became a trope in the *vitae* of learned men.

Aside from the importance of innate brilliance, the role of teachers and supervisors is frequently emphasised. This attests to the value humanists assigned to teaching and intellectual development. In the corpus, teachers (*magistri*) are credited to be sources of scholarly inspiration and the instigators of the cultivation of *ingenium*. Melchior Adam often stresses this in his *Vitae*, for example in characterising Guidus Pancirolus's (1523–1599) teacher as someone who made sure that 'the power of the ingenious mind in the boy had started to shine' during the latter's childhood.[65] It is a trope in the collections of *vitae* that the predestined schoolboy is suddenly enlightened by a teacher. Johann Crato (1519–1585), in whose youth 'something suddenly enlightened the light of his soul and innate brilliance in him', is a case in point.[66] Elsewhere in Adam's *Vitae*, it is said that during Johannes Borcholten's (1535–1593) youth, 'suddenly the light of his innate brilliance plainly manifested itself'.[67] In the process of becoming a man with a cultivated mind, the light of *ingenium* needed to be ignited and make itself known and visible to the world. It is not uncommon for a *vita* to start with a long exposition on the education enjoyed by the learned man in question, with specific reference to all the important teachers in their lives. For the light of *ingenium* to ignite, however, teachers were deemed essential and their names worth remembering, as well the teachers' teachers, thereby creating a sense of lineage and collective identity.

65 '[V]is ingenii in puero elucere coepisset'. Giacomo Filippo Tomasini, *Illustrium virorum elogia Iconibus exornata* [...] (Padua: Donatum Pasquardum, & Socium, 1630), 'Guidus Panricolus', p. 187.
66 '[U]t statim aliquod animi et ingenii lumen in eo eluxerit'. Adam, *Vitae Germanorum Medicorum* [...], *Vitae* II (Heidelberg: Jona Rosa, 1620), 'Ioannes Cratus', p. 262.
67 '[S]tatim manifestum ingenii lumen ostendit'. Adam, *Vitae Germanorum iureconsultorum et politicorum*, *Vitae* III (Heidelberg: Jona Rosa: 1620), 'Ioannes Borcholten', p. 320.

It was crucial to this ignition to surround oneself with erudite men. The first place to find them was in the schools, according to Melchior Adam. He encouraged parents to send young men to school, so that they could associate both with the teachers and their fellow students. In Laurentius Ludovicus's (1536–1594) *vita*, Adam presents an exceptional case to underscore his general rule:

> He was sent to elementary school by his father at age seven. He remained in the city school till he was twenty-two. To him happened the opposite of most other learned men. For often the beginnings of the training of learned men manifests itself as follows: that immediately some light of spirit and innate brilliance started to shine in them.
>
> Missus in ludum litterarum a parente aetatis anno septimo, in scholis patriae ad annum vigesimum secundum substitit. In quo contra illi ac plerisque aliis viris doctis accidit. Sic enim plerunque initia disciplinae eruditorum hominum se dederunt; ut statim aliquod animi et ingenii lumen in iis eluxerit.[68]

The message is clear. If you have any hope of fostering *ingenium* yourself, schools are the first place to start for the light of innate brilliance ('ingenii lumen') to start shining. The next requirement is to travel to meet fellow *viri docti*. While writing about Johannes Goropius (1519–1573), Melchior Adam stressed the importance of travelling to meet ingenious minds: 'at a more advanced age for a more exquisite cultivation of his innate brilliance, he travelled through Italia, Spain, France, and other provinces'.[69] This is the well-known and often often promoted ideal of the *peregrinatio* (literally, meandering journey) to polish one's own crude mind (see Chapter 4). After having been polished by a teacher for some years, it was time to meet learned men beyond the confines of one's school or city.

It is noteworthy that Adam mentioned schools and teachers so often when describing illustrious scholars. Although *ingenium* was an indispensable requirement for young scholars with a predilection for learning, it also needed to be cultivated and polished by excellent teachers as well as meet fellow learned men during adolescence. Adam's characterisation of the essential virtues of *ingenium*, *doctrina*, and *eruditio* are the quintessential attributes of

68 Adam, *Vitae*, I, 'Laurentius Lodovicus', p. 383.

69 'Aetate grandior ad exquisitiorem ingenii cultum, Italiam, Hispaniam, Galliam, aliasque provincias peragravit.' Adam, *Vitae* II, 'Ioannes Goropius', p. 190.

the early modern scholar. When we look at an Italian counterpart, however, different minor interpretations of these core characteristics emerge.

The philologist Gian Vittorio Rossi (1577–1647) also maintained the imagery of enlightened brilliance and polishing one's talents in his three-volume *Pinacotheca*. In this work Rossi explicitly mentioned that these illustrious men were 'praised on account of their learning and innate brilliance'.[70] In several instances Rossi stressed the emergence of *ingenium* in someone due to personal character traits. For example, Antonius Bosius's (1575–1629) innate brilliance took the form of docility, since 'he had *ingenium* teachable by nature'.[71] Elsewhere, Fulvius Testi (1593–1646), praised for his new style of sermons and verses, was credited for gaining his brilliance from his community and travels:

> An incredible and almost divine strength of his innate brilliance diverted him, who was hasting towards the highest glory of poetical accomplishment, but had already left the common path trodden by the itineraries of almost everyone, and pushed him to a place where no one before him had set foot and he invented for himself an utterly laudable way of writing.
>
> Eum igitur, ad summam poeticae facultatis gloriam properantem, incredibilis quaedam, ac prope divina, vis ingenii, a communi atque omnium fere contrita itineribus via abreptum, eo abstraxit, ubi nemo, ante ipsum, vestigium impresserat, ac novam, et in primis laudabilem, sibi scribendi rationem invenit.[72]

In this passage about Testi's life, Rossi makes the strength of innate brilliance (*vis ingenii*) a third-person protagonist, reminiscent of the accompanying muse of spirit originally captured by *genius*. There is no mention of teachers or schools, since Testi's excellence was seen as *prope divina*, almost divine. Rossi's attestation to the divine is in line with the interpretation of *ingenium* by predecessors such as Giorgio Vasari (1511–1574), who considered *ingenium* to be a mysterious, animating, inborn gift for visual artists. According to Vasari, it was a vocation one must answer although one should always stay true to one's particular gift of innate brilliance.[73] Where Vasari maintained

70 '[D]octrinae vel ingenii laude [viri]'. Giovanni Vittorio Rossi, *Pinacotheca Imagnium Illustrium* [...], vol. I (Cologne: apud Jodocum Kalcovium et Socios, 1645), frontispiece.

71 'Habuit ingenium a natura docile'. Ibid., 'Antonius Bosius', p. 233.

72 Ibid., vol. III, 'Fulvius Testa', p. 213.

73 Douglas Biow, *Vasari's Words: The* Lives of the Artists *as a History of Ideas in the Italian Renaissance* (Cambridge: Cambridge University Press, 2018), pp. 52–53.

an ideal of a School of Athens for artists, Rossi generally described something as *commune*, varying between something communal of a city (*commune urbis*), of churchmen, or of all the good (*commune bonorum omnium*). In this context, the darkness of an urban community can limit the possibilities of *ingenium*, as in Martha Marchina's (1600–1647) life:

> in the broad daylight of her town, the great light of her innate brilliance was oppressed by the ignorance of the masses, as if by darkness, preventing her from spreading her beams widely.
>
> in media eius Urbis luce, tantum ingenii lumen, omnium ignoratione, tanquam tenebris, opprimi, ne radios suos late diffunderet.[74]

In general, it is fair to say that the Italian notion of *ingenium* is more linked to divinity and excellence, as opposed to the northern interpretation of brilliance as part of humanist development and teaching.

When we look at the relative frequency of the virtue of *ingenium* in the corpus, we see there is some disparity in its use between the different compendia (see Figure 1.2). Surprisingly, in the collections of *vitae* published on the Italian peninsula, such as Paolo Giovio's *Elogia*, Tomasini's *Elogia Virorum Literis*, and Rossi's *Pinacotheca*, the relative frequency of the virtue of *ingenium* is almost twice as high as in the remaining volumes in the corpus, most of which were published in the Germanic lands on the northern side of the Alps.[75] This might be explained by the fact that the meaning of *ingenium* was more broadly interpreted as innate brilliance on the Italian peninsula in the wake of Vasari's popular *Vite*, whereas *ingenium* was one of the first stages in the narrative of humanistic development in the life writing from the North. Importantly, the term features more in eulogies, such as Giovio's *Elogia*, where the virtue is applied to almost every person eulogised, as a distinction of excellence.

The gift of *ingenium* lived on as a legacy for posterity, especially in Giovio's *Elogia* and Rossi's *Pinacotheca*. Rossi explicitly refers to it in the full title of his work, which he called a 'Pinacotheca of images of men, illustrious due to the praise of their learning and innate brilliance' ('Pinacotheca imaginum, illustrium, doctrinae vel ingenii laude, virorum').[76] Giacomo

74 Rossi, *Pinacotheca*, vol. III, 'Martha Marchina', p. 238.

75 They also appear as keywords in the keyword analysis in Giovio, Rossi (vol. I), and Tomasini.

76 Rossi, *Pinacotheca*, vol. I, frontispiece.

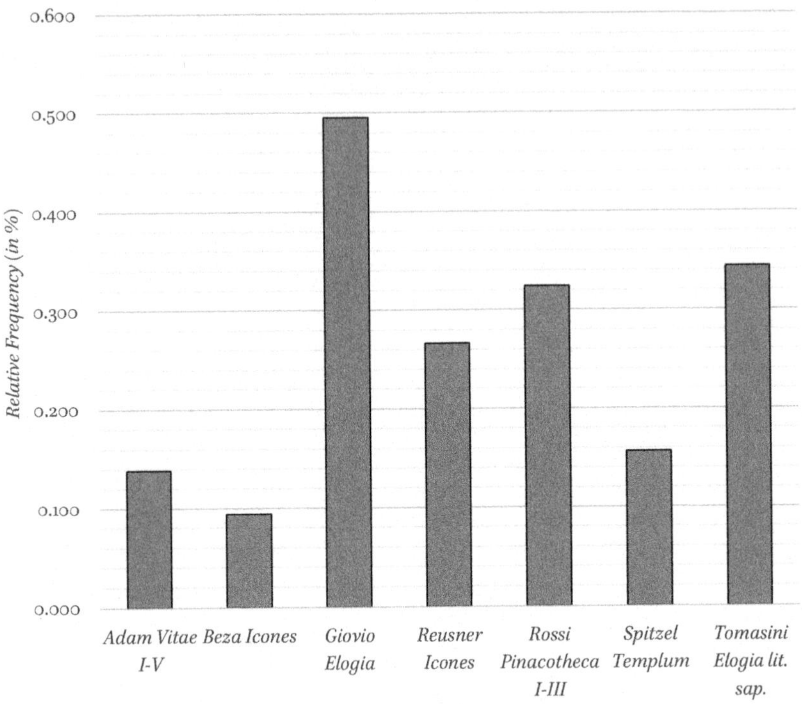


Figure 1.2. The relative word frequency (as a percentage of the total words in the corpus) of the virtue of 'ingenium' in the different collections in the corpus

Filippo Tomasini expressed a similar interest in a learned man who leaves something 'to posterity due to the highest praise of his labour and innate brilliance'.[77] The tomes and work the learned man left behind are his monuments. Or, as Tomasini remarked on Jordanus Moscatellus: 'He left us with these monuments of innate brilliance'.[78] Praise comes to those who served the learned world well and their monuments of innate brilliance became reminders and examples for the reader.

These results are partly in line with an extensive study of the term *ingenium* in early modern European dictionaries.[79] According to the authors of *Logodaedalus*, the Latin and polyglot dictionaries reveal 'a sophisticated account of *ingenium* as "inborn nature" made manifest through a variety of identity markers: one's wit, moral discernment, behaviour, and social

77 'Summa laboris ac ingenii laude posteritati reliquit.' Tomasini, *Elogia*, p. 65.

78 'Relictis hisce ingenii sui monumentis'. Tomasini, *Elogia*, 'Iordanus Moscatellus', p. 256.

79 Alexander Marr et al., *Logodaedalus: Word Histories of Ingenuity in Early Modern Europe* (Pittsburgh, PA: University of Pittsburgh Press, 2018).

status.'[80] In Italian dictionaries, the word *ingegno* was also associated with practical men well versed in the mathematical arts, attesting to Vasari's concept of the artist and maker and his relation to *ingegno*.[81] Yet the discourse on *ingenium* has two dimensions that are expressed more vividly in the scholarly lives: first, as a stage in the learned man's ideal humanistic development, and second, together with learning (*doctrina*) as a twofold characteristic of a learned man.

However, *ingenium* was far from the only virtue associated with learned men. As we can see in Figure 1.1, cardinal and Christian virtues such as courage, justice, wisdom, and prudence also prevailed.

### 1.2.2 Cardinal and Christian Values—*fortitudo, iustitia, sapientia, temperantia* and *prudentia*

Traditional Christian and cardinal virtues also emerge prominently from the quantitative word analysis of the corpus. Traditionally, these consist of the virtues of prudence (*prudentia*), wisdom (*sapientia*), courage (*fortitudo*), temperance (*temperantia*), and justice (*iustitia*).[82] The overall relative frequencies of the virtues in the entire corpus in Figure 1.1 seem to point to the virtues as an integral part of the discourse on early modern scholars. On average, temperance has the lowest relative frequency, at 0.004%, although courage, justice, wisdom, and prudence all emerge in the corpus with a relative frequency between 0.039% and 0.052%. This is comparatively high and means that on average these virtues individually emerge in one out every three *vitae*.[83] Yet when we take a closer look at the distribution of these virtues in the different works in the corpus, we see a more complex distribution of the virtues, as can be seen in Figure 1.3.

Temperance is not a virtue that is commonly associated with learned men. When temperance is used as a virtue to describe a learned man in the *vitae*, the writer generally wants to stress the Christian aspect of someone's life. As such, temperance is often mentioned in the same sentence with virtues of modesty (*modestia*), Christian love (*caritas*), and moderation

80 Ibid., p. 50.

81 Ibid., pp. 84–85.

82 For a general account of virtues, see Timpe and Boyd, *Virtues and Their Vices*, esp. Section I on cardinal virtues.

83 On average, one work contains 170 *vitae*, with an average of 674 words in one *vita*. These four cardinal virtues (*fortitudo, iustitia, sapientia,* and *prudentia*) emerge between 58 and 74 times on average in each individual work. This means that the chance of encountering the virtues in one *vita* is roughly between 34% and 44%, an estimation of one in three (1/3 ≈ 33%).

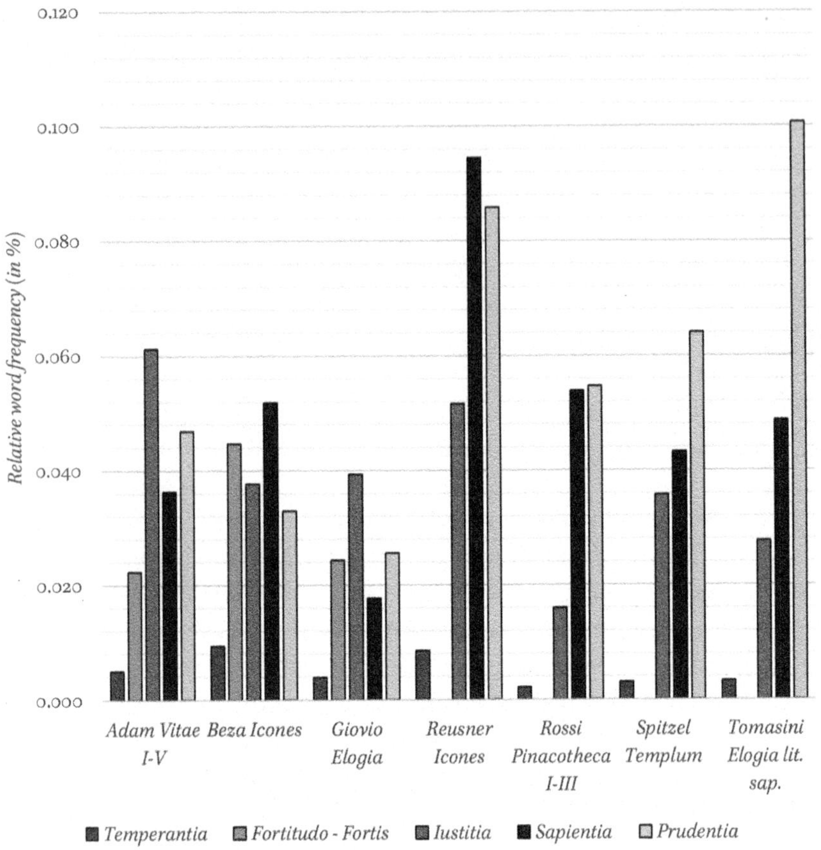


Figure 1.3. The relative word frequency (as a percentage of the total words in the corpus) of the cardinal virtues (*temperantia, fortitudo, iustitia, sapientia*, and *prudentia*) in different books in the corpus

or sobriety (*sobrietas*). Adam presented Lambert Canter (1513–1553), for instance, as a 'rare example of temperance and modesty'.[84] Temperance was also associated with a rustic, isolated life of contemplation, abstinence, and seclusion.[85] Of course, this was an appealing virtue to compilers of collections of vitae of learned men during the Reformation, such as Theodore Beza (1519–1605) and Nicolaus Reusner (1545–1602), although they use it relatively sparingly. Melchior Adam employed the virtue of

84 '[T]emperantiae et modestiae rarum exemplum'. Adam, *Vitae* III, 'Lambertus Canterus', p. 90.

85 See, e.g., Adam, *Vitae* II, 'Levinus Lemnius', p. 100: 'simulque de frugalitate victusque temperantia, et vitae rusticae dignitate.'

temperance to describe Philip Melanchthon (1497–1560), who even during his childhood

> lived according to excellent order, in temperance and sobriety, and in this way, he strengthened his otherwise weak body; so as to be strong enough for the toughest labours.
>
> sed temperantia, sobrietate, ac optimo vivendi ordine, ita corpus alioqui debile roboravit; ut maximis etiam laboribus sufficeret.[86]

Temperance is slightly more popular in the reformed life writing from Northern Europe, such as Adam's *Vitae*, Beza's *Icones*, and Reusner's *Icones*. The virtue of temperance is often employed to present the learned reformers as tempered and modest, in implicit contrast to the abundance, extravagance, and greed they associated with Catholicism.

Much like the virtue of temperance, both courage (or fortitude) and justice (or fairness) tend to appear slightly more often in the *vitae* of Melchior Adam, Theodore Beza, and Nicolaus Reusner compared to the rest of the compendia in the corpus. This suggests that these virtues were part of a Reformation or a northern discourse of humanism. When we take a close look at the virtue of courage, we see a linguistic coincidence that renders the relative frequency less reliable especially in comparison to the other virtues: a full search on *fortitudo* (including *fortis*) includes the ablative *forte*, which is also an adverb (meaning 'perhaps' or 'by chance') that Melchior Adam happens to employ rather frequently. Despite this unfortunate conflation, a glance at the occurrences in the corpus shows a few common collocations: a strong will or mind (*fortis animus*) as example of courage (*exemplum fortitudinis*) and the stout breast (*forte pectus*). The latter is a rendition of a line by Catullus, where Achilles is said to be 'known not by his back, but by his stout breast'.[87] By invoking this imagery, Melchior Adam created a discourse of conflict. Adam evokes a Reformation discourse with the trope of courage in the eulogy of Erasmus Sarcerius (1501–1559), a theologian and clergyman:

> With sincere doctrinal faith, he cares for his sheep.
> What skilful lies and threats of the mighty has he not endured, persevering the hardship with a stout breast?

86 Adam, *Vitae* I, 'Philippus Melanchthon', p. 200.

87 '[H]ostibus haud tergo, sed forti pectore notus'. Catullus, LXIV, l. 339.

Sincero fidei dogmate pascit oves.
Quas non concinnas fraudesque minasque potentum
Sustinuit: forti pectore dura ferens?[88]

It is no exaggeration to say that the Germanic compendia of lives, in particular Melchior Adam's *Vitae*, Nicolaus Reusner's *Icones*, and Theodore Beza's *Icones*, employ militant language to describe the Reformed Protestant theologian as someone who had to care for his flock, be strong, and have courage. This stoic endurance through suffering and adherence to the 'true faith' are quintessential martyrological virtues employed in the wake of the Reformation by its early chroniclers. This conflation of the ideal of a Protestant martyr and the ideal of learned men can be seen even more clearly in the use of the religious virtues of piety and faith.

### 1.2.3 Piety and Faith—*pietas* and *fides*

The virtue of piety was central in the European collections of *vitae* of learned men from the outset. After Johannes Trithemius had compiled a collection of lives of ecclesiastics, he argued that learned men deserved the same respect and treatment. In his *Cathalogus Illustrium Virorum* (*Catalogue of Illustrious Men*, c. 1495), Trithemius explained the importance of remembering learned men in a prefatory letter to Jacob Wimpfeling (1450–1528):

> For what the most holy fathers laudably did for the benefit of posterity, no intelligent person will forbid me to do, motivated as I am by the same intention. Finally, there are many erudite men alive today, and their memorialization for posterity is an act of reverence and of no mean utility. If antiquity had not been motivated to do this by the same reverence, all those famous writers of yore would not have been passed down to us. There are also some people who think that our age boasts no great writers, or only very few. For them I trust that, by bringing to their attention only a few of the many, I will have done something pleasing and deserving of thanks. For even if I have not been able to make mention of all the men of outstanding intelligence, I have nevertheless provided our studious Germans the occasion in turn to add still others and to augment our work with their industry.
>
> Enimvero quod sanctissimi patres ad utilitatem posteritatis laudabiliter fecerunt, mihi eadem intentione facere cupienti nemo sapiens interdicit.

88 Adam, *Vitae* I, 'Erasmus Sarcerius', p. 327.

> Denique multi sunt his temporibus eruditissimi viri: quorum memoriam posteris commendare pietatis est: et non minoris utilitatis. Nisi hoc fecisset pari pietate antiquitas: omnes dudum veteres clarissimi scriptores ad nostram memoriam non venissent. Sunt etiam, qui aut nullos aut paucos hac tempestate illustres scriptores in humanis haberi existiment: quibus me rem iucundam gratamque fecisse confido: si vel paucos de multis ad eorum noticiam praeduxero. Nam etsi non omnium praestantium ingenio virorum mentionem facere potui: studiosis tamen nostris germanis occasionem prestiti: ut his deinceps etiam alios possint adiungere: et opus nostrum sua diligentia ampliare.[89]

From this collection of Christian learned men from *Germania*, a history of northern humanism emerged. The *Cathalogus* aimed to serve as a bibliographic guide, but certainly also argued for monastic reform through the implementation of a humanist education. Trithemius's work also has patriotic overtones and can be considered as the first attempt to erect a paper pantheon of 'great' men of learning.[90] Trithemius's collection is foundational for this northern tradition of collecting and celebrating learned men, which was continued in the sixteenth century by the likes of Johannes Fichard, Theodore Beza, and Nicolaus Reusner in the corpus. Trithemius presented a developing northern strand of humanism in the second half of the fifteenth century called monastic humanism, in opposition to the Florentine, civic humanists.[91] As the case of Trithemius's *Cathalogus* shows, the virtue of piety was closely related to the sense of history and identity of the early compilers of lives of learned men. Monastic and Christian ideals of piety that saw learning as an act of devotion were central for generations of scholars to come, and it is in this context that we can understand the high relative frequencies of the virtues of piety and faith in the corpus.

The relative word frequency of religious virtues can be seen in Figure 1.4. The disparity between the volumes stands out. While the average relative frequency of the virtue of piety in the entire corpus is already high, at 0.155%, Figure 1.4 shows how in both Beza's *Icones* and in Reusner's *Icones* the virtue of piety almost doubles the average, with 0.386% and 0.275% of the

89 Trithemius, *Cathalogus illustrium virorum*, sig. A-ii$^r$. English translation from Johannes Trithemius, 'Johannes Trithemius: *Catalogue of the Illustrious Men Who Have Adorned Germany in Every Way with Their Talents and Tireless Studies* (selections)', ed. Johannes Helmrath, trans. Patrick Baker and Johannes Helmrath, in Baker, *Biography*, pp. 209–239 (pp. 212–213).

90 Helmrath, 'Perception of the Middle Ages', pp. 183–186.

91 Noel L. Brann, *The Abbot Trithemius (1462–1516): The Renaissance of Monastic Humanism* (Leiden; Boston: Brill, 1981).

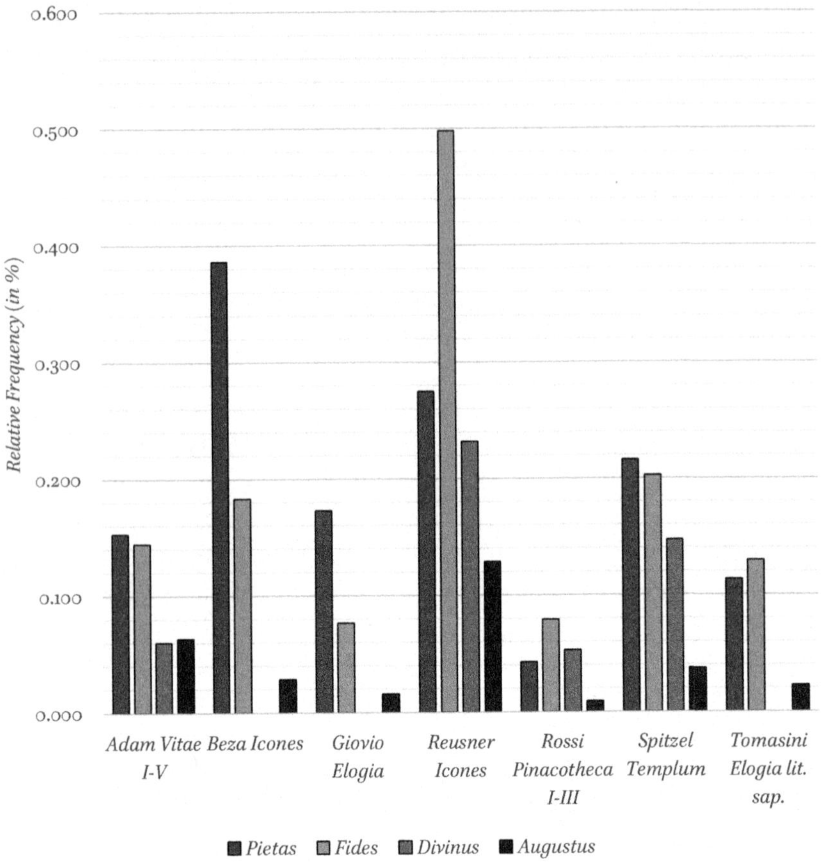


Figure 1.4. The relative word frequency (as a percentage of the total words in the corpus) of religious virtues (*pietas*, *fides*, *divinus*, and *augustus*) in different books in the corpus

total words, respectively. Effectively, this means that the *Icones* genre in the sixteenth century with a reformed twist is unique in its frequent use of the virtue of piety in its discourse. Both *Icones* tomes explicitly referred to piety and faith in their subtitles as well. The first part of the subtitle of Beza's *Icones* reads: 'true representations of illustrious men on account of both their learning and piety' ('verae imagines vivorum doctrina simul et pietate illustrium').[92] Nicolaus Reusner made the link to the Reformation even more unequivocal in his subtitle: 'representations of illustrious men of letters whose faith and doctrine, and the study of religion and *bonae litterae*, in our and our forefathers' time, especially in *Germania*, are completely

92 Theodore Beza, *Icones, id est Verae Imagines Virorum Doctrina Simul et Pietate Illustrium* [...] ([Geneva]: Jean de Laon, 1580).

restored' ('imagines virorum literis illustrium quorum fide et doctrina religionis & bonarum literarum studia, nostra patrumque memoria, in Germania praesertim, integrum sunt restituta').[93]

The Reformation had a lasting impact on the sense of history and identity of both Protestant and Catholic scholars. The *Icones* compiled by Beza and Reusner were based as much on the tradition of remembering learned men as on the cult of saints and confessional martyrology and hagiography. Lutheran and Calvinist scholars adopted the format of the lives of saints and applied it to learned men. As Helen Parish argues, 'Reformation martyrology used the lives of the early martyrs as a precedent for the suffering of the "true church" in the present.'[94] Consequently, Beza and Reusner set out to write an ecclesiastical history which envisaged the learned men and learned martyrs as those who paved the way for a long overdue reformation of the church. The ideals of martyrdom were combined with the emerging discourse on learned men set forth by Trithemius's *Cathalogus* and Fichard's *Vitae*. As such, Beza's *Icones* is an eclectic mix of contemporary Protestant propaganda, an early attempt at a collective history of the Reformation, and the general discourse on learned men.[95]

Beza and Reusner both changed learned men into the main actors of the Reformation. Beza associated piety with learning as foreshadowed in the subtitle of the *Icones*. Take, for example, this opening passage of the life of Wolgang Fabricius Capito (1478–1541), a Protestant reformer connected to Martin Luther (1483–1546) and Huldrych Zwingli (1484–1531):

> The name of Wolfgang Fabricius Capito, his singular erudition, especially in Hebrew letters, and his exceptional piety will never be allowed to die. He has, perhaps, published only few testimonies thereof, but they are brilliant, and posterity would have many more if Capito would not have preferred to confirm what he taught not by writing, but by practising.

> Wolfgangi Fabricii Capitonis nomen eruditio singularis in hebraicis praesertim literis, & eximia pietas, nunquam interire patientur: quarum testimonia pauca quidem, sed luculenta edidit, plura multo habituris

93 Reusner, *Icones sive imagines virorum literis illustrium* (1590).

94 Helen L. Parish, *Monks, Miracles and Magic: Reformation Representations of the Medieval Church* (London; New York: Routledge, 2016), p. 157.

95 Christophe Chazalon, 'Les *Icones* de Théodore de Bèze (1580): Entre mémoire et propagande', *Bibliothèque d'Humanisme et Renaissance* 66, no. 2 (2004): 359–376.

posteris, nisi quae verbis docebat, agendo potius quam scribendo confirmare maluisset.[96]

Capito excelled in his erudition of Hebrew as well as exceptional piety (*eximia pietas*). Piety, it seems, was a prerequisite virtue on the road to erudition.

Reusner presented a similar narrative in his *Icones* several years later. This can be observed perfectly in the motto adorning the portrait of the theologian and philologist Conrad Lycosthenes (1518–1561):

*If piety and the rule of divine Wisdom please you,*
*Then read the sayings of the wise, brought to life by my labour.*

*Si pietas, Sophiaeque placet tibi regula sanctae:*
*Perlege dicta Sophum, viva labore meo.*[97]

In his *vita*, Reusner pointed out that his works of philosophy and theology were brought forth 'with great faith and industry'), once again asserting link between piety and learned works.[98] In the short motto accompanying Simon Grynaeus (1493–1541), we find a similar collocation of piety and learning:

*As I am pious and learned, so I stand out in pleasant manners*
*Of a sound mind, a saintly life, and of serious honour.*

*Quam pius et doctus, tam suavis moribus exsto:*
*Mente probus, vita sanctus, honore gravis.*[99]

These short motto's speak volumes because they summarise the *vita* with clear references to one's virtues. In both the mottos of Lycothenes and Grynaeus we observe that the virtues of piety and learning were deemed essential to learned men in Reusner's *Icones* and certainly in Beza's *Icones*.

This characterisation of the learned man, pious and learned at the same time, continued to resonate in collections of *vitae*, especially in Adam's *Vitae* and Spitzel's *Templum honoris reseratum*. The terms *vir pius* and *vera pietas* occur almost exclusively in the works from the northern regions of

96 Beza, *Icones*, sig. H.j [p. 61].
97 Reusner, *Icones*, p. 98. Original italics.
98 '[M]agna cum fide et industria'. Ibid., p. 99.
99 Ibid., p. 68. Original italics.

Beza, Reusner, Adam, and Spitzel. When Adam sums up the main virtues of Philipp Nicodemus Frischlin (1547–1590), professor of poetics in Tübingen, he stresses his inheritance of piety, erudition, and courage:

> Since it was not the father Jakob or the son Nicodemus who first brought praise on account of their piety, erudition, courage, and other virtues to their paternal family, but these were largely received as a sort of inheritance from their forefathers.
>
> Non enim primus pater Iacobus, aut filius Nicodemus, laudem pietatis, eruditionis, fortitudinis atque aliarum virtutum, in paternam familiam intulerunt, sed eas a maioribus quasi hereditario acceperunt.[100]

Adam effectively brings together the martyrological virtues of piety and courage and adds the scholarly virtue of erudition. Another clear example of this is when Adam considers the merit of the printed works of the humanist theologian Johannes Rivius (1500–1553):

> In these books, his diversity of material, the weight of the arguments, and the truth of his learning, reveal the great erudition, his dedication to the name of God, and a singular evidence of his exceptional conscience. Finally, Rivius was honoured by such many gifts from God.
>
> In his libris et materiae diversitas, et argumentorum gravitas, et doctrinae veritas, ostendunt eruditionem magnam, studium nominis divini, singulare testimonium optimae conscientiae: denique quantis donis ornatus a Deo Rivius fuerit.[101]

Rivius's 'great erudition' made itself known through the weight of his arguments and the truth of his learning. In these volumes, it is piety and erudition that allow a learned man to receive both honour and fame.

The virtue of piety is not only central to the Protestant side of the confessional divide, although Protestants did tend to employ it more often when compared to collections from the Italian peninsula, such as Paolo Giovio's *Elogia* and Giacomo Rossi's *Pinacotheca*. From Figure 1.4, we can deduce that the average relative frequency of the virtue of piety in Beza's *Icones*, Reusner's

100 Adam, *Vitae* I, 'Nicodemus Frischlinus', p. 357.
101 Ibid., 'Ioannes Rivius', p. 158.

*Icones*, Adam's *Vitae*, and Spitzel's *Templum* is 0.243%.[102] In contrast, the relative frequencies of the virtue of piety in Giovio's *Elogia* (0.173%), Rossi's *Pinacotheca* (0.042%), or Tomasini's *Elogia* (0.114%) are significantly lower. The virtue of piety appeared roughly half as often with an average relative frequency of 0.110%. Despite this lower relative frequency, piety remained a central characteristic of learned men in the corpus, although it played a different role in the rhetoric and discourse of these compilers from the Italian peninsula.

Paolo Giovio's *Elogia* was a firm proponent of the idea that men of letters were just as much worthy of glory and fame as men of action, that is statesmen and military leaders.[103] As a result, his conception of virtue was closer to the original Latin meaning of 'manliness' (*vir-tus*) and 'courage', rather than to the more modern meaning of the word. Another factor that significantly changes our interpretation of his conception of piety is that Giovio's adherence was not so much to religion as to his own ecclesiastical career.[104] The collocation of erudition and piety is barely present in Giovio's eulogies.

In sum, the archetypical man of learning must possess innate brilliance, whether it comes from divine sources in Italy or a proper humanistic education in the northern regions of Europe. The Christian virtues of temperance (*temperantia*) and strength (*fortitudo*) were more popular in the northern parts of Europe, especially in the collections that presented the learned man as part of the Reformation movement. Beza and Reusner's tomes presented the man of learning as a strong participant in the conflict with Catholicism. The ideal of the man of learning was greatly influenced by the models of Reformed sanctity and saints after the Reformation.[105] Modesty and Christian virtues were central to the conceptualisation of the learned man throughout Europe, and after the Reformation the virtue of piety became a contested virtue (more on this in Chapter 2.2). However, until now we have mainly looked at the *man* of learning, and the learned *woman* has remained out of sight. The next section will show how female scholars were

102 This is an average of the relative frequencies of Beza, *Icones* (0.386%), Reusner, *Icones* (0.275%), Adam, *Vitae* (0.153%), and Spitzel, *Templum* (0.156%).

103 T. C. Price Zimmermann, *Paolo Giovio: The Historian and the Crisis of Sixteenth-Century Italy* (Princeton, NJ: Princeton University Press, 1995), pp. 271–273.

104 Ibid., p. 281.

105 Jürgen Beyer et al., eds., *Confessional Sanctity (c. 1500–c. 1800)* (Mainz: Philipp von Zabern, 2004); John Exalto, *Gereformeerde heiligen: De religieuze exempeltraditie in vroegmodern Nederland* (Nijmegen: Uitgeverij Vantilt, 2005); Robert Kolb, *For All the Saints: Changing Perceptions of Martyrdom and Sainthood in the Lutheran Reformation* (Macon, GA: Mercer University Press, 1987).

not associated with the same scholarly virtues as men. They were in fact excluded from the works on men and positioned not as peers of the men of learning, but rather as 'different' and as a class of their own.

## 1.3 The Alterity of Female Scholars

The tradition of collective life-writing traditionally focused on the lives of pre-eminent men, especially rulers, emperors, princes, and military leaders.[106] While Petrarch wrote his *De viris illustribus* (*On Illustrious Men*), his contemporary, Giovanni Boccaccio moved away from the strict focus on men and masculinity in collective life-writing with his *De mulieribus claris* (*On Bright Women*). This volume ignited the dissemination of, and attention to illustrious women, first in fourteenth- and fifteenth-century Italy, and later throughout early modern Europe. As a whole, this tradition challenged the common notion of women based on Scripture and Aristotelian teachings.[107] In the Bible, Genesis dictated that woman was created as man's helper and subject. Additionally, the Pauline Epistles provided the foundation of the idea that women were spiritually unequal to men. Aristotle's biology conveyed the sense that women were sexually vulnerable and physically frail, and therefore naturally dependent and deficient.[108] Despite this prevailing doctrine, both men and women voiced non-traditional views of women and their relations to men.[109] As a result, the collections of lives helped shape and reshape identities, social scripts, and roles for women in society.

Boccaccio's *De mulieribus claris*, composed between 1361 and 1362, is often considered to be a foundational text, starting an early modern tradition of life-writing on women.[110] This tradition of men compiling the

106 For the representation of princes in life-writing, see Patrick Baker et al., eds., *Portraying the Prince in the Renaissance: The Humanist Depiction of Rulers in Historiographical and Biographical Texts* (Berlin; Boston: Walter de Gruyter, 2016).

107 Stephen Kolsky, *The Ghost of Boccaccio: Writings on Famous Women in Renaissance Italy* (Turnhout: Brepols, 2005).

108 For an overview of the scriptural and Aristotelian doctrine on women, see Constance Jordan, *Renaissance Feminism: Literary Texts and Political Models* (Ithaca, NY; London: Cornell University Press, 1990), pp. 21–34.

109 Pamela Joseph Benson, *The Invention of the Renaissance Woman: The Challenge of Female Independence in the Literature and Thought of Italy and England* (University Park: Pennsylvania State University Press, 1992).

110 Giovanni Boccaccio, *Famous Women*, ed. and trans. Virginia Brown (Cambridge, MA: Harvard University Press, 2001).

lives of women continued due to the many copycats and evolutions in the second half of the fifteenth century and the first half of the sixteenth century, such as Vespasiano da Bisticci's *Libro delle lodi delle donne* (Book of the praise of women, c. 1479–1480), Giovanni Sabadino degli Arienti's *Gynevera de le clare donne* (Gynevera on bright women, c. 1489–1490) and Giacomo Filippo Foresti's *De plurimis claris selectisque mulieribus* (On many and selected bright women, 1497).[111] In the corpus at hand, Johannes Textor Ravisius's (1470–1542) *De memorabilibus et claris mulieribus* (On memorable and bright women), printed in 1521 in Paris, stands in that Boccaccian tradition. Ravisius's *De mulieribus* is a combination of the work of Plutarch on women titled *De claris mulieribus opusculum* (A little work on bright women) and the Augustine monk Giacomo Filippo Foresti's supplemented and revised edition of Boccaccio's work titled *De plurimis claris selectisque mulieribus* (On many bright and selected women).[112] Ravisius's *De mulieribus* is eclectic in its contents. It describes women from the Bible, such as Mary, Eve, and Magdalene; queens and poets from antiquity, such Dido and Sappho; as well as many early Christian martyrs and examples of piety. In total, the collection contains twenty-six lives written by Plutarch, and 184 originally compiled by Boccaccio and Foresti. Added to these lives are general works and epitaphs on illustrious women, as well as some lives of more recent female philosophers, poets and oratrices.

It is important to note that the adjective *clara* is difficult to translate literally, since it referred to a combination of clear, bright, renowned, and respected. Although learned men are commonly referred to as *viri docti* (learned men), learned women were collectively often referred to as *mulieres clarae* (bright women). In this section, Ravisius's collection is taken to represent the discourse on learned women in the first half of the sixteenth century, especially in relation to the lives of learned men.

In the sixteenth century, traditional values dictated that women should be examples of religion and piety, and men of social success. Desiderius Erasmus (c. 1469–1536), for example, placed high value on education for both men and women after having been impressed by Thomas More's (1478–1535) education of his daughters, although this did not mean he acknowledged

111 Kolsky, *Ghost of Boccaccio*; Carolyn James, 'In Praise of Women: Giovanni Sabadino degli Arienti's *Gynevera de le clare donne*', in *The Intellectual Dynamism of the High Middle Ages*, ed. Clare Monagle, 297–316 (Amsterdam: Amsterdam University Press, 2021).

112 Giacomo Filippo Foresti, *De plurimis claris selectisque mulieribus* (Ferrara: Lorenzo de' Rossi, 1497).

an intrinsic equality between men and women.[113] In his eulogy of Bertha van Heyen († 1490), Erasmus did not present her as an exemplary woman, but rather as an aberration and someone who attained the male virtues. In a matter-of-fact tone, Erasmus wrote:

> Certainly, there is nothing else womanly about her, if you ignore the one difference—that of gender. Why should I not with justice call her a maiden [*virgo*], since she scorned death itself, which terrifies all mortals, with a manly spirit?[114]

Through the eyes of Erasmus, Bertha van Heyen represented not an *exemplum* of a bright woman, but rather a female oddity who possessed male virtues. In general, exemplary women were often seen as a *virgo* of manly spirit.[115]

Learned women were often mythologised into genderless achievers to avoid discussions about the relation between male and female scholarship.[116] As Lisa Jardine explained, the humanist Politian (1454–1494) could not bestow praise on Cassandra Fedele (1465–1558) as a woman: 'In a period which afforded no power to a woman in her own right, a woman's achievement in a sphere which supposedly stood in some active relation to power could not be allowed to stand *as* woman's achievement.'[117] In a similar

113 Erika Rummel, ed., *Erasmus on Women* (Toronto; Buffalo, NY; London: University of Toronto Press, 1996), pp. 9–10.

114 Ibid., p. 239. The full passage reads: 'Furthermore, the frailty of the weaker sex is an addition to the full measure of her glory. For who would not deem these characteristics, which we consider worthy of admiration in men accustomed to warfare from the very cradle (as they say) or in philosophers with their universal knowledge, even more glorious in a woman? Who will be able to find a woman who has such courage? Such a woman has definitely been found—unless perhaps we think she should be termed a maiden rather than a woman. Certainly, there is nothing else womanly about her, if you ignore the one difference—that of gender.'

115 See also Christine Battersby, *Gender and Genius: Towards a Feminist Aesthetics* (Bloomington: Indiana University Press, 1989).

116 I use the word *gender* rather than *sex* to denote the cultural identities that established ideas of male and female, since I refer here to social and cultural differences rather than biological ones. The precarious position of male gender roles in the male/female dichotomy is often referred to as 'anxious' or 'fragile' masculinity. This is because any corrosion of the male/female gender dichotomy incites a decrease of male dominance. This means that commentary on the gender dichotomy often remained unaddressed in the early modern period as well, since it destabilised the prevalent view of male dominance and outstanding virtue. See, e.g., Mark Breitenberg, *Anxious Masculinity in Early Modern England* (Cambridge: Cambridge University Press, 1996).

117 Lisa Jardine, '"O decus Italiae virgo", *or* The Myth of the Learned Lady in the Renaissance', *The Historical Journal* 28, no. 4 (1985): 799–819 (p. 804).

way, a century later, Anna Maria van Schurman (1607–1678) was largely praised on account of her learning and mastery of language, but she was also characterised by learned men as a wonder of nature (*monstrum naturae*) or a miracle (*miraculum*).[118] Throughout the early modern period, virtue, or *virtus* in Latin, was considered to be the quality of a man (*vir*) obtained through a humanist education, making it unattainable for women *as* women. In praise by men, female scholars were often considered eccentric, sexless, or divine muses.

However, scholarship was carried out by women, and they did have a sense of community. Laura Cereta (1469–1499), for example, considered herself to be part of a commonwealth of women, a *muliebris respublica*.[119] In this period, noble and patrician daughters were educated by fathers and teachers in a humanist tradition with a focus on literature, history, poetry, and philosophy, although they were mostly shunned and rejected by universities.[120] In a similar vein, Carol Pal has argued that, although female scholars were often described as oddities and miracles, a respected and functional community of female scholars existed in the seventeenth century.[121] This means that the discourse on learned women did not accurately reflect their societal position and achievements, but can more fruitfully be seen as a way of 'othering' women to strengthen the narrative

118 Mirjam de Baar and Brita Rang, 'Minerva of savante, heilige of dweepster: Receptiegeschiedenis van Anna Maria van Schurman sedert de zeventiende eeuw', in *Anna Maria van Schurman (1607–1678): Een uitzonderlijk geleerde vrouw*, ed. De Baar and Rang, 9–28 (Zutphen: Walburg Pers, 1992), p. 12.

119 In a letter to Bibolo Semproni (likely a fictional character for the sake of polemics), dated 13 January 1488, Laura Cereta forcefully argued against the suppression of women: 'I shall not fail to obstruct tenaciously their treacherous snares [of 'men who are garrulous and puffed up with false pride']. And I shall strive in a war of vengeance against the notorious abuse of those who fill everything with noise, since armed with such abuse, certain insane and infamous men bark and bare their teeth in vicious wrath at the republic of women [*muliebris respublica*], so worthy of veneration.' Laura Cereta, *Collected Letters of a Renaissance Feminist*, trans. and ed. Diana Robin (Chicago; London: University of Chicago Press, 1997), p. 80. I have translated *muliebris respublica* here as 'commonwealth of women'.

120 On humanist female scholars, see Margaret L. King, 'Book-Lined Cells: Women and Humanism in the Early Italian Renaissance', in *Beyond Their Sex: Learned Women of the European Past*, ed. Patricia H. Labalme, 66–90 (New York; London: New York University Press, 1980), esp. pp. 71–75. On the participation of female students in Italian universities, see Paul Oskar Kristeller, 'Learned Women of Early Modern Italy: Humanists and University Scholars', in ibid., pp. 91–106, esp. pp. 99–106. More generally on female education, see Merry E. Wiesner-Hanks, *Women and Gender in Early Modern Europe*, 4th ed. (Cambridge: Cambridge University Press, 2019), ch. 4, 'Learning and Letters', pp. 159–194.

121 Carol Pal, *Republic of Women: Rethinking the Republic of Letters in the Seventeenth Century* (Cambridge: Cambridge University Press, 2012).

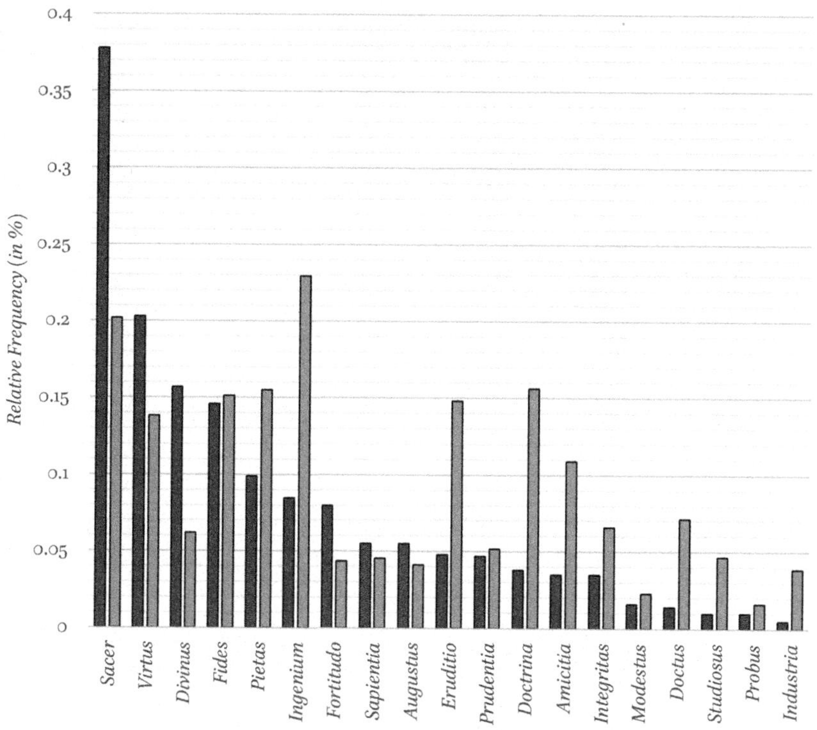


Figure 1.5. Use of virtues in Ravisius's *De claris mulieribus* (1521) (in black) compared to the average of the rest of the corpus on learned men (in grey)

of masculine virtue. As such, the discourse does give us an insight into how learned men negotiated and considered the role of learned women in the learned world, especially in relation to their own role.

When we look at the discourse in Ravisius's *De mulieribus*, we have to take into consideration that the texts he collected date back to the fourteenth and fifteenth centuries in the wake of Boccaccio's influential *De claris mulieribus*. Ravisius collected learned men's views about women which he deemed to be relevant in 1521. In the rest of the corpus, the inclusion of women in the collections of *vitae* is almost non-existent, with a few incidental exceptions. This attests to the fact that learned men considered learned women to be in a category of their own, not to be compared with and included in works on masculine scholarly virtue. That is why this analysis of the discourse in Ravisius is a study of how the alterity of learned women was constructed in a world where the male scholarly archetype was the norm.

The language used in Ravisius's *De claris mulieribus* exemplifies the discourse of alterity to describe learned women. The praise women received focused on their social and religious virtues rather than on the intellectual merit, innate brilliance, or excellent work their male counterparts were praised for. Figure 1.5 shows the relative frequency of virtues in Ravisius's tome compared to the average relative frequency in the works in the corpus on learned men. In general, Ravisius's collection stressed sanctity, virtue, divinity, faith, and piety. The relatively high frequency of the word 'virtue' may seem ironic given its connotation of manliness, although the word is used more broadly in Ravisius's text to praise the esteem of virtue by bright women. By comparing the results with the results from the rest of corpus, the virtues typically associated with male scholars—such as innate brilliance (*ingenium*), erudition (*eruditio*), and learning (*doctrina*)— are noticeably absent.

Let us first take a closer look at the virtues that are overly represented, that is, at the extremely high relative frequency of the words *sacra* (sacred) and *divina* (divine) compared to the rest of the corpus.[122] It must be noted, however, that the use of the word 'sacra' (and all its possible declinations) was used in very different contexts throughout Ravisius's text. The most common collocation is that of 'sacra virgo' ('sacred virgin')—a reference to the virgin Mary—that was commonly attributed to pious learned women. The term *sacred* is universally applied to martyrs, queens, and bright women, usually followed by their name and marital status. For example, Brigid of Scotland is referred to as 'sancta Brigida virgo Scotica'.[123] The adjective 'sacra' does not seem to have a profoundly religious meaning, since in the very first chapter, the Biblical Mary is referred to as 'the perpetual virgin Mary' ('Maria virgo perpetua') without the adjective *holy* or *saintly*.[124] Others are described as 'nobilissima' (the most noble) or 'diva Catharina' (divine Catharina). Ginevra Sforza (1440–1507) is referred to as the wife of Giovanni Bentivoglio (1443–1508) and Cassandra Fedele is simply called a 'Venetian virgin, orator and philosopher'.[125] Regardless of such deviations, the most common way to refer to learned women was as a 'sancta virgo' (sacred virgin/maid). In this context, *sacred* referred to three different aspects of a bright woman: dedication to religion, divinity, and finally moral and spiritual excellence. Other instances of the word *sacra* are related to the holy scriptures (*sacrae*

122 I translate *sacra* here as sacred, rather than holy, since in my view holiness has strong Christian overtones and relations to excellence, dominance, and hierarchy, while the word sacred can be more universally applied.

123 Johannes Ravisius, *De memorabilibus et claris mulieribus* [...] (Paris: Simon de Colines, 1521), fols. 123$^{r}$–124$^{v}$.

124 The title of the chapter reads: 'De Maria virgine perpetua, deique genitrice gloriosissima, coelorum atque angelorum regina'. Ibid., fols. 16$^{v}$–27$^{r}$.

125 'De Cassandra Fideli, Veneta virgine, oratrice & philosopho [*sic*]'. Ibid., fols. 156$^{r}$–157$^{v}$.

*literae*), actual saints, Christ, and holy sacraments. Regardless of how the word *sacra* was used in the text, bright women were more closely associated with divinity and sacredness either directly with the use of the specific adjective or indirectly with references to sainthood, sacraments, and sacred texts.

We do not find the quintessential narrative of the development of *ingenium* for learned men in the *vitae* of bright women. The essential virtues for learned men such as *ingenium* and *doctrina* appear three and five times less frequently, and other words associated with learned men (such as *doctus*, *amicitia*, or *erudition*) are largely absent, too, as can be seen in Figure 1.5. When we take a look at the keyword analysis, the characteristic absence of the words 'doctrina', 'studia', and 'opera' confirm the suspicion that the standard life path for a male scholar (from *ingenium* to polishment by *doctrina* to works of erudition) does not apply and was not used as a model for a female life or *exemplum*. Learned women were not judged by their ability to study or work (*operam dare*). We see this also reflected in the virtue of *industria* (productivity and diligence), which is almost eight times less likely to be associated with learned women compared to men. Religiosity and familiarity with sacred texts was stressed in the developmental phase of the lives of bright women instead of humanistic learning. Moreover, most of the bright women were martyrs of some sort and excelled in their perseverance and adherence to the word of God. Juliana of Nicomedia (c. 285–c. 304), for example, was praiseworthy because of her religiosity and adherence to the Christian faith:

> For a few days she took the time for prayer, and while frequently visiting the house of God, she listened to the holy scriptures with an incredible intensity. When the days came that the marriages were being settled, Juliana denied wanting to be wedded to a man without a reason, unless he himself would first acknowledge the father, the son, and the holy spirit.
>
> Unde singulis diebus orationi vacabat, deique templum frequentando sacras scripturas incredibili ardore auscultabat. Adventantibus igitur diebus quibus nuptiae conficerentur, negavit Iuliana nulla ratione velle sponso coniungi, nisi ipse prius patrem & filium, & spiritum sanctum confiteretur.[126]

It is not surprising that her life as a martyr ends in a dungeon and fiery death. What makes her a bright woman, however, is her 'unique devotion' in the

126 Ravisius, *De mulieribus*, fol. 82$^{r}$.

face of adversity.[127] While this seems to be a typical narrative of martyrdom, it is important to keep in mind that her exclusive devotion to God, and thus not to marriage, is a general aspect of a bright woman. Although this topos is particularly visible in the lives of intellectually gifted martyrs, bright women who were not destined for martyrdom were also characterised by an adherence to God and the unmarried state through devotion.

The very high relative frequency of the word *virgo* (virgin) attests to the centrality of the unmarried state of bright women. In general, early modern women were defined by their marital status, so they were described as a virgin (*virgo*), married woman (*uxor* or *mulier*), or a widow (*vidua*). Learned women's unmarried status was a clear and important marker of their literal and social integrity and allowed them a place in the learned world. Additionally, women are often characterised by a tension between female youth and a mature wisdom in Ravisius's text. For example, Agnes of Rome's (c. 291–c. 304) *vita* describes this patron saint of girls and chastity by referring to this dichotomy:

> Her infancy was counted in years, yet she was very old in mind, with the body of some young girl, but with a grey [i.e., old] soul; with a beautiful face, but even more beautiful of faith. So large was her faith from her childhood onwards that she believed without a doubt that she would be able to fulfil the things God had promised. For this reason, she placed her hope in the future. For her very faith put the universally invisible thing to her before the eyes, and ignoring the visible things of this world, she decided to approach God with trust.

> Infantia eius computabatur in annis: sed erat senectus mentis immensa: corpore quidem iuvencula, sed animo cana, pulchra facie, sed pulchrior fide. Tanta enim ab ipsa pueritia fuit eius fides, ut crederet indubitato q[uod] deus ea quae promiserat, potens foret ut adimpleret: & iccirco spem suam in futurum transtulit. Ipsa enim fides universa invisibilia eidem, iamiam ante oculos posuit, & universa huius mundi deferens, cum fiducia deo se proximare constituit.[128]

She is portrayed as an embodiment of a *virgo* in possession of an old mind. The *vita* praises her external beauty, but her faith even more. The resemblance with the *ingenium* narrative for learned men is striking, yet here it concerns

127 Ibid., fols. 82$^{v}$–83$^{r}$.
128 Ibid., fols. 77$^{v}$–78$^{r}$.

the development of *fides* (faith). Faith came to her as a gift from God giving her fulfilment and a purpose. These female *exempla* of Juliana of Nicomedia and Agnes of Rome show how important martyrdom and virginity were for the archetypical bright woman. Intellectual curiosity was often thought of as immoral and dangerous, and women were therefore dissuaded from engaging in intellectual endeavours. The only way to avoid their immoral and uncurbed curious ventures was to enact a strict social scheme of decorum and restrain.[129] The majority of female examples in Ravisius's work are early Christian examples and the infrequent accounts of contemporary female scholars are placed in a female Christian tradition of piety.

The education of women usually consisted of a form of moral education incorporating the use of *exempla*, such as martyrs. The *vita* of Cassandra Fedele (1465–1558) is a good example of how female learned women were generally described:

> She is bright on account of her studies of philosophy and theology, for through her sharp and very prompt disputations she brings great honour to the female sex as well as to the town of Venice. And for that reason, she is addressed by many pre-eminent people, in both poetry and letters, in this esteem of virtue, for which she is much celebrated. Since she was endowed with very good manners as a young woman, and graced with a beautiful appearance, and not unaware of or without (as said above) Latin speech, which clearly seems to bring to many disciplines not little honour. All are in awe and derive great pleasure from all the learning [*literatura*] of the girl. Her conversation is cultured and easy, while she excels in the gravity of her modesty.

> Est clara philosophiae ac theologiae studijs: acutis enim & promptissimis disputationibus, generi muliebri & orbi Veneto magnum affert decus. Et idcirco a multis praestantissimis, & versibus atque epistolis salutatur, hac opinione virtutis, qua maxime celebratur. Haec namque cum sit adolescens optimis moribus praedita, honestaque forma decorata, non ignaraque aut expers (ut praemissum est) romani sermonis, qui videlicet omnibus disciplinis non parvum decus afferre videtur. omnes mirantur, magnamque ex tanta puellae literatura voluptatem capiunt. Est & illi sermo perhumanus & facilis, ac omni gravitate modestiae excellens.[130]

129 Neil Kenny, *The Uses of Curiosity in Early Modern France and Germany* (Oxford; New York: Oxford University Press, 2004), pp. 386–395.
130 Ravisius, *De mulieribus*, fol. 156$^{v}$.

Fedele is presented as a beautiful curiosity of learning, which contrasts starkly with the archetypical narrative for learned men in two ways. First, there is a strong emphasis on the importance of modesty and good manners. Secondly, the reaction of peers is not one of empathy, where they relate to her on the same level, but rather one of sympathy, which stresses her alterity since they react with *awe* as well as pleasure derived from her learning. It should be said here, however, that the word *literatura* could be translated as both 'reading' and as 'learning', but it is likely that here it refers to her learning in general. Fedele is presented as the quintessential bright woman: modest, awe-inspiring, cultured, and respected by pre-eminent men and women. Her learning and beauty are not presented as a source of inspiration or hard work, but rather as something entertaining and an adornment of the Venetian state.

Apart from martyrs and female *exempla* of modest learning, monasteries are also considered as places for the instruction of female learning and virtue. In a chapter on the ideal scholastic virgin originally written by the Giacomo Filippo Foresti, the moral implications of these lives become apparent.

> Then, she taught also to crush the adversities to be endured in our age, and the storms of this raging sea, with all their devilish crafts and disturbances of demons, through the examples of the holy martyrs. Furthermore, she formed their minds, by being obedient to holy patriarchs and prophets. She herself, however, always held the Book of Gospels in her hand like a mirror.
>
> Cuncta denique huius saeculi adversa perferenda, et huius procellosi maris tempestates, cum omnibus diabolicis versutiis, et infestationibus daemonum, per exempla sanctorum martyrum etiam docebat conculcare. Informabat insuper earum mentem, ex sanctorum patriarcharum prophetarumque obedientia. Ipsa autem quum evangeliorum codicem semper in manibus velut speculum teneret.[131]

Foresti presents the scholastic virgins as living in an age of adversity, withering storms and devilish powers. They became successful through their continuous moral education based on the *exempla* of martyrs. Foresti stresses that spiritual women should study moral examples and obey their patriarchs. At the end we find a trope that will persist throughout the early modern period: women should hold the Bible close for spiritual as well as moral guidance.

131 Ibid., cap. cxxxvi, 'De sancta Scholastica virgine', fols. 122$^{r}$–123$^{r}$.

The eclectic discourse in Ravisius's *De memorabilibus et claris mulieribus* reveals that women were not considered learned like their male counterparts. While men of learning were referred to as *viri docti* (learned men), the term *clarae mulieres* (bright women) was used as a heading for exemplary women of all calibres. The sixteenth century saw the rise of the idea of the 'learned man', with his ingenuity, learning, and erudition. The meaning of the category of 'bright woman' was not constructed in relation or opposition to the category of the 'learned man'. Rather, the former built on centuries of archetypical exemplary women with desirable female traits: religiosity, sanctity, faith, virginity, and docility. When bright women were exhibiting virtues that were traditionally considered masculine, they were praised not *as* women, but rather as a curiosity or an example *despite* their sex. While the early modern period certainly saw a plethora of learned women, their achievements and place in the learned world was not seen in opposition or relation to the male world of learning, but rather in a succession of female *exempla* such as saints, martyrs, and spiritual examples. Thus, the female community was imagined to be more deeply historical than the male community, or at least with comparatively longer chronology as opposed to greater synchronicity. For learned men to praise these women as miracles and examples of faith and modesty was the highest position they could attain. Through these processes of identity formation, women were excluded from participating *as* women in the learned world on an equal level with men.

## 1.4 Conclusion

Collections of *vitae* and *elogia* are rich sources to analyse the collective representation of scholarly identity. In comparison with studies of *vitae* that used close reading methods, the digital text analysis—also taken as a form of distant reading in this chapter—allows us to see larger trends and recurring themes and tropes. This digital textual analysis has acted as a divining rod to point to passages where the tropes and themes that were relevant for the present study could be closely read and analysed. This novel method of analysing historical texts has helped identify archetypical ways of describing the life of a scholar throughout early modern Europe.

The most common virtues and characteristics that were associated with scholars are a combination of moral, social, and epistemic virtues. The essential characteristics of a man of learning are *ingenium* (innate brilliance), *doctrina* (learning), *pietas* (piety), *fides* (faith), *eruditio* (erudition), and *amicitia* (friendship). The virtues of *ingenium*, *doctrina*, and *eruditio* were often used in a context of the development and coming of age of a man of letters. First, he should possess innate brilliance, to be developed and polished into learning and erudition. Christian and cardinal virtues such as piety, prudence, wisdom, and courage were also widely used in northern and southern European compendia of *vitae*. Additionally, late sixteenth-century Reformed collections—especially the *Icones* collections of Beza and Reusner—stress courage and strength as virtues. Because of their influential books, the learned man was increasingly seen as a courageous and pious actor who adhered to the 'true faith', despite challenges and suffering.

The tradition of life-writing was strongly gendered: there were books of *men* of learning as well as those featuring *women* of learning. Female scholars were excluded from the male world of learning, because they were not assessed and talked about in the same language as their male counterparts. Carol Pal has argued for the existence of a female Republic of Letters, which she terms a Republic of Women.[132] The present chapter builds on Pal's research on the praxis of learned women and suggests that female scholars were represented in a female succession and historical community of pious and religious female *exempla*. Female scholars were not imagined in relation to the male learned world and had their own collective identity and sense of history. There were plenty of learned women in the early modern period, yet there were hardly any women in the representations of the learned world (such as the collections of *vitae*), because female scholars were excluded on the basis of identity and the alleged inability of women to attain male scholarly virtues.

Scholarly identity always had to be imagined in a local community. The collections of *vitae* appealed to a European audience, while the ideals of a scholar became more diversified throughout Europe. Local imaginings of the Republic of Letters in the form of collections of *vitae* appealed to a universal ideal of scholarship. That ideal, however, changed considerably throughout the early modern period and depended on geography. The next chapter therefore examines the geographical spread of the content of the *vitae*, and the chronological development of the ideal of the learned man. The symbol of the Republic of Letters—the universal man of learning—and a collective history of the advancement of learning was appropriated and amended by specific learned communities. Quite ironically, the Republic of Letters persisted because learned men all over Europe continued to appropriate its ideal and scholarly virtues and adjusted them to national,

132 Pal, *Republic of Women*.

regional, or even local contexts. Patriotic and confessional imaginings of the Republic of Letters kept the universal ideals afloat. The next chapter continues to focus on collections of *vitae* but looks at the different selection strategies and selection criteria that determined whose *vitae* deserved a place in a compendium. Rather than asking which virtues were associated with scholars, it poses the question: who was remembered where?

# 2. Collective History and Geographical Inclusion in *Vitae* and *Elogia*

**Abstract:** This chapter explores how early modern biographies and commemorations (*vitae* and *elogia*) created a collective history of scholars throughout Europe. It highlights the importance of geography, showing how scholarly identity was shaped by local, national, and transnational factors. The chapter examines how scholars and chroniclers negotiated regional identities alongside the ideal of a universal Republic of Letters. Key examples demonstrate how confessional and political shifts influenced the inclusion of certain scholars, revealing that the concept of a pan-European scholarly community was sustained, even as national identities emerged.

**Keywords:** Digital text analysis, scholarly commemoration, life-writing, national consciousness

> Who knows their own story? Certainly, it makes no sense when we are living in the midst of it. It's all just glamour and confusion. It only becomes a story when we tell it and retell it. Our small precious recollections that we speak again and again to ourselves or to others. First, creating the narrative of our lives, and then keeping the story from dissolving into darkness.[1]
>
> – Nick Cave

With the publication of collections of *vitae* and eulogies, early modern scholars presented not only their sense of a shared past, but also their perception of the learned world, its main members, and its history. By telling and retelling the lives of exemplary scholars in these collections, and cherishing those histories, these writers imbued readers with a sense

1 Nick Cave in *20,000 Days on Earth*, directed by Iain Forsyth and Jane Pollard (Corniche Pictures and British Film Institute, 2014).

Scholten, Koen. *Rethinking the Republic of Letters: Memory and Identity in Early Modern Learned Communities*. Taylor & Francis Group, 2025
DOI 10.5117/9789048559855_CH02

of a shared goal and a shared past. Their volumes offer a glimpse into how early modern compilers imagined the learned world, showing who were thought of as its central members, and to what extent these imaginations of the learned world differed and changed throughout early modern Europe. My diachronic and geographical approach in this chapter aims to answer the question: who is remembering whom and where in these texts? We will assess the content of around twenty collections of lives and eulogies from the sixteenth century up to the early eighteenth century, which were published throughout Europe.

The most straightforward way to gauge a compiler's sense of the learned world is to look at the geographical spread of the learned men in the compilations. Almost every *vita* or eulogy mentioned the *patria*, or 'fatherland', of the learned man in question. The notions of state and *patria* were varied and changed considerably throughout the early modern period.[2] Additionally, political and confessional borders shifted constantly.[3] From a modern standpoint, thinking in nations and in clear-cut borders between nations is self-explanatory, but this is not the case for early modern states and their political and cultural borders. What in an early volume from the sixteenth century could be referred to as *Belgium* could become *Batavia* or *Hollandia* in a later collection. Accuracy is even more problematic: one collection could mention *Italia* or *Germania* as a scholar's *patria*, while others specifically mention a state, such as the Tuscan or Venetian state, or a city such as Augsburg or Basel. Others mention regions, such as Lower Saxony, along the Rhine, or even more vaguely described territories, such as 'north of the Arno'. Up until the seventeenth century, most of the *patriae* in the *vitae* refer to the place or region where one was from in a neutral and apolitical manner. However, later the term *patria* could refer to a political state with a shared history and identity. This is visible, for example, in the increasing frequency of expressions such as *nostra patria* ('our homeland') and *Germania nostra* ('our German lands').[4]

To be able to meaningfully quantify the *patriae* in the texts, I have used the geographical indicators that were employed in the *vitae* themselves

2 See, e.g., N. C. F. van Sas, ed., *Vaderland: Een geschiedenis van de vijftiende eeuw tot 1940* (Amsterdam: Amsterdam University Press, 1999).

3 See, e.g., Benjamin J. Kaplan, *Divided by Faith: Religious Conflict and the Practice of Toleration in Early Modern Europe* (Cambridge, MA; London: Belknap Press of Harvard University Press, 2007), esp. pp. 150–152.

4 This seems to be a development in northern Europe; see, e.g., Mieke B. Smits-Veldt, '"Het vaderland" bij Hollandse rederijkers, circa 1580–1625: grondgebied en identiteit', in Van Sas, *Vaderland*, pp. 83–107.

to represent the general early modern geographical indicators: *Germania* (including *Bavaria*), *Italia* (the entire peninsula), *Helvetia* (Swiss confederacy, including Geneva), *Polonia* (kingdom of Poland), *Hungaria*, *Prussia*, *Austria* (including *Bohemia*, Archduchy of Austria, Tyrol, *Carinthia*, *Moravia* and *Silesia*), *Anglia* (sometimes also *Britannia*), *Scotia*, *Gallia*, *Dania*, *Hispania*, *Frisia*, *Batavia* (the northern Netherlands or the Dutch Republic, including *Hollandia*), *Belgium* (the southern Netherlands or Spanish Netherlands), *Graecia* (or more generally *Byzantium*), unless mentioned otherwise. These early modern categories are largely remnants from a classical Roman geographical and political imagination. A stated *patria* such as *Francia orientalis* or *Westphalia*, but also terms such as *Germanicus* have all been standardised to *Germania*, referring to the German lands in general. This means that detail is lost in the geographical accuracy, but it does enable generalisation and a comparative analysis between the different collections.

This chapter will analyse a variety of collections, mainly from the German lands and the Italian peninsula. In four sections, it will position these collections of *vitae* in their respective cultural context in chronological order. The foundational works of the sixteenth century, especially by Paolo Giovio (1483–1552), were seen as a model upon which other compilers built for at least a century. Section 2.1 will explore the early printed compilations of learned men made by Paolo Giovio and Johannes Fichard (1512–1581). In the early sixteenth century, these compilations imagine the learned world as a transnational and transhistorical community. Section 2.2 considers how *vitae* of learned men became instruments to forge a distinct identity in the wake of the Reformation. The works of Nicolaus Reusner (1545–1602) and Theodore Beza (1519–1605) reflect these contemporary attempts to rewrite and reimagine the learned community as well as its collective history. Section 2.3 moves into the seventeenth century and addresses the trend of praising learned men in the context of patriotic and civic pride. Throughout Europe, the learned man was increasingly seen as an adornment of the state who could bestow fame on a 'nation' on account of his worldliness. Finally, Section 2.4 looks at the developments of the lives of learned men at the end of the seventeenth century and on into the eighteenth century. At this point in time, the learned man became a cultural icon. Popular editions offered an overview of the great scholars of yore to a wider audience. The point of this chapter is not to imply a chronological development of *vitae*, but rather to show how the figure of the learned man was appropriated in different contexts and parts of Europe. The imagination of the learned world took on different forms of history and identity in response to political and cultural events and sensibilities.

## 2.1 Early Ideals of a Transnational Learned Community

While it is hard to pinpoint when the tradition of chronicling the lives of learned men and women began, printed examples of collective scholarly life-writing began to flourish in the sixteenth century and were largely inspired by the chronicling of illustrious men and women in the Italian Renaissance, as well as Roman antiquity (see Chapter 1.1). Two volumes stand out in the revival and appropriation of this tradition in the sixteenth century: Johannes Fichard's *Virorum Vitae* (Frankfurt a.M., 1536), and Paolo Giovio's *Elogia virorum literis illustrium* (Basel, 1577). These collections have been interpreted as the foundational works for the increasing number of collections of *vitae* that followed.[5] Many later volumes refer to both Fichard and especially Giovio as revivers of a classical genre (as explained in the previous chapter). These works cemented the early modern tradition of chronicling learned men with their portraits and lives.

Johannes Fichard (also Johann von Fichard, 1512–1581), an eminent Frankfurt jurist and politician, gave his slim volume of eighteen lives of learned men the title *Virorum qui superiori nostroque seculo eruditione et doctrina illustres atque memorabiles fuerunt, Vitae* (Lives of men who during preceding and our centuries were illustrious and memorable on account of their erudition and learning).[6] This volume was printed in quarto format in Frankfurt in 1536 and became a model for later collections of *vitae*. In the preface, written as a letter to Georg Furster, Fichard elaborated on the necessity of compiling the lives of learned men and subsequently transmitting such 'monuments of learning' (*monumenta literarum*). The value of such a history is to teach conduct.

> Those highest and most prudent men have therefore observed that history is clearly the teacher of life and conduct. The minds of men can overall not be persuaded more suitably in another way when it comes to common interests, or otherwise be stimulated better than when they have to read examples of other superior men, and when these are as it were placed before their eyes to look at.

5 Irena Backus, 'What is a Historical Account? Religious Biography and the Reformation's Break with the Middle Ages', *Archiv für Reformationsgeschichte* 101, no. 1 (2013): 289–304.

6 Johannes Fichard, *Virorum qui superiori nostroque seculo eruditione et doctrina illustres atque memorabiles fuerunt, Vitae* [...] (Frankfurt a.M.: Christianus Egenolphus, 1536). For a short biography of Fichard, see Heinz F. Friederichs, 'Johann von Fichard', *NDB*, vol. 5 (1961), p. 120.

> Viderunt enim viri summi ac prudentissimi illi, historiam plane magistram vitae ac morum esse, hominumque animos non alio pacto de rebus communibus, in omnem partem commodius admoneri, aut alioqui excitari magis posse, quam si iis aliorum ac superiorum exempla legenda & quasi ob oculos spectanda exhiberentur.[7]

In a typical early modern obsession with honour and conduct outlined in the previous chapter, Fichard believed history could teach moral lessons and strengthen the readers' sense of purpose; here he drew on the Ciceronian concept of *historia magistra vitae* and Livy's history.[8] As Fichard reminded his epistolary friend, to be successful one needed to take into consideration the learned men of old and of the present.

> The dignity of all history is amongst all writings and letters by far the best and most outstanding genre. This was not only the opinion of the learned and wise authors in Antiquity, but also held always very true in our time and is confirmed by experience. Likewise, I [Fichard] trust that this labour of ours will be most pleasing to you [Georg Furster], who himself also takes utter delight in learning about this variety of things and examples not only from antiquity but also from today, but also that it will bring some of the highest pleasure and utility for all others who burn with the same desire, affected by the same zeal.

> Atque ut omnis historiae dignitas, inter omnes scriptorum ac literarum genus longe maximum atque praestantissimum est, idque non modo veteres docti ac sapientes senserunt, verum etiam nostra aetate pro verissimo semper habitum, usuque confirmatum est, ita sane confido hanc nostram operam cum tibi, qui ipse quoque in hac rerum atque exemplorum non modo veterum rerum etiam recentium varietate cognoscenda, cum primis delectaris, quam maxime gratam fore, tum vero caeteris omnibus, eodem studio eodemque desyderio incensis & affectis, summam quandam & voluptatem & utilitatem allaturum esse.[9]

Fichard saw his *Vitae* as a history of excellent men of learning in the Plutarchian tradition of moral life-writing. This is somewhat remarkable, since

7 Fichard, *Vitae*, 'Praefatio', sig. a-iij$^{v}$.

8 See Cicero, *De oratore*, 2.36; Livy, *Praefatio*, reads: 'exempli documenta in inlustri posita monumento intueri'.

9 Fichard, *Vitae*, 'Praefatio', sig. a-iij$^{v}$.

the book is a somewhat makeshift compilation of *vitae* of learned man written by other learned men without notes or commentary. It contains short *vitae*, some a mere two pages, but also lengthy expositions that can be up to forty pages.[10] Fichard included a wide-ranging selection of lives, such as Gerolamo Squarzafico's (1501–1576) life of Petrarch, as well as the lives of John Colet (1467–1519) and Thomas More (1478–1535) written by Desiderius Erasmus (1466–1536).

Seen as a whole, the book offers a coherent historical narrative of the development of the European learned world. The origin of erudition started in Italy with luminary scholars such as Petrarch and Pope Pius II (Enea Silvio Bartolomeo Piccolomini, 1405–1464). This learning then spread to the north. First, to *Germania* and *Frisia*, where excellent men like Rudolphus Agricola (1443/1444–1485) and Wessel Gansfort (1419–1489) picked up where the Italian humanists left off at the end of the fifteenth century. Second, these teachings flourished in the works of the humanist scholar Christophe de Longueil (Longolius, 1490–1522) and theologian Johannes Oecolampadius (1482–1531) in *Germania*, and in the works of John Colet and Thomas More in *Anglia*. Fichard's collective portrait of illustrious men on account of their erudition and learning thus provided an expansive view of the learned world, both in terms of geography and fields of knowledge. Since Fichard included theologians, churchmen, poets, philosophers, and humanist writers from three parts of Europe (*Italia, Germania,* and *Anglia*), the volume impressed its readers with a European and multidisciplinary vista of the learned world. As such, Fichard's collection of lives is a representation of a learned world that is European as well as pluri-confessional. Fichard conflated the geography of the humanists, neatly putting them in three categories in the table of contents: *Itali, Germani,* and *Angli*. As can be seen in Figure 2.1, the respective *patriae* of the included scholars reflect Fichard's idea that scholarship emerged in Italy and spread to the north.

Fichard was ambivalent with regard to rising religious tensions, and generally avoided explicit or implicit references to a 'confessional other', Italians, or southerners in general. It is worth mentioning that Fichard did not include lives of reformers such as Martin Luther (1483–1546) or Huldrych Zwingli (1484–1531), but made room for Erasmus, whom he had personally met in Basel, and Pope Pius II. Fichard appears to be impartial about confession, tending to select men on the basis of their scholarly merit,

10 The shortest *vitae* are two or three pages long: 'Wesselus Gansfortius', fols. $87^{v}$–$88^{r}$, and 'Iacobus Papiensis Cardinalis', fols. $80^{r}$–$81^{r}$. The longest *vitae* are thirty-two and forty pages long: 'Pius II, Pont. Max', fols. $10^{v}$–$26^{v}$, and 'Ioannes Ant. Campanus', fols. $27^{v}$–$48^{r}$.

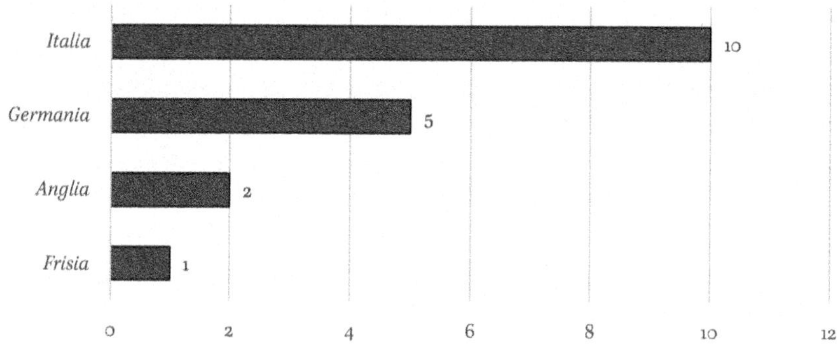


Figure 2.1. The affinity to region and state in Johannes Fichard, *Virorum qui superiori nostroque seculo eruditione et doctrina illustres atque memorabiles fuerunt, Vitae* (Frankfurt, 1536)

ending up with a collection of humanists without provocative reformers.[11] It is reasonable to interpret Fichard's *Vitae* as the start of a genre of scholarly collective biography, as Irena Backus has done.[12] Although Fichard's sense of history and a pan-European learned community (knowledge from the Italian peninsula that spread to *Germania*) would underpin many collections of *vitae*, Fichard's inclusiveness and geographical balance would prove to be more an outlier than the norm.

The second volume that deserves attention is Paolo Giovio's *Elogia*. Giovio was a pre-eminent humanist of the Italian peninsula, well known for his histories and other writings. He wrote numerous lives of popes, military leaders, and learned men and was thus a chronicler of the virtuous behaviour of many Italian Renaissance figures.[13] Giovio started collecting portraits around 1531 and he acquired a villa in Borgo Vico near Lake Como to house his portrait museum with pictures of many great and illustrious men and women, which he referred to as his temple of virtue (*templum virtutis*).[14]

11 Irena Backus argues that Fichard must have been in favour of moderate reform on the basis of his choice of lives. However, since Fichard expresses no such value judgments and did not explain his criteria for inclusion and exclusion of learned men, this claim is difficult to substantiate directly from the sources. See Backus, 'What is a Historical Account?', p. 295.

12 Irena Backus argues that Fichard's *Vitae* 'provided a model of how to choose the best representatives of humanism, including Christian humanism, from the North and from the South, so as to combine them in a collective portrait of best churchmen, theologians, poets, rhetoricians, philosophers, and so on.' Backus, *Life Writing in Reformation Europe: Lives of Reformers by Friends, Disciples and Foes* (Aldershot; Burlington, VT: Ashgate, 2008), p. xxviii.

13 T. C. Price Zimmermann, *Paolo Giovio: The Historian and the Crisis of Sixteenth-Century Italy* (Princeton, NJ: Princeton University Press, 1995).

14 Ibid., pp. 159–162. For details of Giovio's museum, see Eugène Müntz, 'Le musée de portraits de Paul Jove: Contributions pour servir à l'iconographie du Moyen Âge et de la Renaissance',

The portraits and lives Giovio accumulated first found their way into print in 1546 in the collection titled *Elogia veris clarorum virorum imaginibus apposita* (Eulogies placed next to the true portraits of famous men), which was a compendium of the portraits in his villa.[15] A second volume printed in Basel and Florence in 1551 contained only men of war.[16] Although Giovio worked in the vicinity and under the patronage of major ecclesiastical figures, he showed considerable interest in the ideas of ecclesiastical and religious reformers.[17] Franci Minonzio characterised Paolo Giovio as a pluralist intellectual, who approached 'others' such as Lutherans with curiosity.[18] We will come back to this later in this section.

Two types of men were worthy of glory and praise: men famous on account of their virtue of warfare (*viri virtute bellica clari*) and second, learned men (*viri docti*). The two editions featuring learned men are *Elogia doctorum virorum* printed in Antwerp in 1557 and *Elogia virorum literis illustrium* printed in 1577 in Basel.[19] Both editions are compilations of earlier print editions, published long after Giovio's death in 1552. That Giovio's eulogies and portraits were printed throughout the sixteenth and even in the seventeenth century attests to Giovio's influence as a chronicler of learned men, as well as to the increasing interest in histories of learning throughout Europe.[20]

*Mémoires de l'Institut national de France, Académie des Inscriptions et Belles-Lettres* 36, no. 2 (1901): 249–343; Linda S. Klinger, 'The Portrait Collection of Paolo Giovio, 1545–1553' (PhD diss., Princeton University, 1991); R. Pavoni, 'Paolo Giovio, et son musée de portraits à propos d'une exposition', *Gazette des Beaux-Arts*, 105, no. 1394 (1985): 109–116.

15 Paolo Giovio, *Elogia veris clarorum virorum imaginibus apposita* [...] (Venice: Michael Tramezinus, 1546).

16 Paolo Giovio, *Elogia virorum bellica virtute illustrium veris imaginibus supposita* [...], 7 vols. (Florence: Laurentius Torrentinus, 1551).

17 This is also hinted at by T. C. Price Zimmermann: 'His numerous friendships with reformers and Giberti's esteem for him are difficult to explain absent any sympathy on his part for reform.' Price Zimmermann, *Paolo Giovio*, p. 150.

18 Franco Minonzio, *Studi Gioviani: Scienza, Filosofia e Letteratura nell'opera di Paolo Giovio*, vol. 2 (Como: Presso la Società a Villa Gallia, 2002), pp. 307–308.

19 Paolo Giovio, *Elogia doctorum virorum ab avorum memoria publicatis ingenii monumentis illustrium* (Antwerp: Jean Bellère, 1557); Giovio, *Elogia virorum literis illustrium, quotquot vel nostra vel avorum memoria vixere* [...] (Basel: Petrus Perna, 1577). This book has an outdated English translation; see Paolo Giovio, *An Italian Portrait Gallery; Being Brief Biographies of Scholars Illustrious Within the Memory of Our Grandfathers For the Published Monuments of Their Genius*, trans. Florence Alden Gragg (Boston: Chapman & Grimes, 1935).

20 For a consideration of Giovio's portrait books and the representation of humanist learned men, see Paul Ortwin Rave, 'Paolo Giovio und die Bildnisvitenbücher des Humanismus', *Jahrbuch der Berliner Museen* 1 (1959): 119–154.

Giovio divided the learned men in his museum into four types, according to the preface entitled 'The description of his museum of portraits'.[21] This preface conveys Giovio's wide interpretation of who is part of the learned world. The first type of men 'are those who after they died, left monuments of fruitful works [*monumenta felicium operum*] to posterity, because they flourished in the fruitfulness of their innate brilliance.'[22] The second type resembles the first type but describes the living instead of the dead. Or, as Giovio put it: 'those who live today, and who, on account of publishing the gifts of their innate brilliance, enjoy an illustrious fame, as the most certain fruit of their nocturnal labour'.[23] The third type includes visual and literary artists:

> This third type will follow the artists of excellent works. This will be explained in a delightful little book, which brings back the memory of the adornment of picture and engraving, shown by accurate monuments of noble craftsmen. Additionally, it restores the memory of the wittiest men who comfort our troublesome minds by making us laugh with their sayings or writings.
>
> Tertia porro classis praecellentium operum artifices excipiet. Haec periucundo libello explicabitur, quum praeter picturae, caelaturaeq[ue], decus, ex certis nobilium artificum monumentis demonstratum, facetissimorum etiam hominum, qui dictis aut scriptis excitato risu, aegrorum animorum curas allevarint, memoria renovetur.[24]

Finally, the fourth type consists of historians and virtuous men who provide the people with *exempla*. 'The fourth will be the class of popes, kings, and dukes who attained glory in peace and war, and gave posterity the famous examples of deeds to imitate and avoid'.[25] According to Giovio, these four types deserved fame, and thus a place in the memory of posterity. The

21 Giovio, *Elogia virorum literis illustrium*, 'Musaei Ioviani Descriptio', pp. 3–6. This description of the museum can be found in other printed editions as well.

22 'Prima eorum est qui fato functi, quum ingenii foecunditate floruerint, felicium operum monumenta posteris reliquerunt.' Ibid., p. 5.

23 'Secunda classis horum erit, qui hodie vivunt, & publicatis ingenii dotibus, illustri fama, tanquam certissimo vigiliarum fructu perfruuntur.' Ibid., p. 6.

24 Ibid.

25 'Quarta erit [classis] maximorum Pontificum, Regum & Ducum, qui pace & bello gloriam consecuti, praeclara ingentium facinorum exempla imitanda, aut vitanda posteris tradiderunt.' Ibid.

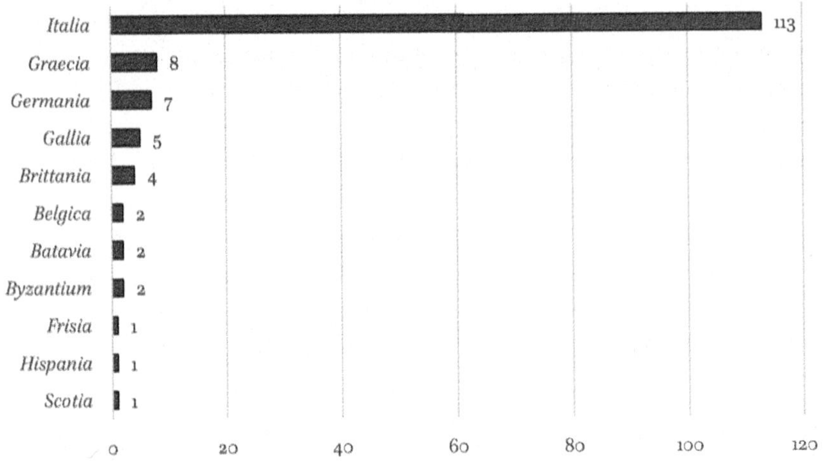


Figure 2.2. The affinity to region and state in Paolo Giovio, *Elogia virorum literis illustrium* (Basel: Petrus Perna, 1577)

classification reveals Giovio's broad fascination with virtue and with men of learning, as well as craftsmen, popes, kings, and commanders.

Giovio's *Elogia* largely contained learned men from the Italian peninsula and some from *Germania*, *Graecia*, *Gallia*, and to a lesser extent from other parts of Europe (see Figure 2.2). Since Giovio was the first to collect and print the portraits of learned men on such a large scale, the inclusion or exclusion does not simply represent Giovio's interest in the person, but also the availability of a printed portrait. One can imagine that obtaining a portrait from a learned man from the Italian peninsula must have been much easier than finding a likeness from across the Alps. This might partly explain the skew towards learned men from the Italian peninsula, but it also denotes that Giovio's circle consisted largely of Italian humanists. The overwhelming part of the volume was filled with men from Florence, Rome, and Venice. To analyse Giovio's sense of history and collective identity, it is meaningful to look at the organisation of the book, paying particular attention to the quarter of the portraits dedicated to learned men from *Graecia* (eight men), *Germania* (seven men), and *Gallia* (five men).

The *Elogia* is prefaced with a life and portrait of Giovio himself. The composition of the book is not thematic, but strictly chronological according to year of birth. The book opens with a portrait and eulogy of the theologian, doctor of the church and bishop Albertus Magnus (c. 1200–1280), who originated from Cologne and is therefore earmarked as coming from *Germania*. He is followed by a doctor of the church, Thomas Aquinas (1225–1274), and by the Franciscan medieval philosopher Duns Scotus (c. 1266–1308). The book thus starts with Catholic scholars who were central to a Catholic ecclesiastical history. Next are major literary and intellectual Italian figures from the fourteenth and fifteenth centuries: Dante (c. 1265–1321), Petrarch (1304–1374), Boccaccio (1313–1375), Leonardo Bruni (1370–1444), Poggio Bracciolini (1380–1459), Lorenzo Valla (1407–1457), Flavio Biondo (1392–1463), and Bessarion (1403–1472). This order constituted the early history of learning in Giovio's eyes; these are the men who constitute the early examples of learned men. Giovio considered scholastic doctors of the church such as Albertus Magnus and Thomas Aquinas to be part of a lineage and history in which Italian Renaissance humanists such as Petrarch and Bruni were the next historical step. Giovio glossed over ancient philosophers such as Plato and Aristotle, whose writings have inspired the work of many of the scholars included in the book. Giovio thus traced the rise of learning and the learned man to fourteenth- and fifteenth-century Italy.

The majority (113 out of 146, or 77%) of the men contained in Giovio's *Elogia* were from the Italian peninsula. These include many of Giovio's learned friends throughout Italy, but mainly from centres of learning such as Rome, Florence, Venice, and Milan. The ensemble of learned men in the *Elogia* reflect Giovio's preoccupations: *cavaliere* of Pope Leo X (1475–1521), companion of Pope Clement VII (1478–1534), an active career ecclesiast, and also an avid humanist writer and energetic member of the *Academia Romana*. The eulogy of the first Italian in the volume states:

> Dante Alighieri is with good reason shown as the first of the Italians [in this book], for he deserves the first place among their portraits not only because of its chronological order, but also because of the exceeding fruitfulness of his grave innate brilliance.
>
> Primus Italorum Danthes Algiderius, non instituto vetustatis ordine tantum, sed praecellenti gravis ingenii foecunditate primum locum inter imagines meritus, optimo iure conspicitur.[26]

Dante was presented as the start of a learned tradition in *Italia*, followed by Petrarch, who received from Dante 'the ease of the Tuscan tongue, clearly defined and adorned with certain metres',[27] followed by Boccaccio, Leonardo Bruni, Lorenzo Valla, and Flavio Biondo, amongst others. All the *elogia* from

26 Ibid., p. 10.

27 'Franciscus Petrarcha eodem cive magistroque suo Danthe, Etruscae linguae facultatem constitutam plane & certis adornatam numeris flagranter excepit.' Ibid., p. 12.

these men from *Italia* convey a sense of the institution and restoration of knowledge from the ancient Greeks and Romans, which happened first and foremost in *Italia*, according to Giovio. Leonardo Bruni, for example, is referred to as the man who 'was the first man in Italy to raise and restore the glory of Greek letters which had been trampled underfoot by many centuries of barbarians with monstrous tyranny.'[28] The generations of barbarians Giovio was referring to are probably the warlords and scribes in monasteries, whom humanists criticised for their ignorance of ancient philosophers. Yet, since Giovio does include scholastic theologians Albertus Magnus and Thomas Aquinas, it shows that he does not discredit medieval scholars outright. In a similar regard for the revival of classical knowledge, Pietro Leoni of Spoleto (c. 1445–1492) is said to have been 'among the first physicians to open the true door to medicine, after publishing almost all of Galen'.[29] Giovio thus invokes the traditional story of the Renaissance with the revival of knowledge and learning in fourteenth- and fifteenth-century Italy.

A group of Byzantine scholars are key in Giovio's narrative of the revival of knowledge. This group of Byzantine scholars stand out, because they are the largest group of non-Italians in Giovio's work. All these learned men fled from *Graecia* to Italy due to the fall of Constantinople to Sultan Mehmed II (1432–1481) in 1453, the loss of the Byzantine empire, and the death of its Christian leader, Constantine XI (1405–1453), and the ensuing persecution of Christians.[30] *Graeci* thus refers to Byzantine Greeks, and most came from Constantinople specifically. Many of these scholars were responsible for the translation of ancient Greek authors into Latin, such as Aristotle and Plato, but also Ptolemy and Euclid.[31] Giovio included George of Trebizond (1395–1486), Theodorus Gaza (c. 1398–c. 1475), as well as émigré scholars John Argyropoulos (c. 1415–1487), Michael Tarchaniota Marullus (1458–1500), Laonikos Chalkokondyles (c. 1430–c. 1470), Marcus Musurus (1470–1517), and Janus Lascaris (1445–1535). Some of these Byzantine emigrant scholars were remembered as Italian citizens, such as the eminent Byzantine bishop and scholar Bessarion (1403–1472) who is not

28 'Primus in Italia Leonardus Aretii natus, Graecarum literarum decus a multis seculis Barbarorum immani tyrannide proculca tum erexit, atque restituit.' Ibid., p. 19.

29 '[I]nter medicos primus fere prolato Galeno, verum medicinae limen aperuit'. Ibid., p. 67.

30 For an overview of Greek emigrants, the causes of emigration, and the knowledge and technology they transmitted, see Jonathan Harris, *Greek Émigres in the West, 1400–1520* (Camberley: Porphyrogenitus, 1995).

31 For a detailed study of the transmission and translation of Greek texts by Byzantine scholars, see N. G. Wilson, *From Byzantium to Italy: Greek Studies in the Italian Renaissance* (London: Duckworth, 1992); see also John Monfasani, *Byzantine Scholars in Renaissance Italy: Cardinal Bessarion and Other Emigrés* (Aldershot; Brookfield, VT: Ashgate, 1995), esp. ch. XII, 'L'insegnamento universitario e la cultura bizantina in Italia nel quattrocento'.

included as *Byzantius* or *Graecus*, but as *Florentinus*.[32] Thus, Giovio presented a history of learning in which Greek Byzantine scholars played a central role in the transmission of texts and knowledge.

In contrast, the scholars from *Germania* in Giovio's tome are much more scattered and diffuse. These include the previously mentioned Albertus Magnus, Henricus Cornelius Agrippa (1486–1535), Georg Sauermann (c. 1492–1530),[33] Jacob Ziegler (c. 1470–1549), Johann Reuchlin (1455–1522), and Regiomontanus (Johannes Müller von Königsberg, 1436–1476). What they all have in common are their travels through Europe and the Italian peninsula, as well as their dedication to the study of the Latin, Greek, or Hebrew languages, except for Regiomontanus, who was primarily concerned with astronomy. Giovio distinguished between different northern states, such as *Brittannia* (British Isles), *Batavia* (Holland), *Belgium* (the southern Low Countries), *Scotia* (Scotland), *Hispania* (Iberian Peninsula), and even *Frisia*. His selection of learned men is eclectic. From *Brittannia* we encounter martyrs under Henry VIII (1491–1547) such as cardinal John Fisher (1469–1535) and Thomas More, as well as the humanist scholar Thomas Linacre (c. 1460–1524). The rest of the learned men from Northern Europe are renowned northern humanists, such as the 'Batavian' Desiderius Erasmus, his 'Frisian' mentor Rudolph Agricola, Christophe de Longueil from *Belgium* (specifically, the Duchy of Brabant), and Juan Luis Vives (c. 1493–1540) who married in *Belgium* but was originally from Valencia. The latter was more commonly associated with *Belgium*, rather than his hometown Valencia. All of these are very well-known figures who were in intense contact with each other. More importantly, all of these scholars were part of a contemporary memory culture. For example, More and Fisher were venerated as Catholic martyrs during the English Reformation,[34] Vives was part of the memory culture of the *Collegium Trilingue* in Leuven with a memorial,[35] and Erasmus tried to secure his own memory as a European man of learning and invented himself as a successor to Rudolph Agricola and Saint Jerome.[36] Their representations in the gallery reflect Giovio's

32 Giovio, *Elogia*, pp. 29–30.

33 Gustav Bauch, 'Georg Sauermann', *ADB*, vol. 30 (1890), p. 417.

34 Anne Dilon, *The Construction of Martyrdom in the English Catholic Community, 1535–1603* (London; New York: Routledge, 2016), pp. 36–48.

35 Gilbert Tournoy et al., eds., *Vives te Leuven: Catalogus van een tentoonstelling in de Centrale Bibliotheek te Leuven, 28 juni–20 augustus 1993* (Leuven: Leuven University Press, 1993), no. 14, pp. 79–80.

36 Lisa Jardine, *Erasmus, Man of Letters: The Construction of Charisma in Print* (Princeton, NJ: Princeton University Press, 1993), esp. ch. 2, 'The In(de)scribable Aura of the Scholar-Saint in

pan-European interest and reception of these learned cultures. The implicit message of Giovio's work is clear: learned men flourish all over Europe. The learned man was represented as a man of vocation and the rise of learning as a pan-European phenomenon.

The learned men from *Gallia* in Giovio's tome show a similar inclination to fame. He included portraits and eulogies of the botanist Jean Ruelle (1474–1537), of Guillaume Budé (1467–1540), Jacques Lefèvre d'Étaples (c. 1455–c. 1536), Peter Martyr d'Anghiera (1457–1526), and Germain de Brie (1490–1538). All these men were well-known scholars of Greek and Latin languages in early modern Europe. The Italian point of view of this book comes to the fore, for example, in the short *vita* of Robert Gaguin (1433/1434–1501), in which Giovio points out that 'France owes a great deal to this man, but those who are concerned with the transmission of an unflawed memory of things to posterity, should be on their guard as to his trustworthiness'.[37] According to Giovio, Gaguin was an unreliable writer, especially in Italian matters. Significantly, this shows that the criterium for inclusion was not primarily the quality of scholarship, but simply the amount of fame and renown someone had acquired. Giovio's *Elogia* was collected on account of fame, rather than to present a coherent intellectual agenda.

T. C. Price Zimmermann argued that Giovio's *Elogia* of illustrious men helped to spread a rhetoric of individuality that posited individual character as the driving force of action.[38] Giovio's book of eulogies of learned men strengthened the link between fame, virtue, and learning. Much like Johannes Fichard, Giovio also provided a view of learning as a pan-European affair, with representative and illustrious examples of learned men from all over Europe. While it is true that the larger portion of the learned men in the *Elogia virorum doctorum* were considered Italians, Giovio highlighted the centrality of Greek Byzantine scholars in the revival of ancient knowledge. As such, Giovio presented a history of learning in which knowledge originated from ancient authors, was transmitted and translated by Byzantine scholars to Italian scholars, and from which the fruits were then reaped throughout Europe.

I would argue that both Fichard and Giovio's collections represent the earliest early modern imaginations of the learned world in print. Together,

His Study: Erasmus's *Life* and *Letters* of Saint Jerome', and ch. 3, 'Inventing Rudolph Agricola: Recovery and Transmission of the *De inventione dialectica*'.

37 'Gallia huic plurimum debet: sed de fide illi viderint, quorum interest incorruptam rerum memoriam ad posteros transmitti.' Giovio, *Elogia*, p. 138.

38 T. C. Price Zimmermann, 'Paolo Giovio and the Rhetoric of Individuality', in *The Rhetorics of Life-Writing in Early Modern Europe: Forms of Biography from Cassandra Fedele to Louis XIV*, ed. Thomas F. Mayer and Daniel R. Woolf, 39–62 (Ann Arbor: University of Michigan Press, 1995).

they are representative of the increasing cultural visibility of the learned man as a pan-European, virtuous, and exemplary figure. Both Fichard and Giovo's selections reflected their own idiosyncratic sensitivity to different developments in Europe, such as Giovio's inclusion of Greek Byzantine émigré scholars, and Fichard's geographical division of Europe into *Brittannia*, *Germania*, and *Italia*. Although Fichard and Giovio did not write history in the modern sense of the word, that is, as a cohesive narrative, they certainly imagined a pan-European learned community which had a collective history. Both collections impressed on the reader that learning and scholarship was thriving throughout Europe. Fichard traced the development and spread of knowledge from Italian Renaissance figures to northern humanists. Giovio's collection is certainly less programmatic, but nevertheless invokes the sense that learned men can be found everywhere throughout Europe, all dedicated to the cause and transfer of knowledge.

This universal ideal of the learned man as someone who can be found anywhere in Europe remained popular in learned circles throughout the sixteenth century. At the same time, however, ecclesiastical and religious reforms between roughly 1520 and 1550 had a profound impact on the perception of the learned world as a pan-European community with a universal history. This in turn significantly changed the role and exemplary function of the learned man.

## 2.2 The Learned Man Amidst Confessionalisation

In the second half of the sixteenth century, learned men figured prominently in the propaganda and rhetoric in the wake of religious reforms throughout Europe. In the previous chapter we have already established that the virtue of piety took on different meanings in the Catholic and reformed parts of Europe. In this section we will examine the geographical and confessional sensibilities of compilers of *vitae* in the second half of the sixteenth century. Specifically, we will take a closer look at the work of two compilers, Theodore Beza (1519–1605) and Nicolaus Reusner (1545–1602), who both compiled books of icons and eulogies of learned men. The tone and representation of the learned world are different from the previously mentioned collections by Johannes Fichard and Paolo Giovio, but show continuities as well. The *Icones* of Beza adopted the visual aspects from martyrology and combined them with the *vitae* tradition of learned men to forge a new icon of the learned man, as an instrument to propagate his ideal of religious reform, the learned world, and its history. Reusner, on the other hand, appropriated Beza to make the first steps to rewrite the canon of learned men of his homeland, *Germania nostra*. Both show how malleable the learned identity was and how it could be adjusted to fit a Reformed sense of history and identity. These works show that the church was not an antagonist of the learned world, but that the perception of the learned world changed in response to a fragmentation of religious sensibilities.

Wolfgang Reinhard and Heinz Schilling introduced the concept of confessionalisation to denote the confession-building undertaken by early modern states and rulers, often by means of social disciplinary measures based on a confession of faith.[39] In the context of the Reformation, confessionalisation is often taken to mean a cultural, political, and religious shift in European religious communities and cultures as a result of deliberate policies. Ute Lotz-Heumann identified seven aspects of the top-down political process of confessionalisation in Europe that can help us to understand the breadth of this development: the establishment of 'pure doctrine'; the distribution and enforcement of these norms; propaganda and censure; internalising of norms through education; disciplining the population and ensuring homogeneity of religion and rites; and finally, the regulation of language.[40] Lotz-Heumann points to the cultural aspects of confessionalisation that are representative of the developments of many regions, states, and confessions in sixteenth-century Europe. From the second half of the sixteenth century up to the first half of the seventeenth century, regional and religious leaders solidified their confessions. The Catholic (Council of Trent, 1545–1563), Gallican (1559), Belgic (1561), Helvetic (1561–1562), and English (Church of England, 1571) confessions were influential due to religious and state intervention. Historians have nuanced this top-down view and argued that small regions sometimes deviated and prescribed their own confessions. Additionally, historians argued that confessionalisation entailed more than the state's enforcement of pure doctrine, rites, and practices, but also involved the development of a regional and confessional sense of history and identity.[41] As we shall see in the books of icons by Beza and Reusner,

39 See, e.g., Heinz Schilling, 'Die Konfessionalisierung im Reich: Religiöser und gesellschaftlicher Wandel in Deutschland zwischen 1555 und 1620', *Historische Zeitschrift* 246 (1988): 1–45; Wolfgang Reinhard, 'Konfession und Konfessionalisierung in Europa', in *Bekenntnis und Geschichte: Die Confessio Augustana im historischen Zusammenhang*, ed. Reinhard, 165–189 (Munich: Vögel, 1981).

40 Ute Lotz-Heumann, 'The Concept of "Confessionalisation": A Historiographical Paradigm in Dispute', *Memoria y Civilización* 4 (2001): 93–114.

41 This is explored in Bruce Gordon, ed., *Protestant History and Identity in Sixteenth-Century Europe*, 2 vols. (Aldershot: Ashgate, 1996).

the cultural icon of the learned man changed to adhere to the confessional interpretations of doctrine, rites, and virtues.

Confessionalisation had myriad consequences for the collective perceptions of the learned world.[42] The focus on confession led to a reconfiguration of both a political and a nascent national identity (more on that in Section 2.3).[43] In this section, we will focus on how Beza and Reusner appropriated the figure of the learned man to play a central part in their confessional history of learning. Their compilations of icons and eulogies of learned men represent the malleability of cultural identity as well as the divergent perceptions of learned men and European learned communities from the late sixteenth century onwards.

Fichard's apparently ambivalent stance with regards to the religious fidelity of scholars was untenable for Reformed scholars such as Theodore Beza and Nicolaus Reusner. Their compilations of lives and portraits of learned men, Beza's *Icones* (Geneva, 1580) and Reusner's *Icones* (Strasbourg, 1587), are a case in point.[44] These two volumes are adorned with plates of learned men, each followed by a short biography of the subject and a description of how they contributed to instituting *vera religio*. As such, the volumes present a community of a particular type of learned men: those who paved the way to the Reformation.

Beza is known as a reformer, as the right hand and successor of John Calvin (1509–1564), as the spiritual leader of the Republic of Geneva, as well as a theologian, poet, and professor.[45] He is also known for providing the

42 Erika Rummel, *The Confessionalization of Humanism in Reformation Germany* (New York: Oxford University Press, 2000); Matthias Pohlig, *Zwischen Gelehrsamkeit und konfessioneller Identitätsstiftung: Lutherische Kirchen- und Universalgeschichtsschreibung, 1546–1617* (Tübingen: Mohr Siebeck, 2007); Ian Maclean, *Scholarship, Commerce, Religion: The Learned Book in the Age of Confessions, 1560–1630* (Cambridge, MA; London: Harvard University Press, 2012).

43 Heinz Schilling, 'Nationale Identität und Konfession in der europäischen Neuzeit', in *Nationale und kulturelle Identität: Studien zur Entwicklung des kollektiven Bewußtseins in der Neuzeit*, ed. Bernard Giesen, 192–252 (Frankfurt a.M.: Suhrkamp, 1991); Schilling, 'Konfessionelle und politische Identität im frühneuzeitlichen Europa', in *Nationale und ethnische Minderheiten und regionale Identitäten in Mittelalter und Neuzeit*, ed. Antoni Czacharowski, 103–123 (Toruń: Wydawnictwo Uniwersytetu Mikołaja Kopernika, 1994); Irena Backus, *Life Writing*, pp. 229–231.

44 Theodore Beza, *Icones, id est verae imagines virorum doctrina simul et pietate illustrium* [...] (Geneva: Jean de Laon, 1580); Nicolaus Reusner, *Icones sive imagines virorum literis illustrium* [...] (Strasbourg: Bernard Jobin, 1587). For the digital analysis a digitised version of a later edition was used, namely Nicolaus Reusner, *Icones sive imagines virorum literis illustrium* [...] (Frankfurt a.M.: Balthasar Diehl, 1719).

45 For a complete survey of Theodore Beza's contributions to religion, politics, theology, exegesis, literature, and philology, see Irena Backus, ed., *Théodore de Bèze (1519–1605): Actes du Colloque de Genève (septembre 2005)* (Geneva: Librairie Droz, 2007); Alison Adams, *Web of Allusion: French*

Calvinist reformation movement with a new translation of the Greek New Testament with his *Novum Testamentum* (1557), as well as a Reformed history with his *Histoire ecclésiastique des Eglises réformées*.[46] After Calvin's death in 1564, Beza succeeded him as the leader of the Calvinist movement and worked hard to purify and spread Calvin's theology and religious reform movement.[47] The *Icones* we will discuss here are part of Beza's programme to consolidate the Calvinist Reformed religious movement with its own doctrine, history, religious text, theology, and true icons whose exemplary behaviour and printed portrait inspired pursuance. Within Beza's theology, the *Icones* denote the centrality of good works in his theology, where one's deeds define a pious Christian life.[48] More generally, the *Icones* and the included emblems (*emblemata*) are there to inspire religious followers to contemplate and emulate a good Christian life in the Protestant sense of the term.[49]

The *Icones* thus represent a reformulation of the learned man and the genre of collective life-writing to match confessional sensibilities.[50] Beza edited a collection of lives of learned men who contributed to the development of the Reformation, veering between genuine history and Reformed religious writing. The full title suggests as much: *Icones, id est verae imagines virorum doctrina simul et pietate illustrium, quorum praecipue ministerio partim bonarum literarum studia sunt restituta, partim vera Religio in variis orbis Christiani regionibus, nostra patrumque memoria*

*Protestant Emblem Books of the Sixteenth Century* (Geneva: Librairie Droz, 2003), ch. 2, 'Théodore de Bèze: Theologian and Poet', pp. 119–153.

46 See Bernard Roussel, 'Le *Novum Testamentum* de Théodore de Bèze: L'édition, la traduction et l'annotation de l'*Epître de Jude*', in Backus, *Théodore de Bèze*, pp. 185–194; and Marianne Carbonnier-Burkard, 'L'*Histoire ecclésiastique des Eglises réformées*...: la construction bézienne d'un "corps d'histoire" ', in ibid., pp. 145–161. For an account of the role of history in the French Reformation, see David Watson, 'Jean Crespin and the Writing of History in the French Reformation', in Gordon, *Protestant History and Identity*, vol. 2, *The Later Reformation*, pp. 39– 58.

47 See, e.g., Robert M. Kingdon, *Geneva and the Consolidation of the French Protestant Movement, 1564–1572* (Madison: University of Wisconsin Press, 1967).

48 John S. Bray, 'The Value of Works in the Theology of Calvin and Beza', *The Sixteenth Century Journal* 4, no. 2 (1973): 77–86 (esp. pp. 84–85).

49 On Theodore Beza's emblems, see Alison Adams, 'The *Emblemata* of Théodore de Bèze (1580)', in *Mundus Emblematicus: Studies in Neo-Latin Emblem Books*, ed. Karl A. E. Enenkel and Arnoud S. Q. Visser, 71–99 (Turnhout: Brepols, 2003); Adams, *Web of Allusion*, pp. 119–153.

50 Patricia Eichel-Lojkine, for example, has rightfully commented on Beza's *Icones* that it was not so much a continuation of the Plutarchan tradition of life-writing as an adoption of the popular genre of the lives of martyrs, see Eichel-Lojkine, *Le siècle des grand hommes: Les recueils de vies d'hommes illustres avec portraits du XVIème siècle* (Leuven; Paris; Sterling, VA: Éditions Peeters, 2001), pp. 277–281, 345.

*fuit instaurata* (Icons, that is true images of men illustrious on account of both their learning and piety, whose ministry of faith in particular partly revived the study of good letters, and partly restored the true Religion in various regions of the Christian world, in our and our forefathers' time). In the preface, a dedicatory epistle to James VI, King of Scotland (1566–1625), Beza described his ideal of a complete renewal and revival of the true practice of religion. The learned men in his volume are referred to as 'the advanced martyrs of the renewed Christianity from our and our forefather's time, from very diverse regions of the world, from England, Bohemia, Italy, selected by God',[51] including men such as John Wycliffe (1328–1384) and Jerome of Prague (1379–1416). In Beza's view, these men gave their lives to pave the way for the Reformation and are thus essential to his sense of history and community. The *Icones* also contain 'pre-eminent Christians of our and our forefathers time, who built the tools for the Germans', including figures such as Erasmus, Luther, and Philip Melanchthon (1497–1560).[52] Similar parts are dedicated to the religious reformers in France and the Swiss states, for instance reformers such as Calvin and Zwingli. At the same time, Beza included men of letters renowned for their textual scholarship, such as Julius Caesar Scaliger (1484–1558) and Robert Estienne (1503–1559), but also the printer Johann Froben (c. 1460–1527) from Basel. With their work, learning, and piety, Beza contended, these men built the tools for a successful Reformation. Beza used his *Icones* to introduce a new historical consciousness and a teleological history of progress towards religious reform in which reformers, theologians, scholars, and printers were crucial actors.

Despite Beza aiming to show that religious reforms were a pan-European development with the inclusion of men such as Girolamo Savonarola (1452–1498) and Jerome of Prague (1379–1416), his volume is strongly geared towards the German lands, France, and the Swiss states (see Figure 2.3). Savonarola was commonly included in Protestant histories in the sixteenth century, because he was seen as a true witness of Christ, much like the later reformers.[53] As such, Beza appropriated learned men from the past to fit his historical narrative: they were described as part

51 'Antesignani patrum memoria instaurandi Christianismi martyres, ex diversissimis orbis terrarum regionibus, Anglia, Bohemia, Italia, divinitus selecti.' Beza, *Icones*, sig. A-i$^{r}$.

52 'Praecipua Christianismi, nostra patrumque memoria, inter Germanos instaurati organa'. Ibid., [sig. B-v$^{r}$].

53 Bruce Gordon, '"This Worthy Witness of Christ": Protestant Uses of Savonarola in the Sixteenth Century', in Gordon, *Protestant History and Identity*, vol. 1, *The Medieval Inheritance*, pp. 93–107.

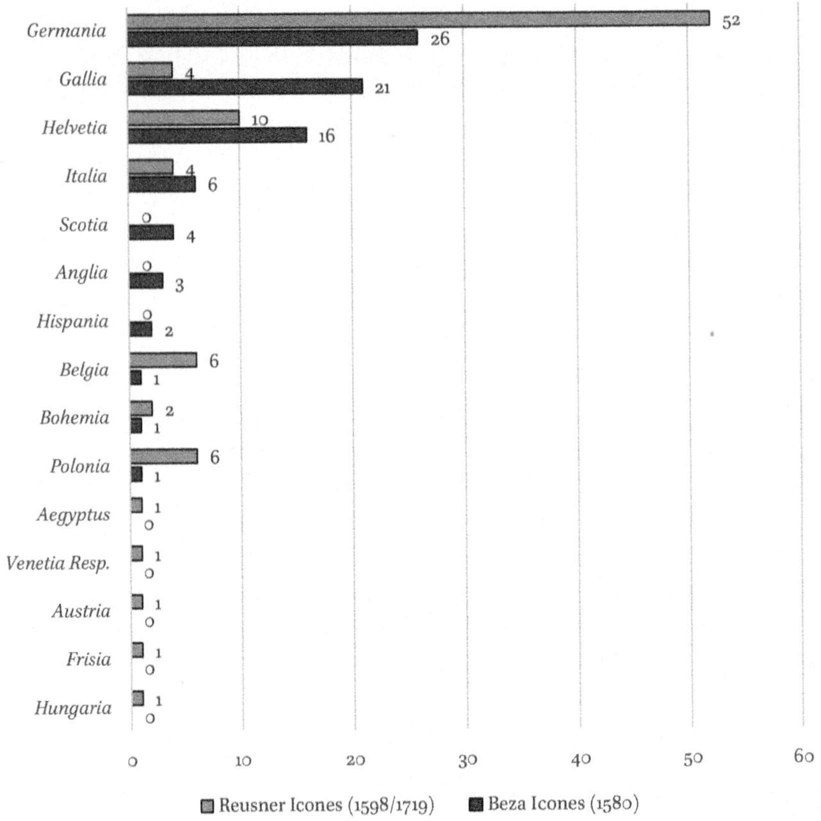


Figure 2.3. The affinity with state and region in Nicolaus Reusner, *Icones* (1719; first published in 1587) and Beza, *Icones* (1580)

of a long line of witnesses of Christ in order to give their own religious movement a sense of history and identity.[54] Religious reform was bound to happen and witnesses of the *vera religio* could be found in every place and period.

This broad scope not only reflects the historical reality of religious reforms taking place primarily in the northern parts of Europe, but also Beza's allegiance to John Calvin, whose supporters could primarily be found in France and the Swiss states. Yet to characterise Beza's *Icones* as 'Reformed propaganda', as Christophe Chazalon did, misses the point that the *Icones* are part of a sincere attempt to reach out to religious followers and offer inspiration for a Christian life, as well as provide the religious reforms in

54 This appropriation is also described in Evan Cameron, 'Medieval Heretics as Protestant Martyrs', *Studies in Church History* 30 (1993): 185–207.

Europe at the time with a history and cultural icons.[55] While Beza's *Icones* were certainly influential in establishing the Reformed learned man as a cultural icon for decades to come, its example was appropriated rather than copied.

Nicolaus Reusner came from a wealthy and respected family in Silesia and made a name for himself as a prolific author, publisher, and professor of law in Strassbourg (1583–1588) and Jena (1589–1602).[56] He was a member of the Diet of Augsburg in 1555 when the peace of Augsburg was concluded, received the noble title of *Comes Palatinus* (Count Palatine) from Emperor Rudolf II in 1594, and was in favour of the Holy Roman Emperor. Reusner's *Icones sive imagines virorum literis illustrium* (Icons or images of men illustrious on account of learning, 1587) has been described—when it ever has been described—as Reformed propaganda and as a copy of Beza's *Icones*. However, the comparison of this *Icones* to other biographical collections compiled by Reusner reveals a more nuanced account.[57] The *Icones* of reformers and theologians was certainly more popular, both with contemporary readers and modern historians. A comparison with Reusner's second collection of learned men, the *Icones virorum literis illustrium*, published two years later, shows that the identity of the learned man was clearly malleable. While a modern reader would expect friction, Reusner seemed to see no contradiction in publishing a book about Reformed learned men as well as one concerning learned men from all over Europe.

The first book of icons was published as *Icones sive imagines virorum literis illustrium quorum fide et doctrina religionis & bonarum literarum studia, nostra patrumque memoria, in Germania praesertim, in integrum sunt restituta* (Icons or images of men illustrious on account of learning, who with their faith and teaching completely restored the study of religion and good letters, especially in Germania, in our and our forefathers' time, 1587).[58] The title already foreshadows a Reformed historical consciousness. The composition in terms of the geographical origins of the men included in

55 Christophe Chazalon, 'Les *Icones* de Théodore de Bèze (1580): Entre mémoire et propagande', *Bibliothèque d'Humanisme et Renaissance* 66, no. 2 (2004): 359–376.

56 August Ritter von Eisenhart, 'Nicolaus von Reusner', in *ADB*, vol. 28 (1889), pp. 299–303.

57 Chazalon, 'Les Icones', p. 361.

58 This collection of icons from 1587 had several reprints in the sixteenth and eighteenth centuries: Nicolaus Reusner, *Icones sive imagines virorum literis illustrium* [...] (Strasbourg: Bernhard Jobin, 1590); Reusner, *Icones sive imagines virorum literis illustrium* [...] (Strasbourg: Bernard Jobin, 1593); Reusner, *Icones sive imagines virorum literis illustrium* [...] (Frankfurt a.M.: Johann Balthasar Diehl, 1714); Reusner, *Icones sive imagines virorum literis illustrium* [...] (Frankfurt a.M.: Johann Balthasar Diehl, 1719).

the volume is very similar to that of Beza's *Icones*, as can be seen in Figure 2.3, although its focus is somewhat broader with the inclusion of theologians and reformers from various Reformed Protestant regions all over Europe, such as the Swiss Confederacy, England, Scotland, and the Kingdom of Poland. In the preface, Reusner positioned himself as a successor to Theodore Beza:

> That work of icons by Theodore Beza published a few years ago has been favourably compared to this [Giovio's Museum]; when it concerns the standing of our fatherland it deserves far higher praise. In this [work by Beza], next to the most elegant eulogies, one can see icons and effigies of the same type of men illustrious of account of both their learning and piety, especially from our *Germania*, and France and England. Chiefly due to their help the study of good letters was revived on one hand; on the other, true religion in various regions of the Christian world was brought back, in our time and the time of our forefathers.
>
> Cum quo non immerito comparatur, & quod ad patriae dignitatem attinet, longe etiam maiore meretur laudem, opus illud Iconum Theodori Bezae, ante paucos annos in lucem editum: in quo & ipso, praeter Elogia elegantissima, conspiciuntur eiusdem generis Icones & effigies virorum doctrina simul & pietate illustrium: Germaniae praesertim nostrae, & Galliae & Angliae: quorum praecipue ministerio partim bonarum literarum studia restituta, partim vera religio in varijs orbis Christinae regionibus, nostra patrumque memoria fuit instaurata.[59]

Reusner wanted to continue this tradition of providing the fatherland ('Germania nostra') with the respect it deserved, in the form of a book of icons of men who had revived the study of letters in the German lands, and thereby helped and shaped religious reforms throughout Europe. While Reusner's *Icones* may at first glance seem confessional, much like Beza's *Icones*, it is important to also identify the differences between them. Both Beza and Reusner accommodate learned men from any non-Catholic confession, thus providing a non-Catholic, multiconfessional history of the Reformation, where theologians and reformers such as Luther and Calvin play equal parts. But Reusner was more concerned with learning and its revival in the Reformed parts of Europe from the 1500s onwards. Beza wanted to present a coherent set of characters who played a major role in the religious reforms in northern Europe (including 'dissident' Catholic voices such as Savonarola and Erasmus).

59 Reusner, *Icones* (1587), 'Praefatio', fol. 3$^{r}$.

While it is true that Reusner copied much from Beza, he also added seemingly random inclusions, such as the ancient astronomer Ptolemy (c. 100 –c. 170), the Hungarian physician Georg Wirth (1470–1524), and the anatomist Andreas Vesalius (1514–1564) and thereby expanded the selection of his learned men.

Beza and Reusner both included four Italian scholars and theologians. These Italians, in the imagination of Beza and Reusner, had contributed to the cause of the Reformation, such as proto-reformer Savonarola and well-known reformer Peter Martyr Vermigli (1499–1562). Although they were Italian, Savonarola and Vermigli recognised and criticised the papal authority, as befitted a pious learned man in the eyes of Beza and Reusner. These two volumes of icons offer a historical representation of the Reformation through the lives of reformers, theologians, and scholars and emphasise learning and textual criticism. As such, these volumes contributed to, and are indicative of, the increasing confessionalisation and cultural visibility of learned memory culture on the northern side of the divide.[60] In short, Beza and Reusner remembered, memorialised, and represented learned men as northern, Germanic, and Reformed theologians and scholars.

With his collection, Reusner did not stress religious reforms and reformers as Beza did, but rather emphasised that learning and erudition were flourishing in his *Germania* in the 1500s. He included mathematicians and astronomers: Ptolemy, for example, presented as the 'prince of mathematical learning',[61] figures in this history of mathematics and is succeeded by the Polish astronomer Nicolaus Copernicus (1473–1543), hailed as 'the most erudite in the studies of mathematical disciplines'.[62] Of course, the reformers from Beza's volume can be found in Reusner, too, but Reusner presents them as learned and erudite men, rather than actors in the religious reformation of Europe. Reusner included many other types of learning besides theology, such as law, medicine, and mathematics (as can be seen in Table 2.1).

**Table 2.1. Professions and occupations mentioned in Reusner's Icones (1587)**

| Profession/occupation | Count |
|---|---|
| *Theologus* (theologian) | 28 |
| *Jurisconsultus* (legal scholar, jurist) | 11 |
| *Medicus* (physician) | 7 |
| *Pastor* (pastor, minister of the word) | 5 |

60 For a consideration of the sixteenth-century confessionalisation of humanism, i.e., education, see Rummel, *The Confessionalization of Humanism*.

61 'Princeps doctrinae Mathematicae'. Reusner, *Icones*, p. 2.

62 'Mathematicarum studiis disciplinarum eruditissimus'. Ibid., p. 116.

| Profession/occupation | Count |
|---|---|
| *Historicus* (historian) | 5 |
| *Mathematicus* (mathematician) | 5 |
| *Theologiae Doctor* (doctor of theology) | 4 |
| *Philologus* (philologist) | 4 |
| *Archiater* (chief physician) | 4 |
| *Ecclesiast* (ecclesiast, church official) | 2 |
| *Theologus & Philologus* (theologian and philologist) | 2 |
| *Philosophus* (philosopher) | 2 |
| *Martyr* (martyr) | 1 |
| *Musicus* (musician) | 1 |
| *Typographus* (printer/publisher) | 1 |
| *Cosmographus* (cosmographer) | 1 |
| *Princeps* (prince) | 1 |
| *Medicus & Historicus* (physician and historian) | 1 |
| *Eques* (count) | 1 |
| *Eques & Poeta* (count and poet) | 1 |
| *Theologus & Philosophus* (theologian and philosopher) | 1 |
| *Philosophus & Medicus* (philosopher and physician) | 1 |
| *Poeta* (poet) | 1 |
| *Medicus Jurisconsultusque* (physician and jurist) | 1 |

Reusner's focus on erudition and learning is even clearer in a collection of icons he published several years after the first *Icones* from 1587. This second book is titled *Icones sive imagines vivae, literis Cl[arorum] virorum, Italiae, Graeciae, Germaniae, Galliae, Angliae, Ungariae* (Icons or images after life of men famous on account of learning, from Italy, Greece, Germania, France, England, and Hungary), and was published in Basel in 1589.[63] As the title suggests, this collection was dedicated to learned men throughout Europe, and was based on Giovio's *Elogia*. It presents a history of learning originating with the ancients, such as Aristotle and Cicero, which then truly flourished in the 1400s in Italy, with luminaries such as Dante and Leonardo Bruni; it also includes the addition of émigré Byzantine scholars, echoing Giovio's *Elogia*. Chronologically, it can be seen as a prequel to the first *Icones* from 1587, since it collects the learned men who came before the revival of learning in *Germania* described in the 1587 *Icones*. Consequently, the 1589 *Icones* has a strong focus on learned men in Italy, as can be seen in Figure 2.4.

The pan-European learned community and the communication between its members are sometimes described as being resistant to the

63 Nicolaus Reusner, *Icones sive imagines vivae, literis Cl[arorum] virorum, Italiae, Graeciae, Germaniae, Galliae, Angliae, Ungariae* (Basel: Conrad Waldkirch, 1589).

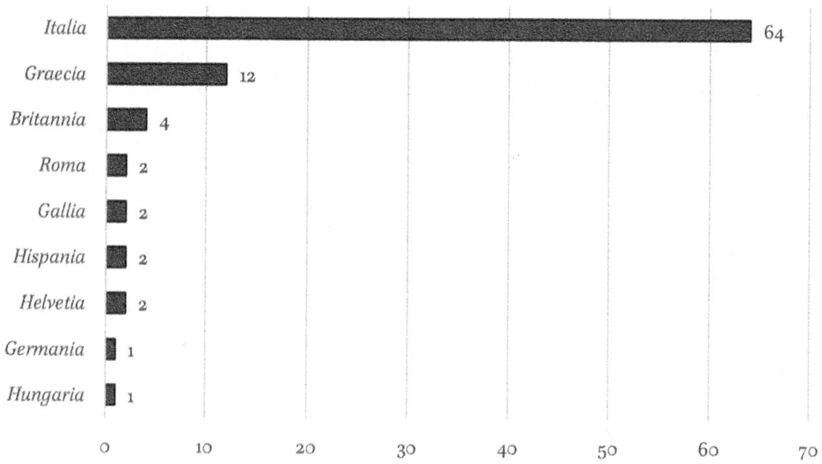


Figure 2.4 The affinity to state and region in Nicolaus Reusner, *Icones sive imagines vivae, literis Cl[arorum] virorum* [...] (Basel, 1589)

force of confessionalisation, for example in the neutral and pragmatic use of Latin.[64] Indeed, in many historical accounts scholars were expected to avoid and resist religious controversy and confessionalism. At the same time, scholars such as Ian Maclean have clearly demonstrated that the spread of knowledge through means such as print was seriously hindered by local confessional issues, such as the expulsion of non-Lutheran publishers Frankfurt am Main around the end of the sixteenth century.[65] However, many scholars were at the forefront of religious reforms themselves. In the context of this historical tension, Beza and Reusner's books show how the cultural symbol of the learned man was adapted to confessional sensibilities. In the *Icones*, Beza and Reusner presented learned men as proto-reformers.[66] While Beza's *Icones* can be,

64 See *Die Territorien des Reichs im Zeitalter der Reformation und Konfessionalisierung: Land und Konfession, 1500–1650*, 7 vols. (Münster: Aschendorff, 1989–1997); this sentiment is also repeated in Ute Lotz-Heumann and Matthias Pohlig, 'Confessionalization and Literature in the Empire, 1555–1700', *Central European History* 40, no. 1 (2007): 35–61 (p. 38); and subsequently cited in Dmitri Levitin, 'Introduction: Confessionalisation and Erudition in Early Modern Europe: A Comparative Overview of a Neglected Episode in the History of the Humanities', in *Confessionalisation and Erudition in Early Modern Europe: An Episode in the History of the Humanities*, ed. Nicholas Hardy and Dmitri Levitin, 1–94 (Oxford: Oxford University Press, 2019), p. 4, n. 14.

65 See Maclean, *Scholarship, Commerce, Religion*, esp. ch. 2, 'In Medias Res: A Literary Agent in Frankfurt, 1606–1615'.

66 For a consideration of the study of collective identities and representations in the context of secularisation, see Matthias Pohlig et al., *Säkularisierungen in der Frühen Neuzeit: Methodische Probleme und empirische Fallstudien* (Berlin: Duncker & Humblot, 2008), pp. 110–123.

and has been, construed as Calvinist propaganda,[67] he had no intention of rewriting and replacing the established history of learning. Rather, Beza wanted to provide his religious followers and reformers with a book full of inspiring figures, and thus broadened the conception of the learned man to include reformers.

The two books of icons compiled by Reusner strengthen the thesis that early modern scholars did not mean to reject and replace the leading history of learning—from the ancients, via Italy, to northern Europe—but rather to emphasise that northern theologians and reformers deserved their respective places in that narrative. That Reusner could publish a book of icons in 1587 featuring largely Reformed theologians, as well as a book primarily of learned men from Italy in 1589, reveals the quest to unite these two distinct imaginations of the learned world. On the one hand, no one was disputing that fourteenth- and fifteenth-century Italian luminaries such as Dante and Bruni deserved their place in a collection of learned men. Yet, on the other hand, by praising the Reformed theologian as a purifier of literature and religion, Reusner as well as Beza excluded Catholic scholars from the history of learning in the northern parts of Europe. Reusner imagined the past of the learned world as pan-European, just as Giovio and Fichard had. Together with Beza, however, Reusner imagined the present state of the learned community as fragmented along confessional and political lines. Their work contains different attempts to imagine a new sense of a learned community, its identity, and collective history. Throughout Europe, different regions and confessions imagined and formed a distinct sense of a Protestant identity.[68] While the learned man became a cultural icon in future histories of the Reformation, this did not erase nor replace the historical consciousness of learning put in place in tomes such as Giovio's *Elogia* and Fichard's *Vitae*.

67 Chazalon, 'Les Icones'; Catharine Randall, 'A Protestant Poetics of Process: Reformation Rhetorics of the Self in Sponde, de Bèze, and d'Aubigné', in *Rhetorics of Life-Writing*, ed. Mayer and Woolf, pp. 243–282.

68 Bruce Gordon, 'The Changing Face of Protestant History and Identity in the Sixteenth Century', introduction to Gordon, *Protestant History and Identity*, vol. 1, pp. 1–22, with the important observation on pp. 6–7: 'It was a struggle not only to establish the primacy of the Word, but to control words: to formulate a language which could be used to legitimate the Protestant cause and bring people into the new communities through the use of propaganda. In this evolution it became manifestly evident that there was no Protestant view of history, but several, and each acquired its voice and character from the local circumstances out of which it arose.' Also see Backus, *Life Writing*.

## 2.3 The Learned Man and National Consciousness

Historical thought and writing gained considerable popularity at the end of the sixteenth century. History and *vitae* not only became rhetorical weapons in the wake of the Reformation, but states and regions increasingly started to compose regional histories which were part of the reconfiguration of collective identities following confessional and political changes.[69] Before we discuss these developments in collective life-writing any further, we have to address the historical debate about the existence and development of such a national framework in the early modern period.[70] Although we have already touched on the ambiguity of the term *patria*, a further review of the historiography of nationalism and historical consciousness provides a background to our understanding of the changes in the collections of *vitae* of learned men throughout the seventeenth century.

Put crudely, historians are divided into two camps on the question what nationalism is, and consequently, when it emerged and flourished. On the one hand, the modernist interpretation views nationalism as a by-product of the nation state, with its roots in the nineteenth century. In this interpretation, nationalism is a product of the nineteenth-century nation state and cannot exist without it. The nineteenth-century nation state and nationalism are thus inextricably linked.

On the other hand, the traditionalist interpretation of nationalism holds that nationalism is inherent to *any* form of state—modern or premodern—and the form in which it manifests itself is historically contingent. In the latter, admittedly broader, interpretation of nationalism, nineteenth-century nationalism is understood as a historically specific form of nationalism. Although this is a rather crude dichotomy, it clarifies that to be able to discuss nationalism in an early modern context, we need to subscribe to the traditionalist interpretation. Nationalism is thus taken to be, as Azar Gat noted, 'the doctrine and ideology that a people is bound together in solidarity, fate, and common political aspirations'.[71] This includes the common early

69 Thomas Fuchs, 'Reformation, Tradition und Geschichte. Erinnerungsstrategien der reformatorischen Bewegung', in *Protestantische Identität und Erinnerung: Von der Reformation bis zur Bürgerrechtsbewegung in der DDR*, ed. Joachim Eibach and Marcus Sandl, 71–89 (Göttingen: Vandenhoeck & Ruprecht, 2003).

70 For a clear introduction to this debate and history, see Azar Gat, with Alexander Yakobson, *Nations: The Long History and Deep Roots of Political Ethnicity and Nationalism* (Cambridge: Cambridge University Press, 2012); Herman Roodenburg, ed., *Forging European Identities, 1400–1700* (Cambridge: Cambridge University Press, 2007).

71 Gat, *Nations*, p. 26.

modern sentiment of patriotism, which referred to the perceived kinship and commonality of culture on the level of the state, region, or *patria*.

The concept of nationalism as a shared sense of community, past, and future became copious in seventeenth-century editions of *vitae*. As I will argue in this section, learned men were increasingly remembered in the context of the *patria*, and became part of a national and patriotic historical narrative. As opposed to the representation of learned men in collective life-writing from the sixteenth century discussed above, collections from the seventeenth century often appeal to a nationalistic sentiment and were presented to ecclesiastical and political leaders as gifts. Many learned men framed the learned world as a 'national' affair, characterised by a common culture, history, and future on the level of an imagined *patria*, where the learned men from a specific region or state were presented as the next historical step after the revival of the classics throughout Europe in the fifteenth and sixteenth centuries. This development is part of a larger European trend: states, regions, and communities started to develop a historical consciousness. From the end of the sixteenth century onwards, communities started to write their own histories and hagiographies, with examples such as the rise of papal history and patriotic, national histories.[72]

Thus, in the seventeenth century, learned men were increasingly presented within a national and patriotic framework. While most compilers did acknowledge earlier collections, especially Giovio's *Elogia*, they presented learned men from their own region or state as worthy additions to the established history of learning in the fourteenth and fifteenth centuries. The historical consciousness in which learned men were celebrated and granted a stage was of a national and patriotic kind.

Melchior Adam's five-volume *Vitae* (1615–1620) and Gian Vittorio Rossi's three-volume *Pinacotheca* (1645–1648) are representative of the seventeenth-century national framework of *vitae* of learned men, as well as the growth of collective scholarly life-writing, both in the amount of works published and in their respective sizes. In comparison with the aforementioned books by Beza and Reusner, Adam's and Rossi's multivolume works are compiled along patriotic and disciplinary lines. The collections of Adam and Rossi are meant to represent the learned men of a certain *patria* or region, but also

72 Stefan Bauer, *The Invention of Papal History: Onofrio Panvinio between Renaissance and Catholic Reform*, Oxford-Warburg Studies (Oxford: Oxford University Press, 2020); Sandra Langereis, 'Geschiedenis als ambacht: oudheidkunde in de Gouden Eeuw: Arnoldus Buchelius en Petrus Scriverius' (PhD diss., University of Amsterdam 2001); Gordon, *Protestant History and Identity*.

aimed at a pan-European readership. These massive collections conveyed to the reader: here are the learned men from our region or state that deserve a place in the transnational history of learning.

Little biographical information is known about Melchior Adam (c. 1580s–1622), whose work we already analysed in the previous chapter. He had studied in Heidelberg since 1598 and worked at the Heidelberger Paedagogium, first as a *magister*, and later as a co-rector and rector.[73] I would argue that Adam's *Vitae* in five volumes reflect three trends in scholarly collective life-writing in the seventeenth century: firstly, the increase in size of the projects, where collections easily take up several volumes; secondly, the confinement of a collection to a region or state, which in Adam's case is *Germania*; and thirdly, the organisation of learned men according to disciplines. The titles of Adam's individual volumes reflect this patriotic and disciplinary organisation of lives of learned men:

1. *Vitae Germanorum philosophorum* (Lives of German philosophers)[74]
2. *Vitae Germanorum medicorum* (Lives of German men of medicine)[75]
3. *Vitae Germanorum iureconsultorum et politicorum* (Lives of German jurists and politicians)[76]
4. *Vitae Germanorum theologorum* (Lives of German theologians)[77]
5. *Vitae theologorum exterorum* (Lives of foreign theologians)[78]

Adam adheres to the division of disciplines commonly seen in universities: theology, medicine, law, and philosophy. All the titles stress that the men described in these tomes are from *Germania*, that is, Germanic lands that roughly correspond to the Roman province of *Germania*, enclosed by *Gallia*, the Alps, and *Bohemia*.

A closer look at Adam's attribution of *patria* to these philosophers in the first volume clearly shows his wide interpretation of *Germania* as everything north of the Alps (see Figure 2.5), including the kingdom of Poland, the Low Countries, and even the Swiss Confederacy. In light of the Beza and Reusner's *Icones*, Adam's choice to limit his geographical remit to *Germania* suggests

73 Friedrich Hermann Schubert, 'Adam, Melchior', in *NDB*, vol. 1 (1953), p. 53.

74 Melchior Adam, *Vitae Germanorum philosophorum* [...], *Vitae* I (Frankfurt a.M.: Jona Rosa, 1615).

75 Adam, *Vitae Germanorum medicorum* [...], *Vitae* II (Heidelberg: Jona Rosa, 1620).

76 Adam, *Vitae Germanorum iureconsultorum et politicorum* [...], *Vitae* III (Heidelberg: Jona Rosa, 1620).

77 Adam, *Vitae Germanorum theologorum*, *Vitae* IV (Heidelberg: Jona Rosa, 1620)

78 Adam, *Decades duae continentes vitas theologorum exterorum principum* [...] (Frankfurt a.M.: Jona Rosa's widow, 1623).

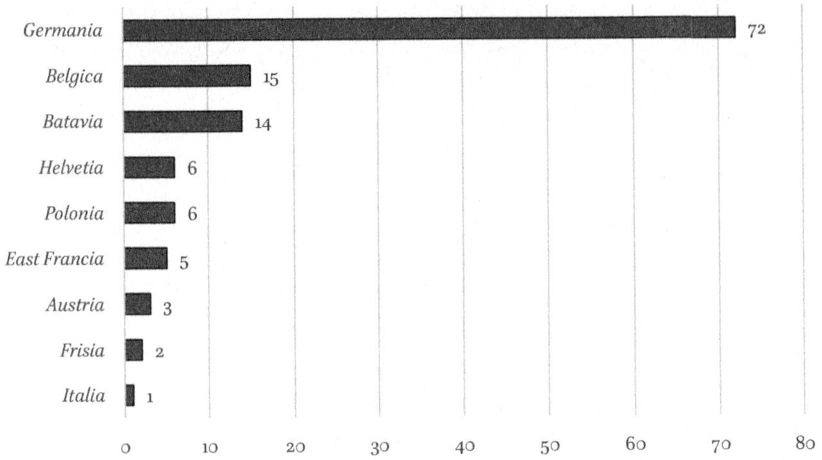


Figure 2.5. Adherence to state and regions in Melchior Adam, *Vitae Germanorum philosophorum* […], *Vitae* I (Frankfurt a.M.: Jona Rosa, 1615)

a continuation of a confessional selection of scholars who contributed to religious reforms. Yet Adam's goal was much more patriotic and nationalistic, rather than confessional. Adam's work can be seen as an attempt to cross a confessional barrier in the scholarly world, since he wanted to expand the history of learning with learned men from a variety of confessions from the Germanic lands.

> Thus, for the not undeserving German, I regarded it was shameful: to know and comprehend in one's mind what some Italian, or Frenchman, or someone across the sea, each in his *patria* will have accomplished in his type of life and learning, but to know very little or nothing about countrymen, born even in the midst of Germany, who served their church or country very well, and to often never have heard of their name, or existence.

> Nam Germano non immerito, opinor, turpe duxerim; nosse, animoque complecti; quid vel Italus aliquis, vel Gallus, vel alius transmarinus, in sua quisque patria, & suo vitae doctrinaeque genere praestiterit: de popularibus vero, & in media Germania natis viris, praeclareque de Ecclesia vel Republ[ica] meritis, parum admodum aut nihil scire, saepe etiam ne nomen quidem, aut natum esse, inaudiisse.[79]

79 Adam, *Vitae* IV, 'Epistola dedicatoria', fol. $3^{v}$.

Adam lamented that Germanic learned men were aware of men of learning from the Italian peninsula, but rarely knew their local colleagues. He had two audiences in mind for this plea: the learned men in Germany who could now get acquainted with their fellow learned men and the history of their community, and also men across the Alps, to ensure that Italian learned men would know of their Germanic counterparts. Adam thus abandoned the narrative of a confessional fragmentation of the European learned world and replaced it with a nationalistic framework.

Adam dedicated most pages in the volume that contains *philosophi* to textual scholars, such as Melanchthon (twenty-one pages), Petrus Lotichius Secundus (1528–1560, seventeen pages), William Canter (1542–1575, seventeen pages), Justus Lipsius (1547–1606, sixteen pages), and Erasmus (fifteen pages). In the fourth volume, Adam presents reformers such as Luther and Calvin as theologians, rather than part of a collective historical process towards the purification of religion and faith. This, too, was a departure from the historical narratives presented by Reusner and Beza, who stressed the centrality of the Reformation.

To help those who were unfamiliar with the Germanic learned men to find their counterparts, Adam divided his overview into disciplines (theology, philosophy, medicine, and law). Adam explains the organisation of his work as follows:

> With this goal in mind for quite a few years now, I fostered for myself amongst other things a desire: to assemble from anywhere I could the lives, accomplishments, writings, virtues, and even vices of Germans, who flourished during the previous century and a little later, known for their erudite learning and who did their heirs fine services; to collect these in certain volumes from wherever I could, placed according to their type of discipline, in the succession of time [chronological]; and to embrace them with simple and concise style, summarily and to not lump them together in a Nocerian[80] leisure of Giovio.

> Hoc fine a complusculis jam annis id mihi studio inter alia habui: ut Germanorum vitas, res gestas, scripta, virtutes, nec non & vitia, qui superiore illo, & quod excurrit, seculo eruditae doctrinae nomine floruerunt; deque posteris nobis bene meriti sunt; unde unde possem, conquirerem; ac certis voluminibus, juxta disciplinarum genera, observata temporum

80 'Nocerian' refers to Paolo Giovio's position as a bishop of the Nocera Diocese of Nocera de' Pagani.

> serie, stylo simplici ac compendioso complecterer, & summatim, non in ocio Jovii Nuceriniano coagmentarem.[81]

Adam refers to the sixteenth century as the age of erudite learning, which is an age of patriotic pride for him. He criticised Giovio's haphazard organisation and clearly positioned his own work as a counterpart to Giovio's elaborate collection of Italian learned men as well as an improvement on his style. Adam did not explain why he chose these respective disciplines as an ordering principle, but his preface does explicate the disciplines. The first volume contains the lives of 'philosophers', which was meant to include philologists, poets, orators, historians, and mathematicians. The second volume included physicians (both doctors and surgeons). The third volume included lawyers and politicians. And finally, the last two volumes contained theologians, 'who have supported and propagated the kingdom of Christ by means of their voice and writings'.[82] Thus, the division of the volumes according to university disciplines moved away from the example set by Giovio. Adam presented knowledge and learned men in a structure and hierarchy that is typical of the early modern university faculties, instead of adopting a discourse on exceeding virtue and exemplarity that had been common since Giovio.

While the confessional aspect is less prevalent in Adam's *Vitae*, the fourth volume with 'German' theologians does have confessional overtones. This tone was immediately set with a quotation from Paul's Epistle to the Hebrews in the preface:

> Remember your leaders who taught you the word of God. Think of all the good that has come from their lives and follow the example of their faith.[83]

In the first of the two volumes on theologians, Adam provided a historical narrative of the learned community in the German lands, in which Reformed theologians and learned men, such as Melanchthon and Calvin figure prominently. Adam presented these theologians as glorious, 'who in our *Germania*, fought strenuously for the church of Christ (*ecclesia Christi*) with their voice and writings'.[84] In Adam's sense of history and collective

81 Adam, *Vitae* I, 'Epistola Dedicatoria', [n.p.], fol. 2$^{v}$.

82 '[Q]ui regnum Christi, voce atque scriptis propugnarunt & propagarunt'. Ibid., fol. 3$^{r}$.

83 Hebrews 13:7. Adam, *Vitae* IV, 'Ad lectorem', [n.p.], fol. 2$^{r}$.

84 '[D]um heic uno in volumine tibi exhibeo vitae Theologorum clarissimorum: qui in Germania nostra nati, ecclesiam Christi strenue propugnarunt, & voce scriptisque late propagarunt.' Adam, *Vitae* IV, 'Lectori benevolo S.P.', [n.p.], fol. 1$^{r}$.

identity, these Protestant theologians—most prominently Luther, Calvin, Melanchthon, and Zwingli—are the most important men of learning, because they fought a theological, political, and intellectual battle against Roman despotism. According to Adam, the lives of these theologians can serve as Germanic 'exemplars of virtue and learning' ('exempla virtutis & doctrinae').

At the same time, Adam saw his work as a considerable improvement of the tradition of life-writing originating from Roman historians, later followed by Giovio. He appropriated Ovid to pay tribute to the Roman tradition of bestowing fame through writing *vitae*, while at the same time implying that the attention should now turn to German learned men:

> and turn to me, *oh*, Overseers of the Muses, and look at the descriptions of these German men:
> *Those whom a first generation has admired, a second generation will also admire*
> *Their applause roars in the learned theatres.*
>
> & tuemini mihi, o Musarum antistites, viros istos Germanos heic descriptos:
> *Quos prior est mirata, sequens mirabitur aetas:*
> *In quorum plausus docta theatra sonant*[85]

Here Adam changed Ovid's 'tota theatra' (all the theatres) to 'docta theatra' (the learned theatres) to convey the idea of a learned circle that received praise and applause from its onlookers.[86] It evokes the image of roaring applause and cheers in a Roman theatre, but the ovations are now directed at the German learned men contained in the volume. According to Adam, these German learned men deserve the highest praise and stand in a tradition going back to Ovid. Adam's message is clear: the next grand step in the history of learning would be made in the Germanic lands, not in the Rome of Ovid or the Pope.

The Italian counterpart to Melchior Adam's *Vitae* was Gian Vittorio Rossi's three-volume *Pinacotheca*. Like Adam's *Vitae*, it was massive, containing 303 lives in total, and it enjoyed increasing popularity, with several reprints

85 Adam, *Vitae* I, 'Epistola dedicatoria', sig. *4 verso. Original italics.

86 The original lines from Ovid, *Ex Ponto*, II, vi, 27–28 read: 'quos prior est mirata, sequens mirabitur aetas, in quorum plausus tota theatra sonant.' My translation based on Ovid, *Tristia; Ex Ponto*, trans. Arthur Leslie Wheeler (Cambridge, MA: Harvard University Press, 1975), pp. 348–349.

throughout Europe after its original publication in 1643. Gian Vittorio Rossi (1577–1647) was a philologist from Rome, who was educated by the Jesuits at the Collegio Romano and afterwards became an important member of the prestigious learned society called the *Accademia degli Umoristi*.[87] Rossi was well connected to fellow European scholars and spent a large part of his career as secretary to cardinal Andrea Peretti (1572–1629), whom he served between 1610 and 1628. After Peretti's death, Rossi returned to Rome, where his house on Gianicolo became a central hub of scholarly exchange. During his visit to Rome in 1629, the French bibliographer, librarian, and scholar Gabriel Naudé (1600–1653) inspired Rossi to construct his vast collection of images and *vitae* of scholars.[88]

*Pinacotheca imaginum illustrium, doctrinae vel ingenii laude, virorum, qui, auctore superstite, diem suum obierunt* (Gallery of images of illustrious men, praised on account of their learning or innate brilliance who died while the author lived) was published in three volumes in 1643, 1645, and 1648 under his Latin pseudonym, Janus Nicius Erythraeus, and a false printer's mark, in both cases to avoid Roman censors.[89] The title is similarly misleading, since the books did not contain a picture gallery (as the word *Pinacotheca* commonly denoted), nor do any of the three volumes contain any visual images, but only textual portraits, mostly written by Rossi. The *Pinacotheca* was probably published outside of Italy, most likely in Amsterdam by Johannes Blaeu (1596–1673), since censors were highly suspicious of Rossi after the publication of his *Eudemia*, satirising the Roman clergy.[90] The *Pinacotheca* had a relatively wide readership in both the Protestant and Catholic parts of seventeenth-century Europe, but especially in the Germanic lands, corroborated by the reprints in Nuremberg in 1676; in Leipzig in 1692, 1693, and 1712; and in Wolfenbüttel in 1729.[91]

87 Luisella Giachino, 'Gian Vittorio Rossi', in *DBI*, vol. 88 (2017).

88 Ingo Herklotz, 'Ianus Nicius Erythraeus und Jean-Jacques Bouchard: Zur schweren Geburt einer neulateinischen Vitensammlung des 17. Jahrhunderts', in *Neulateinisches Jahrbuch: Journal of Neo-Latin Language and Literature*, vol. 10, ed. Marc Laureys and Karl August Neuhausen, 145–176 (Hildesheim; Zürich; New York: Olms-Weidmann, 2008), p. 152.

89 For more details of the publication history, see ibid., pp. 154–156. For the false printers of Rossi's works, see Emil Weller, *Die falschen und fingirten Druckorte: Repertorium der seit Erfindung der Buchdruckerkunst unter falscher Firma verschienenen deutschen, lateinischen und französischen Schriften*, 2 vols. (Leipzig: Wilhelm Engelmann, 1864), 1:261, 262, 263; 2:126.

90 For the false printer of Rossi's *Pinacotheca*, see ibid., 1:261.

91 Gian Vittorio Rossi, *Pinacotheca imaginum illustrium* [...], 3 vols. (Leipzig: Thomas Fritsch, 1712); Rossi, *Pinacothecaimaginum illustrium* [...], 3 vols. (Wolfenbüttel: J. C. Meisner, 1729). For a short consideration of the popularity of Rossi's *Pinacotheca*, see Carlo Dionisotti, 'La galleria degli uomini illustri', in *Cultura e società nel Rinascimento tra riforme e manierismi*,

Although Rossi had a wider European view of the learned world, his *Pinacotheca* focused primarily on learned men from the Italian peninsula (see Figure 2.6). While the content suggests an Italian version of Adam's *Vitae*, the tone is neither patriotic nor nationalistic. The first pages of the *Pinacotheca* contain a description of Rossi's collection and his selection criteria:

> I only included in this *Pinacotheca* the images of those, who, during my age, stood out as worthy of memory, and they who have already passed away while I live. Yet I have not embraced the images of all of them, but only of those, who are either bound to me by friendship or known to me by hearsay or fame. And I do not fear to have angered those of whom I do not speak if they complain to have been ignored, simply because the dead do not know how to be angry. Yet those cannot be said to have been neglected if I have not read their works and did not even know they were born.

> Eorum tantum imagines in hanc Pinacothecam intuli, qui, mea aetate, memoria digni extitere, quique, me vivo, fato functi sunt suo; neque tamen omnium imagines sum complexus, sed eorum tantum, qui mihi, vel amicitia conjuncti, vel auditione & fama noti fuere; eos autem, de quibus tacui, iratos non vereor, quod se fuisse neglectos querantur; quoniam mortui omnino irasci nesciunt: quamquam neglecti non sunt dicendi ii, quorum ego scripta non legerim, quoque ne natos quidem noverim.[92]

This prefatory comment affirms Rossi's defensive stance towards the content of the book: he only included those who were known to him on account of their scholarly fame and those who were within his circle of scholarly friends. Rossi's volume is thus presented as idiosyncratic and not as programmatic as, for example, Adam's *Vitae*.

When we examine Rossi's *Pinacotheca* more closely, we see a clear bias towards Italian learned men. Although Figure 2.6 shows the largest portion dedicated to *Italia*, it is important to note that Rossi often points out more precisely where a certain learned man was either born or active. This

Civiltà Veneziana 32, ed. Vittore Branca and Carlo Ossola, 449–461 (Florence: Olski, 1984), pp. 457–458.

92 Rossi, *Pinacotheca*, vol. 1 ([Amsterdam: Johannes Blaeu], 1642), 'Idem [Ianus Nicius Erythraeus] lectori', fol. 1$^{v}$. The phrase 'fato functi suo sunt' comes from Quintilian, *Institutio Oratoria*, bk. 3, ch. 7, ll. 10.3.

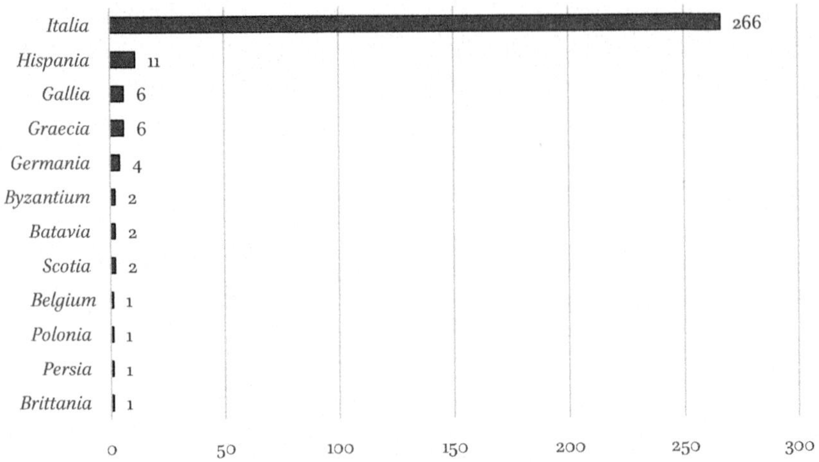


Figure 2.6. Adherence to state and region in Gian Vittorio Rossi's *Pinacotheca*, 3 vols. (1643–1648)

indicates that, unlike Adam, Rossi did not present *Italia* as a consistent geographical category. The *Pinacotheca* is not dramatically skewed to any Italian region and includes an equal number of scholars from Florence, Bologna, Rome, Siena, and to a lesser extent Genoa, Venice, and the regions of Perugia and Urbino. As such, it is a representation of the learned world Rossi was a part of, which centred around Italian cities of learning such as those mentioned above.[93]

Apart from Adam's and Rossi's large-volume works, single-volume lives and eulogies also appeared throughout the seventeenth century. These were most often printed to praise scholars coming from a certain state or academy and can be understood as part of a surge of patriotism throughout Europe, which often manifested itself in regional pride as I mentioned earlier.[94] Examples from Italian compilers are Giovanni Imperiali's *Musaeum Historicum et Physicum* (Museum of historians and physicists, 1640) and Giacomo Filippo Tomasini's *Elogia Virorum Literis & Sapientia Illustrium* (Eulogies of illustrious men on account of their learning and wisdom, 1644). Although the titles of these volumes do not appeal to a nationalistic sensibility, their dedications, prefaces, and contents reveal an appeal to a regional learned world.

93 Mauro Sarnelli, 'Spettacolarità, commercio e scienza nelle raccolte biografiche del medio Seicento', *Proteo* 4, no. 2 (1998): 7–20.

94 Lotte Jensen, ed., *The Roots of Nationalism: National Identity Formation in Early Modern Europe, 1600–1815* (Amsterdam: Amsterdam University Press, 2016).

Imperiali's *Musaeum* was dedicated to a senator of the Venetian Republic, Giovanni Francesco Barbarigo.[95] Dedications were often part of a rhetorical strategy to consolidate patronage, and a place where one can gauge the changing role and position of learned men.[96] Imperiali's book is part of a trend in which collective scholarly life-writing could be presented as a gift to appease and please rulers and statesmen. Learned men and women were increasingly seen as adornments of the state, and a sign of the success of the rulers who governed them. Imperiali thus presented his book as a testament of the success of the state and its ruler.

The *Musaeum* is divided into two books: one dedicated to illustrious men of letters ('illustres literis viri') and the second to ingenious minds of nature ('animi ingenii naturae'). The first book, the *Musaeum historicum* (Historical museum), is a typical collection of portraits and *vitae*, while the second book, the *Musaeum physicum* (Physical museum), is a philosophical exploration of the nature of human brilliance (*ingenium*).[97] In terms of content, Imperiali's volume mostly contains Italian scholars and ecclesiasts, as well as notable Catholic learned men from the perceived periphery: the scholar Justus Lipsius, the Dominican friar Luís de Granada (1504–1588), the poet John Barclay (1582–1621), the anatomist Andreas Vesalius (1514–1564), and Arnaldus de Villa Nova (c. 1240–1311) (see Figure 2.7). When Imperiali addressed the reader, he stressed the continuation of his work with Paolo Giovio's 'well-known volume', including the illustrious learned men from his own age to his *literatorum conspectus*.[98]

In a similar way, Tomasini dedicated his works of collective life-writing to aspiring patrons. His work was stylistically modelled on Rossi's *Pinacotheca*.[99] In the context of the role of prefatory material, it is worth noting that Tomasini's previous volume with lives, *Illustrium Virorum Elogia* (Eulogies of illustrious men, 1630), was dedicated to Giovanni Battista Agucchi (1570–1632), then secretary to Pope Urban VIII (1568–1644) and titular bishop

95 Giovanni Francesco Barbarigo appears to be largely unknown, but the Barbarigo were a well-known patrician family from Venice, bringing forth bishops and cardinals throughout the early modern period.

96 For a consideration of Imperiali's *Musaeum* in the context of the genre of collecting and presenting famous men, see Dionisotti, 'La galleria degli uomini illustri', pp. 460–461.

97 Giovanni Imperiali, *Museaum Historicum et Physicum* [...], 2 vols. (Venice: Giunti, 1640). Vol. 1 is titled *Musaeum Historicum, sive de viris doctrina illustribus*; vol. 2 is titled *Musaeum Physicum, sive de humano ingenio.*

98 Ibid., vol. 1, 'Auctor Lectori S.', fol. 2. The sentence fragment reads: 'praeferente sic nobis facem in percelebri volumine Paolo Iovio, a cuius non abscedentes instituto, continuandam, praestantium literis hominum commemorationem suscepimus[.]'

99 Dionisotti, 'La galleria degli uomini illustri', p. 459.

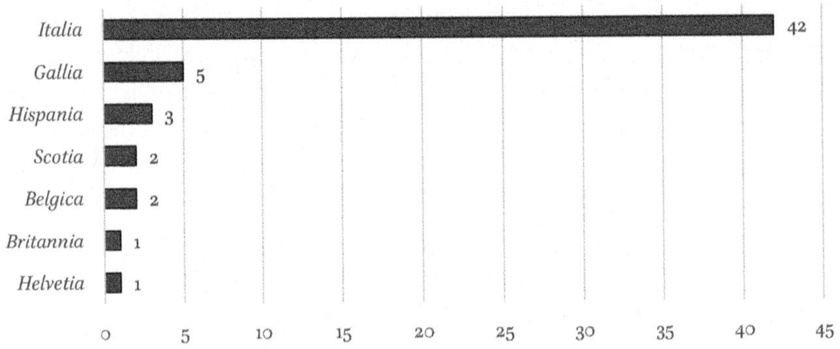


Figure 2.7. Adherence and reference to state and regions in Imperiali's *Museaum* (1640)

of Amasea.[100] By the time Tomasini published his second collection in 1644 on learned and wise men, he himself had taken up Agucchi's position as bishop and became close with high-ranking clergymen. His second work with eulogies, the *Elogia Virorum Literis & Sapientia Illustrium* (Eulogies of illustrious men on account of their wisdom and learning, 1644) was dedicated to Anne of Austria (1601–1666), queen of France. Additionally, another prefatory letter was addressed to cardinal Jules Mazarin (1602–1661), who served as chief minister to the kings of France. Tomasini's dedicatory letters show how life-writing, through dedication and inclusion, could function as an honourable gesture and thus as worth the compiler's while. In a social world of civility, such a gesture could turn the literary genre of life-writing into a social asset.[101] This function of collections of learned men and women as gifts is a development that became the norm, rather than the exception it was in the sixteenth century. Almost every collection of learned men and women was strategically dedicated to a patron or political leader throughout the seventeenth century.

Tomasini's *Elogia Virorum Literis & Sapientia Illustrium* includes Agucchi as a learned and wise man, as well as many other contemporary Italian cardinals and bishops. Tomasini stretched the meaning of a learned man to also include those who have acted wisely in their role as ecclesiastical leaders. The work clearly has a different tone and historical context to the previously mentioned tomes, which positioned themselves in relation to the practice of historical and life-writing in the wake of Paolo Giovio. Tomasini's work as a

100 Giacomo Filippo Tomasini, *Illustrium Virorum Elogia iconibus exornata* [...] (Padua: Donatus Parquardus, 1630), 'Epistola dedicatoria'.
101 On the role of dedications as a social asset to consolidate favours, see Rienk Vermij, 'On the Function of Dedications in Early Modern Scientific Books', *Nuncius* 33, no. 2 (2018): 171–197.

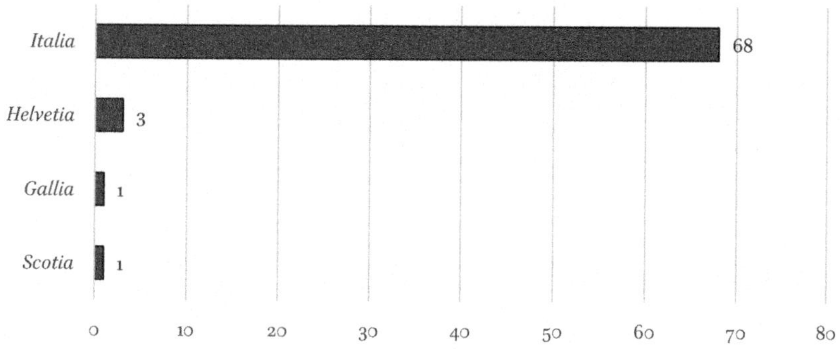


Figure 2.8. Adherence to state and regions in Tomasini's *Illustrium Virorum Elogia* (1630)

chronicler of the lives of 'learned' men is part of a culture of gift-giving and career ecclesiasts. By providing future patrons with a flattering historical narrative of, for example, their ecclesiastical community, Tomasini showed how much of a positive example and cultural icon the man of letters had become in Italy in the seventeenth century.

Together, Imperiali's *Musaeum* and Tomasini's *Elogia* represent a seventeenth-century trend to use collective scholarly life-writing as a means of bestowing praise on friends, patrons, and statesmen, by admiring their state. In these two Italian works, the learned man becomes the pride of states, and a compliment to patrons. The compositions of these collections are in effect more focused on regional peers, who were often active in the same part of society, field of learning, and certainly confession. While it is easy to interpret the change in content as a self-explanatory result of confessionalisation, I would argue that the works of Imperiali and Tomasini show that they are not only representative of confessional learned communities, but also of the increasingly important role of collections as cultural capital.

This trend can also be observed in collections of scholarly lives from the German lands in the seventeenth century. In his *Templum Honoris Reseratum* (Temple of honour unlocked), Gottlieb Spitzel (1639–1691),[102] a Lutheran minister and scholar, turned his attention solely to learned men whom he deemed important to the public, religious, and intellectual success of Leipzig.[103] The frontispiece of the volume shows this imagined temple of honour, its entrance flanked by quotations from the Apostle Paul and Augustine of Hippo, and its halls adorned with portraits of learned men.

102 Herbert Jaumann, 'Gottlieb Spitzel', in *NDB*, vol. 24 (2010), pp. 718–720.
103 Gottlieb Spitzel, *Templum honoris reseratum* [...] (Augsburg: Koppmayer, 1673).

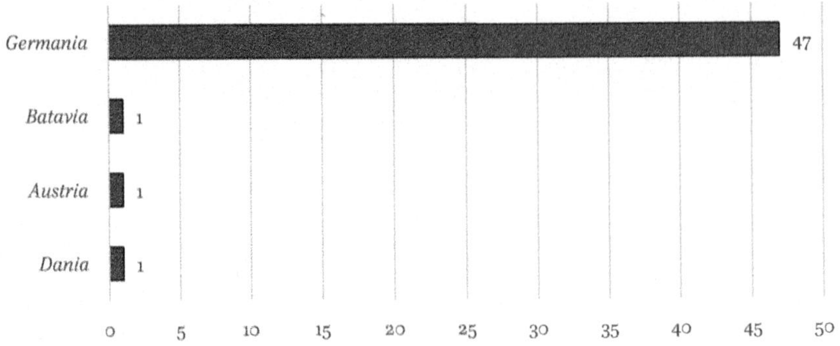


Figure 2.9. Adherence to state and regions in Spitzel, *Templum Honoris Reseratum* (1673)

Spitzel welcomed the reader into a temple that celebrated learned men who, with their voice, writings, or leadership, bestowed honour on the universities and churches with their industriousness, piety, and gravity.[104] Starting with the 'prince of theologians', Martin Chemnitz (1522–1586), Spitzel offered a view of eminent German theologians and philologists from the past century, excluding earlier learned men and Protestant reformers such as Luther. Spitzel's *Templum* honoured those who worked to spread the Lutheran message after Chemnitz. Spitzel's *Templum* is therefore representative of both a confessionalised scholarly memory culture, as well as a regional or 'nationalised' memory culture, in this case from the perspective of Leipzig. The main criteria for inclusion in his compendium were firstly, a scholar's confession and *patria*, and secondly, his merit and prestige. This temple is thus a place where the Lutheran faith and the learning of Lutheran theologians and philologers were celebrated. What is more, it was a temple of worship, an argument that in a certain place or state, in this case *Germania*, virtue was wonderfully nurtured and cultivated.

Scholarly collective life-writing changed considerably throughout seventeenth-century Europe. First, large collections spanning multiple volumes started to emerge both in the Germanic lands and on the Italian peninsula. Both Melchior Adam's *Vitae* and Gian Vittorio Rossi's *Pinacotheca* were attempts to inscribe a new community into the wider history of learning and scholarship. Adam wanted to ensure that Germanic learned men

104 Ibid., 'Epistola dedicatoria', fol. 3$^{v}$. The full sentence reads: 'Horum laudabili excitatus exemplo, de extruendo & adornando quodam Honoris Theologico Literarii Templo sui sollicitus, in quo quinquaginta Virorum imagines sisterentur, qui nostra aetate Academicis pariter ac Ecclesiasticis muneribus singulari cum industria, pietate, gravitate, imo cum bonorum omnium plausi praefuerunt, atque Plus Caeteris Laborantes, rem Theologicam & Literariam non exiguo locupletarunt augmento.'

became part of the pan-European canon of learning; Rossi provided an extensive overview of learned clergymen and scholars throughout Italy. Both acknowledged the influence of Giovio's *Elogia*, which ensured that the archetypical image of the virtuous learned man became part of the cultural and learned landscape throughout Europe.

Most collective scholarly life-writing from the seventeenth century shows an inclination towards a patriotic interpretation of history. The confessional identity that was central in the sixteenth century slowly gave way to a nationalistic context that honoured the achievements of the learned man. In tomes from both sides of the Alps, we observe that collections of learned men were increasingly presented as rhetorical and gestural gifts to rulers and clergy of a separate state. The short collection of lives became a gift that triggered remembrance of learned men in a national context.

## 2.4 Chronicles and Chroniclers of Learned Communities

Until now, we have explored the varieties of collective scholarly life-writing throughout Europe in the early modern period. As we have seen in the previous section, the seventeenth century saw an increase in large multivolume collections of lives, as well as smaller collections that often served as a regional history of learning and a gift to local rulers. In this last section, I want to turn to the larger, encyclopaedic volumes that became prominent in the seventeenth century serving as inspiration for later extensive collections of lives in the eighteenth century.

Learned communities throughout Europe knew the tradition of chroniclers who meticulously recorded those who had passed away in their community. We can see these traditions in the histories of universities, such as the collections of lives of professors at Tübingen or Leiden.[105] At the end of the seventeenth century, we see a move to chronicle the learned world in large tomes. Two large volumes are exemplary of this trend towards encyclopaedia of learned men at the end of the seventeenth century: Paul Freher's *Theatrum Virorum Eruditione Clarorum* (Theatre of men famous

105 See Richard Kirwan, 'Miscellanies of Memory: From Scholarly Biography to Institutional History in the Early German University', in *Memory and Identity in the Learned World: Community Formation in the Early Modern World of Learning and Science*, ed. Koen Scholten et al., 235–254 (Leiden; Boston: Brill, 2022); Joannes Meursius, *Illustrium Hollandiae & Westfrisiae ordinum alma Academia Leidensis: contenta proxima pagina docebit* (Leiden: Jacob Marcus & Joost van Colster, 1614); R. E. O. Ekkart, ed., *Icones Leidenses: De portretverzameling van de Rijksuniversiteit te Leiden* (Leiden: Universitaire Pers, 1973).

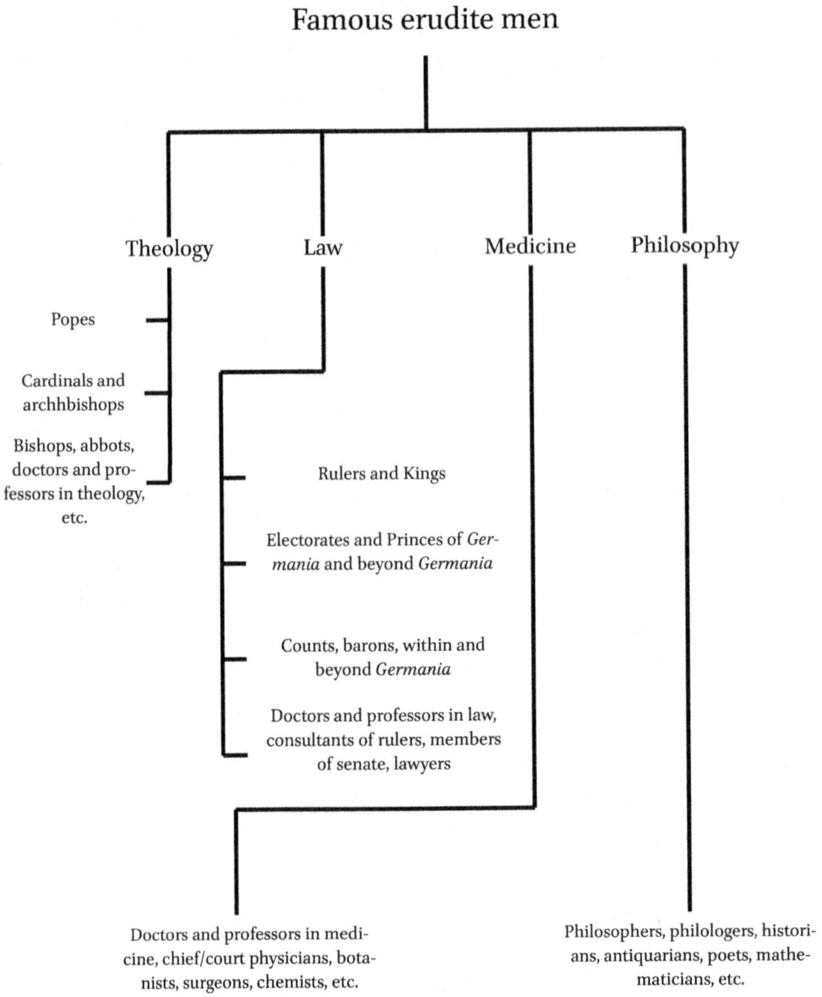


Figure 2.10. Disciplinary classification in Freher, *Theatrum Virorum Eruditione Clarorum* (1688)

on account of their erudition, 1688), and Antoine Tessier's compilation of lives mentioned in Jacques-Auguste de Thou's works.

The *Theatrum Virorum Eruditione Clarorum* by Paul Freher (1611–1682), a distinguished physician from Nürnberg, was an ambitious work that joined theologians, men of law and order, and scholars from the faculties of medicine, and philosophy (liberal arts).[106] Much like earlier works by Melchior Adam and Gian Vittorio Rossi, Freher's *Theatrum* is immense,

106 Paul Freher, *Theatrum virorum eruditione clarorum* [...], 2 vols. (Nuremberg: Johannes Hofmann and Andrea Knorz's widow, 1688).

spanning almost two thousand pages in two volumes. Freher's volume widens the definition of an 'erudite man' considerably. He incorporated a rigid classification scheme that underpinned the structure of the book, as can be seen in Figure 2.10. In essence, Figure 2.10 is a visual display of the table of contents of the *Theatrum.* The book is divided into four parts representing the university faculties in order of importance: theology, law, medicine, and philosophy. Each faculty is then divided into different sections, also according to social and civic status.

The tradition of printed theatres was popular throughout the early modern period. Samuel Quiccheberg's *Inscriptiones vel tituli theatri amplissimi* (Inscriptions, or titles of the highest theatre, 1565) is often seen as the beginning of this tradition, in which a certain topic or collection was displayed to the reader, as if in a theatre. The printed theatre offered a stage for a collection but also showcased the conceptual and organisational framework of the two volumes.[107] In Freher's *Theatrum*, the reader can observe the many men of learning in the four academic disciplines, from 1200 until the 1680s. This *Theatrum* is indicative of a trend that would continue well into the eighteenth century: the publication of large collections of lives of 'erudite' or 'learned' men (and sometimes learned women) in the form of a 'theatre' or encyclopaedia. Such books could be used as a reference work to find the lives and portraits of eminent men as well as introduce the reader to the history of learning, scholarship, and erudition. More importantly, the *Theatrum* provided a stage for the cultural icon that was the learned man. In Freher's *Theatrum* the learned man can take many shapes and forms, but its underlying framework is the university and its four faculties. Freher's *Theatrum* is an example of a systematic, visual encyclopaedia that presents the history of erudition as a chronological progression of eminent men. This history, however, is disciplinary. Each of the four faculties of knowledge—theology, law, medicine, and philosophy—all have their own series of eminent men.

The increasing popularity and visibility of the history of scholarship can also be observed in the work of Antoine Teissier (1632–1712), who at the height of his career was the historian and ambassador of the king of Prussia.[108] In

107 Steffen Siegel, 'Bildnisordnungen: Visuelle Pragmatik in Paul Frehers Gelehrtenlexikon "Theatrum virorum eruditione clarorum" (Nürnberg 1688)', *Archiv für Kulturgeschichte* 90, no. 1 (2013): 79–108.

108 For a more elaborate sketch of Antoine Teissier's life, see Claude Cristin, 'Aux origins de L'histoire Littéraire Française: "Les éloges des Hommes sçavans Tirez de l'histoire de M. de Thou par Antoine Teissier" (1683–1715)', *Revue d'Histoire littéraire de la France* 72, no. 2 (1972): 238–246 (pp. 239–240).

1683, Teissier published the first edition of *Les Eloges des hommes savans* (Eulogies of learned men). Teissier had meticulously read all the historical works of Jacques Auguste de Thou (1553–1617), a sixteenth-century French historian, who had chronicled his lifetime in his *Historiae sui temporis* (Histories of his time).[109] De Thou considered himself to be a successor to Paolo Giovio, who also wrote a work titled *Historiae sui temporis* (Histories of his time) during his own lifetime, which ended in 1545.[110] De Thou started writing immediately after this and his *Historiae* span the years between 1546 up until 1601. De Thou was clearly still inspired by classical Roman historians such as Livy and wrote in elaborate prose, much like Giovio.

Much had changed at the end of the seventeenth century, as can clearly be seen in the adaptation of De Thou's work by Antoine Teissier. He published *Les Eloges des Hommes Scavans Tirez de l'Histoire de M. de Thou* [...] (The eulogies of learned men taken from the history of De Thou), which included eulogies of all the learned men mentioned in De Thou's *Historiae*; the first edition appeared in 1683 and was reprinted in Lyon in 1686.[111] This compilation was hugely popular, as is clear from the appearance of a second edition in 1696 and 1697 in Utrecht, a third edition in Berlin in 1704, and finally, a fourth edition in Leiden in 1715.[112] Over the years, scholars added to Teissier's first volume and the *Eloges* was used as a biographical dictionary and reference work by many.

The *Eloges* are part of the shift of the representation of learned men in narrative to a reference work, usually strictly chronologically ordered. We already saw this in Freher's *Theatrum*, and Teissier's work shows similar traits of organisation and usability. What is even more telling is the clear interest in learned men and women beyond one's own region and *patria*. Both De Thou's *Historiae* and, to an even greater degree, Teissier's adaptation and extension of the lives of learned men from the sixteenth and seventeenth centuries display a pan-European conception of the learned

109 Jacque August de Thou, *Historiarum sui temporis pars prima* (Paris: Ambrose and Jerome Drouart, 1606); De Thou, *Historiarum sui temporis libri CXXXVIII*, ed. Thomas Carte, 7 vols. (London: Samuel Buckley, 1733). For an analysis of De Thou's *Historiae*, see Ingrid A. R. de Smet, *Thuanus: The Making of Jacques-Auguste de Thou (1553–1617)* (Geneva: Librairie Droz, 2006), pp. 201–261.

110 Ibid., pp. 15, 247.

111 Antoine Teissier, ed., *Les Eloges des Hommes Sçavans Tirez de l'Histoire de M. de Thou. Avec des additions Contenans l'Abrégé de leur vie, le jugement, et le Catalogue de leurs Ouvrages* (Geneva: Wilderhold, 1683); Teissier, ed., *Les Eloges des Hommes Sçavans* [...], 2 vols. (Utrecht: François Halma, 1696).

112 Samuel Kinser, *The Works of Jacques-Auguste de Thou* (The Hague: Martinus Nijhoff, 1966), pp. 302–303.

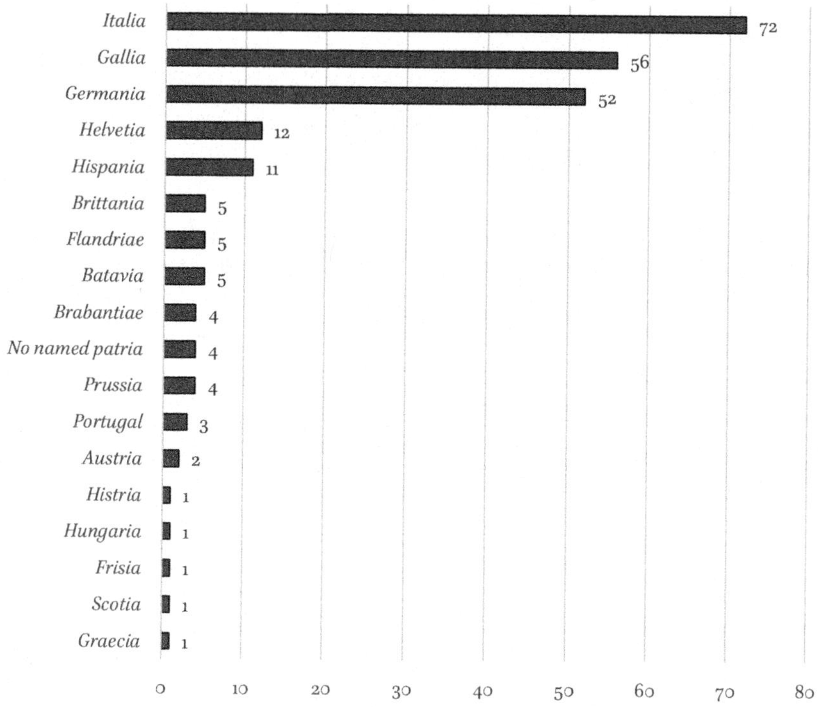


Figure 2.11. Adherence to state and region in Teissier, *Les Eloges des Hommes Sçavans Tirez de Histoire de M. de Thou*, vol. 1 (1696)

world. Figure 2.11 shows that most of the learned men and women come from the Italian peninsula (30 per cent).[113] The next two largest regions by percentage are France (*Gallia*), represented in 23 per cent of lives, and the German lands (*Germania*), appearing in 22 per cent. Teissier reinforced the idea of scholarship thriving throughout Europe on both sides of the Alps, both in the Italian peninsula among cardinals, bishops, professors, and scholars, and in the German lands and its universities as well as the French professors, learned men of government and the king, and clergy. It is important to point out that, although Teissier translated all the lives into French, this did not mean he focused solely on the French-speaking parts of Europe. Quite the opposite: Teissier opened the history of scholarship and its main protagonists, according to De Thou, to a French-reading public, not only for those in France but also for the learned reader beyond

113 The indexation and quantification of Antoine Teissier, *Les Eloges des Hommes Sçavans*, vol. 1 (Utrecht: François Halma, 1696) has primarily been carried out by Nienke van Doeselaar between September 2019 and January 2020.

its borders, since French was slowly overtaking Latin as the language of learning.[114]

Teissier's compilation should thus primarily be seen in the context of an epistemic shift at the end of the seventeenth century. In an attempt to condense and present centuries of accumulated knowledge, scholars started to systematically organise and manage knowledge in reference works, compilations, bibliographies, and catalogues.[115] The epistemic ideal was different for De Thou when compared to Teissier. While De Thou modelled his work on classical Roman historians and wrote in prose, Teissier wanted to compile a systematic reference work. In this quest, Teissier provided the pan-European learned world with a book of its most prominent members from days of yore. In his review of Teissier's *Éloges* in the *Nouvelles de la République des Lettres*, Pierre Bayle (1647–1706) praised the books, since the eulogies 'were spread across a vast corpus' ('répandus dans un vaste corpus') of De Thou's *Historiae* and copies were rare.[116] Bayle wrote:

> Finally, we can assure the reader that he will find a collection of curiosities so full, so diverse, so exact, and so well provided with the necessary citation, that he will have no regret having profited by reading a few hours, a harvest which undoubtedly cost the one who gathered it many years. Besides, the partialities of religion have been wisely banished from this work.

> Enfin on peut assurer le lecteur qu'il trouvera un assemblage de curiositez si plein, si diversifié, si exact, & si bien muni des citations nécessaires, qu'il ne se repentira point d'avoir profité par une lecture de quelques heures, d'une moisson qui a coûté sans doute beaucoup d'années à celui qui l'a recueillie. Au reste les partialitez de Religion ont été bannies sagement de cet Ouvrage.[117]

114 Françoise Waquet, *Latin or the Empire of a Sign: From the Sixteenth to the Twentieth Centuries* (London; New York: Verso, 2001), esp. ch. 3, 'Latin Scholarship'; Dena Goodman, *The Republic of Letters: A Cultural History of the French Enlightenment* (Ithaca, NY; London: Cornell University Press, 1994).

115 Ann Blair, *Too Much To Know: Managing Scholarly Information Before the Modern Age* (New Haven, CT: Yale University Press, 2010), ch. 3, pp. 117–172.

116 The full quote reads: 'Premièrement les exemplaires de cette Histoire sont rares. En second leiu les éloges n'y sont pas de suite, ils sont répandus dans un vaste corps.' Pierre Bayle, 'Article II: *Les Eloges des hommes Sçavans tirez de l'Histoire de M. de Thou, avec des additions contenant l'Abregé de leur vie, le jugement & le Catalogue de leurs Ouvrages, par Antoine Teissier Avocat au Présidial de Nimes.* A Geneve chez Jean Herman Widerhold 1683. 2 vol. in 12.', *Nouvelles de la République des Lettres*, October 1684 (Amsterdam: Henry Desbordes, 1686), pp. 772–773.

117 Ibid., p. 774.

Bayle's praise of the banishment of partialities of religion refers to the fact that both De Thou and Teissier were not partisan towards a single confession or religion in the compilation of the *Éloges*. That Bayle noted this as an advantage reflects the emergence of the eighteenth-century ideal of a pan-European community of learning that disregarded political and confessional borders. As can be seen in Figure 2.11, Teissier presented a compilation of learned men from all over Europe, as well as from different confessions (both Catholic and Protestant confessions). We also know that the passage cited above was not just empty praise, because Bayle cited Teissier in his *Dictionnaire historique et critique* (1697) extensively. Very often one finds 'Mr. Teissier reports ...' ('Mr. Teissier rapporte ...'), but also more critically 'Mr. Teissier who cites him shouldn't have' ('Mr. Teissier qui le cite ne devoit pas').[118] The *Éloges* was a primary example of how the knowledge of learned men of yore was represented from the late seventeenth century onwards: that is, as pan-European and supra-confessional.

Another interesting development that can be observed in the composition of Teissier's *Éloges* is the rise of the liberal arts in the compilations of *vitae*. While earlier tomes often largely focused on theologians and clergy, the eulogies taken from De Thou's *Historiae* were primarily from what we would now call the liberal arts, as can be seen in Figure 2.12.[119] Learned men and scholars from the liberal arts constitute more than 50 per cent of the content of Teissier's *Éloges*. While the prominence of liberal arts scholars in the *Éloges* can partly be explained by De Thou's interests and scientific acquaintances, it also signals an increasing interest and value attached to these fields of knowledge.

The popularity of compilations of *vitae* of learned men and women steadily increased from the late seventeenth century onwards. In the eighteenth century, many of the works mentioned above were reprinted in new editions throughout Europe, suggesting a rise in the interest of the *vitae* of learned men and women. The many editions and reprints of Freher's *Theatrum* and especially Teissier's *Éloges* signify this change. The collections in Latin slowly started to recede in quantity and visibility, making way for French,

118 Pierre Bayle, *Dictionaire historique et critique*, vol. 2, pt. 1 (Rotterdam: Reinier Leers, 1697), pp. 135 and 466.

119 Liberal arts include mathematics, grammar, rhetoric, philology, history, philosophy, antiquarianism, and humanities in general (e.g., 'professeur en humanité'). Clergy includes abbots, bishops, cardinals, and other secretaries of the church. Government includes members of parliament, senate, and other functionaries of regional governments.

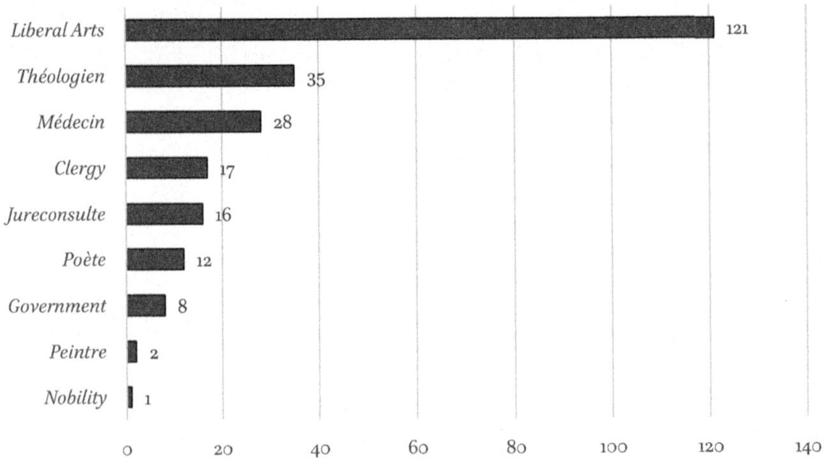


Figure 2.12. Preoccupation and field of knowledge of people described in Teissier, *Les Éloges des Hommes Sçavans* [...], vol. 1 (1696)

the new lingua franca of the learned world.[120] At the same time, the history of scholarship and scholars gained popularity in a more popular and accessible form. Lives of scholars began to appear in multivolume periodical publications written in the vernacular.

A good example of such a popular, vernacular edition of scholarly life is Gerardus Outhof's (1673–1733) compilation of famous scholars. In the 1730s, Outhof, a preacher and rector of a Latin school in the Dutch town of Kampen, instigated the publication of a bi-monthly publication containing 'descriptions of lives of well-known and learned men' in the Dutch language.[121] Outhof is also known as the author of numerous volumes on theology, antiquity, and a history of floods from Noach to the present. The six volumes on learned men constituted a series of thirty-one issues published between 1730 and 1733. Each number could be bought individually.[122] In the preface of the first volume, Outhof expresses his surprise that no one had undertaken a series of volumes in Dutch on the lives of learned men. Furthermore, Outhof used the preface to explain his choice of subjects, starting with 'that time, when in the thick darkness of ignorance and godlessness, some began to resist the scandalous lives of the so-called Roman clergyman, and their

120 Waquet, *Latin*, esp. ch. 10, 'Yearning for the Universal', pp. 257–270; also see April G. Shelford, *Transforming the Republic of Letters: Pierre-Daniel Huet and European Intellectual Life, 1650–1720* (Rochester, NY: University of Rochester Press, 2007).

121 P. J. Blok and P. C. Molhuysen, 'Gerardus Outhof', in *NNBW*, vol. 2 (1912), pp. 1046–1047.

122 Gerardus Outhof, *Levensbeschryvinge van bekende en geleerde mannen*, 6 vols. (Amsterdam: A. Wor and Erve G. onder de Linden, 1730–1733).

wrongful understanding of the sacrament of the Lord's supper'.[123] Readers were encouraged to send written lives of learned men, in this case to the publisher Adriaan Wor in Amsterdam. Outhof's work leaned towards a confessional collection of lives, although it contains a wide variety of learned men, from both the German, Italian, and Greek lands. The reformers took centre stage, with Zwingli's life counting 101 pages, Luther's 168 pages, and Calvin's 107.[124] While the volumes are filled with learned men of varying nationalities (for want of a better term), they are all framed as contributors to the Protestant cause because they prepared and advanced the purification of religion through a critical reading of the Bible. Outhof's collection thus presents the Reformation as a transnational phenomenon: the foundations were established by Agricola and Valla, the first bricks laid by Luther, Calvin, and Zwingli. At the same time, Outhof emphasised textual scholarship that was not blatantly confessional, such as that of the neo-Stoic scholar Justus Lipsius. The inclusion of Lipsius is particularly striking, because he converted back to Catholicism at the end of his life when he moved to Leuven. Regardless, the learned man is presented as one contributing to the 'purification of text and religion' in a typical Protestant fashion. However, when we examine the contents of the individuals in terms of their *patria*, it quickly becomes clear that Outhof maintained a balance between a variety of nationalities (as can be seen in Figure 2.13).

At the same time, Outhof's work placed a lot of emphasis on a local culture of remembrance. When we take a glance at the frontispiece, we can see the symbols of the city of Amsterdam hanging proudly above the allegory (see Figure 2.14). The point of remembering is illustrated here: the industrious youth is eager to feed from the lap of truth, personified by Pallas Athena, with time on her side. The young move books and portraits around in the large library. At the top of the frontispiece, the crown and the weapon of the city of Amsterdam are adorned with the portraits of learned men that the young have hung up. The frontispiece thus invokes the idea that learning and the obtainment of truth go hand in hand with the remembrance of learned men.[125] The remembrance of learned men through these lives is thus depicted as both an educational and an honourable pursuit. This

123 '[E]en begin te maken met dien tydt, wanneer in de dikste duisternisse van ontwetenheidt en godtloosheidt, sommigen zich hebben beginnen te verzetten tegen het schandelyk leven der *Roomsche* zo genoemde *Geestelyken*, en hun verkeert begrip van *het Sakrament des H. Avontmaals*.' Ibid., vol. 1, 'Aan den lezer'.

124 Ibid., vol. 4, pp. 421–522, 559–648; vol. 5, pp. 3–82, 695–799, resp.

125 The page facing the frontispiece explains the allegorical depiction as follows: 'De Jeugt verzamelt al d'Afbeeltsels, hoog geacht, // Der wyze Mannen, om hun waerde hier beschreven.

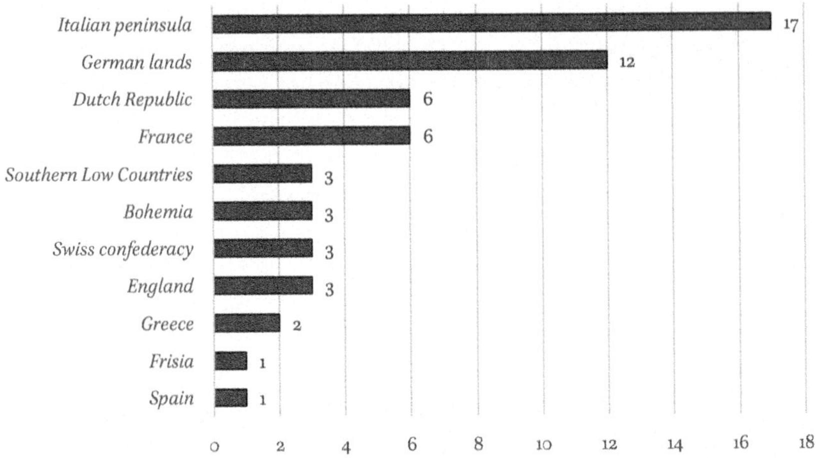


Figure 2.13. Adherence to state and region in Outhof *Levensbeschryvingen van bekende en geleerde mannen*, 6 vols. (1730–1733)

Figure 2.14. Frontispiece of Gerardus Outhof, *Levensbeschryvingen van bekende en geleerde mannen*, vol. 1 (1730). UBU: S OCT 111, DL 1

suggests that the readers Outhof had in mind were young and local, and that Outhof believed that his audience should know the pan-European history and major scholars of the pursuit of textual and religious purity through scholarship. Outhof showed the direction the cultural icon of the learned man was taking in the eighteenth century: it became a way of life that was attainable for studious Dutch young men, who did not necessarily know Latin or French, but who could nonetheless try to follow in the footsteps of great scholars before them.

Outhof's book reflected the rising popularity of science and scholarship in the public sphere.[126] With the rendition of reference works such as Antoine Teissier's compilation of lives from Jean-Jacques de Thou's *Historiae sui temporis*, the life of the learned and famous man reached the new learned French-speaking audience. In the early eighteenth century, vernacular collections of lives of pan-European heroes of scholarship and knowledge started to appear. Outhof's *Levensbeschryvingen* shows how the collective European history of scholarship—based on the idea that knowledge came from Greece, bloomed in Italy, and then spread to the rest of Europe—was appropriated for local heroes (in this case Erasmus), and religious reformers (Luther, Zwingli, and Melanchthon). Outhof's work also shows how the learned man became a more popular cultural icon, since his work was aimed at a popular audience as opposed to a Latinate or French academic or scholarly one. It is a facet of the popular imagination of the learned world, amidst works of critique, parody, and caricature from the early eighteenth-century.[127] The scholar was increasingly seen as a public figure fit for praise, parody, and scrutiny. At the same time, the learned man was increasingly expected to benefit the public good of the region and state.[128]

// Voorts ziet men Amsterdam in't flaeu verschiet vertoont, // Dat Nederlant met dees vergulde lettren kroont.' Ibid., vol. 1, frontispiece.

126 Bernadette Bensaude-Vincent and Christine Blondel, eds., *Science and Spectacle in the European Enlightenment* (Aldershot: Ashgate, 2008); Geoffrey V. Sutton, *Science for a Polite Society: Gender, Culture and the Demonstration of Enlightenment* (Boulder, CO: Westview Press, 1995); Michael Lynn, 'Enlightenment and the Public Sphere: The Musé de Monsieur and Scientific Culture in Late Eighteenth-Century Paris', *Eighteenth Century Studies* 32 (1999): 463–476; Huib J. Zuidervaart, 'Science for the Public: The Translation of Popular Texts on Experimental Philosophy into the Dutch Language in Mid-Eighteenth Century', in *Cultural Transfer Through Translation: The Circulation of Enlightened Thought in Europe by Means of Translation*, ed. Stefanie Stockhorst, 231–262 (Amsterdam; New York: Rodopi, 2010).

127 Sari Kivistö, *The Vices of Learning: Morality and Knowledge at Early Modern Universities* (Leiden; Boston: Brill, 2014).

128 Reinhart Siegert, 'Scientific "Patriotism" between Self-Importance, Self-Recommendation and Camouflage: The Enlightenment Requirement of Public Benefit as Reflected in the Title Pages and Prefaces of Popular Works by Scholars', in *Scholars in Action: The Practice of Knowledge*

## 2.5 Conclusion

Throughout the large timespan covered in this chapter we can see two parallel memory cultures. On the one hand, we detect a collective identity and memory culture of a pan-European learned community that in its imagination transcended both political and confessional borders. Part of this collective identity was a shared history, in which the flourishing of scholarship throughout Europe was seen as part of the same tradition, exemplified most strongly by the sixteenth-century works of Johannes Fichard and by Antoine Teissier's late seventeenth-century compilation of eulogies from Jacques-Auguste de Thou's *Historia sui temporis*. This is a master narrative of the history of scholarship that runs through many of the collections in one form or another. On the other hand, we see a collective identity that was regional and confessional, especially between the end of the sixteenth century up until the seventeenth century. Compilations of *vitae* were idiosyncratic and were strongly tied to their political and cultural contexts: the Reformation *Icones* tomes of Theodore Beza and Nicolaus Reusner; large compilation projects from the seventeenth century, such as Melchior Adam's *Vitae* and Gian Vittorio Rossi's *Pinacotheca;* as well as compilations with a regional character such as Tomasini's *Elogia*. In one way or another, these very different memory cultures all dealt with the master narrative described above: Beza and Reusner turned their back on Catholic books of martyrs and *viri docti*, and tried to change the definition of the illustrious learned man; Adam and Rossi explicitly saw their work as an expansion of Giovio's *Elogia* and this historical master narrative; while regional works such as those of Tomasini and Spitzel stressed the pan-European fame and praise that their local learned men enjoyed and, in their view, deserved. The continuation of the pan-European master narrative thus relied on regional and confessional engagements with this very same narrative. At the end of the seventeenth century, and into the eighteenth century, the collections became larger and more encyclopaedic. The vernacular works of Teissier and Outhof show that the learned man became a more popular cultural icon, attracting interest from outside scholarly circles.

Throughout the early modern period, collections of *vitae* helped establish the cultural icon of the learned man as a virtuous and praiseworthy figure. The collections reflect learned communities with regional, political, and confessional memory cultures. The imagined community of the Republic of

*and the Figure of the Savant in the 18th Century*, ed. André Holenstein et al., vol. 2 (Leiden; Boston: Brill, 2013) pp. 853–876.

Letters was thus an amalgamation of local imaginations that all appropriated a master narrative of pan-European learning and scholarship. These local learned men inscribed themselves into the ideal of the Republic of Letters as a pan-European community of learning beyond political and confessional boundaries.

# 3. Collective Memory and Identity in Hugo Grotius's Correspondence

**Abstract:** Focusing on Hugo Grotius's extensive correspondence, this chapter delves into how scholars constructed collective memory through letters. Grotius's epistolary exchanges, particularly his frequent citations of ancient, religious, and early Christian authors, reflect a shared intellectual framework and historical consciousness. The chapter emphasises how values like friendship, trust, and brilliance shaped scholarly identities and how Grotius positioned himself in the Republic of Letters by fostering historical awareness through citation practices. His letters serve as an example of how early modern intellectuals sustained learned communities through shared values and historical references.

**Keywords:** Hugo Grotius, epistolary networks, collective memory, citation, scholarly virtues, scholarly values

The exchange of letters was a substantial part of the exchange of knowledge in the early modern period. Throughout the entire period, scholars exchanged many epistles (after the Latin *epistolae*). Such epistolary exchanges not only served the purpose of informing fellow scholars of recent developments, but also signified one's importance in the learned world. The role of letters and letter-writing is fundamental in understanding the formation of learned communities in the early modern learned world.

The early modern obsession with letters and their exchange was shaped by the resurgence and popularity of Roman learned culture among scholars and students. Romans wrote a lot, including letters, a selection of which has been passed down to us.[1] The exchange of letters created a certain sense of community, in which the sender and receiver fostered a special

1 Stephanie Ann Frampton, *Empire of Letters: Writing in Roman Literature and Thought From Lucretius to Ovid* (New York: Oxford University Press, 2019).

Scholten, Koen. *Rethinking the Republic of Letters: Memory and Identity in Early Modern Learned Communities*. Taylor & Francis Group, 2025
DOI 10.5117/9789048559855_CH03

bond based on the principles of trust. The practice of ancient letter-writing served to forge and foster relationships between two writers. In their letters, writers would often stress the relationship with their addressees, as well as pointing out that they both belonged to the same community with shared values and beliefs.[2] A clear example of this type of community-building can be found in Cicero's letters. Ingo Gildenhard argued that these approximately three hundred letters from the period between 49 BCE and 44 BCE, just after Caesar's crossing of the Rubicon, were designed to unite a community of readers in a Republican cause, despite Caesar's autocratic rule and the waning power of high-profile Republicans.[3] The letters allowed their authors to create and sustain communities that reached far beyond the relation between sender and recipient. Cicero's letters were clearly aimed at a wider reading public that sympathised with the Republican cause. Thus, Gildenhard argued, Cicero's letters served 'as a means and medium in which Cicero could promote a (nostalgic) sense of civic community.'[4]

The clearly stylised rhetoric of ancient epistles and their potential to forge communities from the Roman period are relevant, because early modern scholars modelled and mimicked the rhetoric and style of these letters and they thus imagined and sustained communities in a similar way. Letter-writing was an essential aspect of being a well-connected, learned man or woman in the early modern period. The writing of letters was an art that had to be taught and learned. In this light, we can understand the rise of manuals on how to write elegant and learned letters from the end of the fifteenth century onwards. The most popular manual was Desiderius Erasmus's (1466–1536) *De conscribendis epistolis* (On the writing of letters).[5] In this enriched school manual Erasmus offers readers types of letters, each with their own style, form, and rhetoric.[6] It contains numerous examples from the letters of Cicero, Pliny, and Politian

2 Paola Ceccarelli et al., eds., *Letters and Communities: Studies in the Socio-Political Dimensions of Ancient Epistolography* (Oxford: Oxford University Press, 2018); see esp. Thorsten Fögen, 'Ancient Approaches to Letter-Writing and the Configuration of Communities through Epistles', in ibid., pp. 43–80.

3 Ingo Gildenhard, 'A Republic in Letters: Epistolary Communities in Cicero's Correspondence, 49–44 BCE', in *Letters and Communities*, ed. Ceccarelli et al., pp. 205–236.

4 Ibid., p. 233.

5 Desiderius Erasmus, 'On the Writing of Letters / De conscribendis epistolis', trans. and annotated by Charles Fantazzi, *Collected Works of Erasmus: Literary and Educational Writings*, vol. 25, ed. J. K. Sowards, 1–254 (Toronto; Buffalo, NY; London: University of Toronto Press, 1985).

6 Charles Fantazzi, 'Vives versus Erasmus on the Art of Letter-Writing', in *Self-Presentation and Social Identification: The Rhetoric and Pragmatics of Letter-Writing in Early Modern Times*, ed. Toon Van Houdt et al., 39–56 ((Leuven: Leuven University Press, 2002).

to teach people how to write an elegant epistle, suitable in form, tone, and rhetoric for various situations and contexts. In other words, the art of letter-writing could be learned from classical authors and followed rules and forms.[7]

The letter form and correspondence in general are often considered to represent a private dialogue between a single author and a single addressee. The characterisation of a letter as a 'dialogue between absent friends' can be found in many letters and letter writing manuals. Erasmus repeated this trope in his *De conscribendis epistolis*:

> The letter is like a mutual talk between absent friends (as the comedian Turpilius aptly wrote). It should not be harsh or rough, also not beautified nor solely about one topic, nor tediously long. This type of letter favours simplicity, clarity, joy, and wit. Therefore, we shall strive for gracefulness of expression, except when topics or persons require otherwise.

> Est enim (quod scite scriptum est a Turpilio comico) epistola absentium amicorum quasi mutuus sermo, quem neque agreste, aut asperum, neque fucatum esse opportet neque una de re, neque ad fastidium usque multum. Eoque simplicitate, candore, festivitate, argutiam amat hoc epistolarum genus. In his igitur omnia leporum genera consectabimur, nisi quum res, aut personae diversa postulabunt.[8]

Erasmus also compared the character of the letter with an intimate, soft-spoken conversation between friends:

> For this ought to be the character of the letter: as if you were whispering in a corner with a dear friend, not shouting in the theatre, or otherwise somewhat more freely. For we commit many things to letters, which would be shameful to express in public.

7 For an overview of letter-writing manuals, see Dirk van Miert, 'Letters and Epistolography', in *The Classical Tradition*, ed. Anthony Grafton et al., 520–523 (Cambridge, MA: Belknap Press, 2010).

8 Erasmus, *Opus de conscribendis epistolis recognitum denuo ab autore & locupletatum* (Cologne: apud Heronem Alopecium, 1523), fol. 10$^{r-v}$, sig. B-ij$^{r-v}$. For the modern edition see Desiderius Erasmus, *Opera Omnia*, vol. 1, no. 2 (Amsterdam: North-Holland Publishing Company, 1971), pp. 153–579, hereafter referred to as ASD. For an English translation, see Desiderius Erasmus, 'On the Writing of Letters', in *Collected Works of Erasmus*, ed. Sowards, pp. 1–254.

> Is enim debet esse epistolae caracter, tanquam cum amiculo in angulo susurres, non in theatro clames, aut paulo etiam liberior. Multa enim epistolae committus, quae coram pudet expromere.[9]

This view of the letter as an intimate conversation between friends continued to be a model throughout the early modern period. In these settings, such friends can express their common ideals and goals.[10] Erasmus's work on letter-writing was popular throughout the seventeenth century and was enthusiastically republished and consulted.[11]

The letter could truly be a private dialogue *in absentia*, but often had a more public character.[12] Letter-writing was part of the early modern culture of exchange and communication.[13] We can think of the exchange of letters in political, domestic, religious contexts, and in academic or scholarly contexts. The letter was a well-known cultural symbol; early modern readers and writers could gauge the content of a letter based on its material qualities such as the way it was folded, the ink, or the quality of the paper.[14] More importantly and relevantly to the topic at hand, letters were vital in scholarly and scientific exchanges in the early modern period.[15]

9 Desiderius Erasmus, *Libellus de conscribendis epistolis* (Cambridge: Siberch, 1521), fol. 1$^r$, cited in, and translation based on Lisa Jardine, *Erasmus, Man of Letters: The Construction of Charisma in Print* (Princeton, NJ: Princeton University Press, 1993), p. 151.

10 For intellectual friendships, see Kathy Eden, *The Renaissance Rediscovery of Intimacy* (Chicago; London: University of Chicago Press, 2012); Eden, *Friends Hold All Things in Common: Tradition, Intellectual Property, and the* Adages *of Erasmus* (New Haven, CT: Yale University Press, 2001).

11 For the editions of Erasmus's *De conscribendis epistolis* from 1540 onwards, see ASD vol. 1, no. 2, pp. 173–184.

12 Gary Schneider, *The Culture of Epistolarity: Vernacular Letters and Letter Writing in Early Modern England, 1500–1700* (Newark: University of Delaware Press, 2005).

13 Ulrich Johannes Schneider, ed., *Kultur der Kommunikation: Die europäische Gelehrtenrepublik im Zeitalter von Leibniz und Lessing* (Wiesbaden: Harrassowitz, 2005); Hans Bots and Françoise Waquet, eds., *Commercium Litterarium: La Communication dans la République des Lettres / Forms of Communications in the Republic of Letters, 1600–1750* (Amsterdam; Maarssen: APA-Holland University Press, 1994).

14 For a detailed account of material aspects and practices of early modern letters, see James Daybell, *The Material Letter in Early Modern England: Manuscript Letters and the Culture and Practices of Letter-Writing, 1512–1635* (Basingstoke; New York: Palgrave Macmillan, 2012).

15 Dirk van Miert, ed., *Communicating Observations in Early Modern Letters (1500–1675): Epistolography and Epistemology in the Age of the Scientific Revolution* (London: Warburg Institute; Turin: Nino Aragno, 2013); Andrea Rusnock, 'Correspondence Networks and the Royal Society, 1700–1750', *British Journal of the History of Science* 32 (1999): 155–169; J. L. Pearl, 'The Role of Personal Correspondence in the Exchange of Scientific Information in Early Modern France', *Renaissance and Reformation* 8, no. 2 (1984): 106–113.

Scholarly epistolary exchanges could, for example, contain the results of measurements, opinions, and responses to recently published theses, a more intimate exchange of thoughts, a list of the most recently published books, or something as mundane as the weather or the local news. Women could use correspondence to communicate and fashion themselves as learned and well-connected.[16] Thus, the letter served many functions and could be used as a form of self-presentation and community-building. Learned men and women could create and spread their self-image as well as define themselves as part of a community.[17]

In this chapter we will focus on the role of letters and letter-writing in the construction of imagined learned communities by looking at the rhetoric, shared values, and historical consciousness in the letters of Hugo Grotius (1583–1645). With a corpus of letters that spans almost his entire lifetime, we can assess the mechanisms through which epistles served as a vehicle of community formation in the early modern learned world. Four aspects of epistolary correspondence made it an essential vehicle of community-building. First, in contrast with the temporary nature of conversations, disputations, or lectures, the permanent character of the letter made them collectible items, embedded in an early modern practice of collecting and of memory cultures. Secondly, letters can extend communities in space, because the letter allowed writers to reach an audience, both private and public, throughout Europe. The letter also extends communities in time, since a letter can become a document through which a scholar can converse with the past. In the most extreme case, the letter can become a transhistorical conversation, as in the case of Petrarch's letters to Cicero after his discovery of Cicero's lost letters to Atticus in 1345.[18] Thirdly, correspondence collections enabled the publication of popular memorial editions containing letters or specific epistolary exchanges. In the early modern period, letters of pre-eminent figures were often collected and reprinted in an *opus epistolarum* ('work of letters'). These printed correspondence editions further helped to memorialise the correspondent. Furthermore, they elevated both

16 Julie D. Campbell and Anne R. Larsen, eds., *Early Modern Women and Transnational Communities of Letters* (Farnham; Burlington, VT: Ashgate, 2009); James Daybell, 'Letters', in *The Cambridge Companion to Early Modern Women's Writing*, ed. Laura Lunger Knoppers, 181–193 (Cambridge: Cambridge University Press, 2009); Carol Pal, *Republic of Women: Rethinking the Republic of Letters in the Seventeenth Century* (Cambridge: Cambridge University Press, 2012).

17 Van Houdt et al., *Self-Presentation*.

18 Chris Heesakkers, 'Driemaal Cicero: De Neolatijnse Cicero-cultus van Petrarca tot Lipsius', *Lampas* 26, no. 2 (1993): 131–158; Eden, *Renaissance Rediscovery of Intimacy*, ch. 2., 'A Rhetoric and Hermeneutics of Intimacy in Petrarch's *Familiares*', pp. 49–72.

the individual learned letter and extensive correspondences as scholarly achievements in their own right. As a consequence, these collections became part of lively memory cultures surrounding early modern scholars. Fourthly and finally, the archival collections of letters as well as the printed correspondence editions could be used for different intellectual or political purposes. This flexibility of correspondence collections allowed them to be used and reused throughout the early modern period (and beyond) in quite diverse memory cultures.[19]

The extant correspondence of Hugo Grotius consists of 7,459 letters. Molhuysen began work on the edition of the correspondence in 1917; the first volume appeared in 1928, and the final volume came out in 2001. The textual analysis of the letters in this chapter is based on this seventeen-volume collection.[20] The ePistolarium project made these printed and edited letters available through a digital portal.[21] The results and textual analysis are based on the digital text files in the ePistolarium database. Each letter was formatted in a plain text file and subsequently analysed using AntConc software.[22] Additionally, the entire correspondence collection was placed in a database containing all the metadata from the ePistolarium, including the author and addressee of the letter, as well as the place and date of writing. This database was further managed, disambiguated, and analysed with the use of OpenRefine.[23]

The number of letters Hugo Grotius wrote and received each year varied considerably. The number of correspondents increased dramatically throughout his life as well as the number of letters he wrote when he became more successful and better known. At the same time, we must take into

19 Patrick Baker, *Italian Renaissance Humanism in the Mirror* (Cambridge: Cambridge University Press, 2015).

20 Hugo Grotius, *Briefwisseling van Hugo Grotius*, ed. P. C. Molhuysen et al., 17 vols., Rijks geschiedkundige publicaties, Grote serie 64, 82, 105, 113, 119, 124, 130, 136, 142, 154, 179, 197, 213, 222, 238, 246 and 248 (The Hague: Martinus Nijhoff, 1928–2001). This collection of correspondence will be referred to as *BW* in the remainder of this chapter. The correspondence of Grotius can also be accessed online at https://grotius.huygens.knaw.nl. The analysis that follows fully relies upon and follows the transcriptions in this edition. The appendices to the *Briefwisseling* have been excluded from the analysis here.

21 Walter Ravenek et al., 'The ePistolarium: Origins and Techniques', in *CLARIN in the Low Countries*, ed. Jan Odijk and Arjan van Hessen, 317–323 (London: Ubiquity Press, 2017).

22 AntConc 3.5.7 (Macintosh OS X and Windows), developed by Laurence Anthony, Faculty of Science and Engineering, Waseda University, Japan. For a short introduction to AntConc and its use, see Shawn Graham et al., *Exploring Big Historical Data: The Historian's Macroscope* (London: Imperial College Press, 2016), pp. 79–83.

23 For a complete guide to OpenRefine, see Ruben Verborgh and Max de Wilde, *Using OpenRefine* (Birmingham: Packt, 2013).

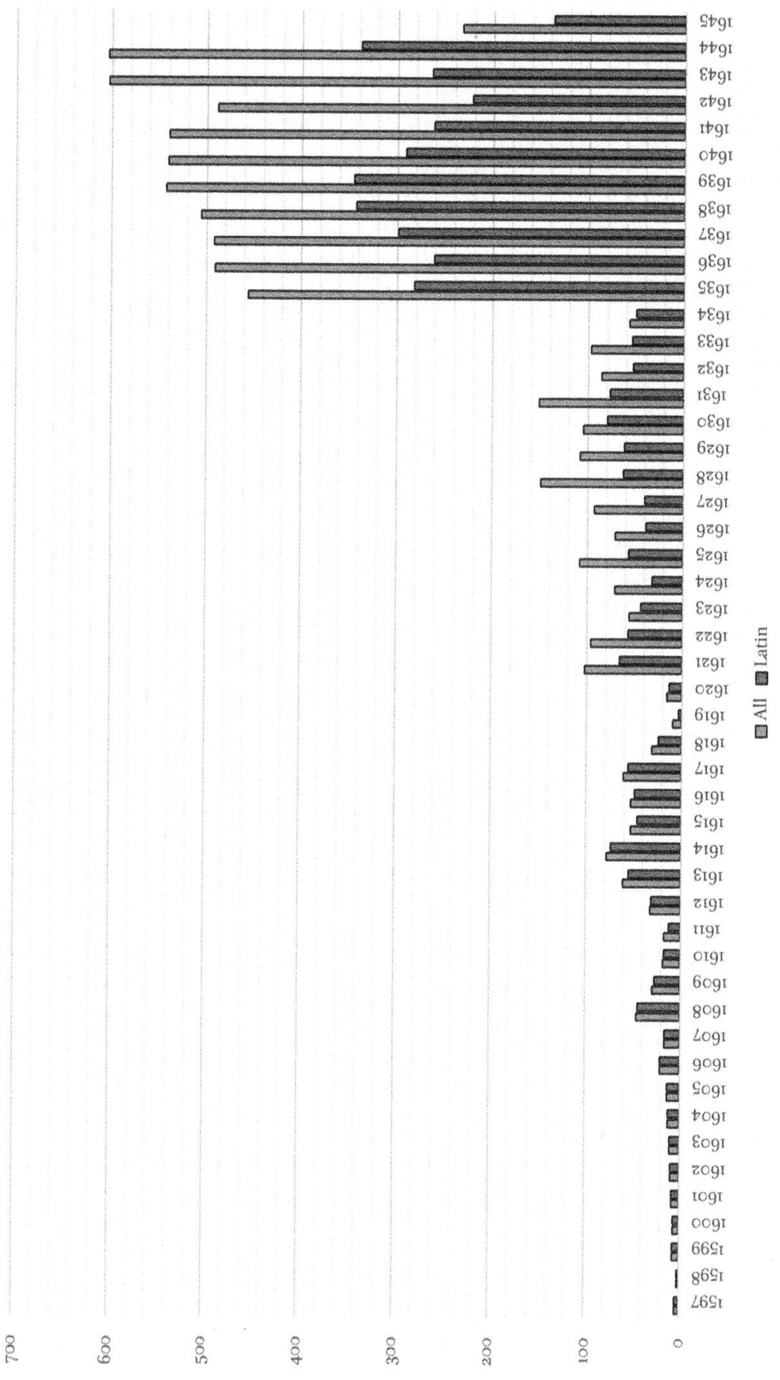


Figure 3.1. Latin letters written to and from Hugo Grotius per year compared to all letters in all languages

consideration that while Grotius wrote considerably fewer letters in his early years, it is also likely that some earlier correspondence was lost. Or, conversely, since Grotius gained prominence in later life, his letters likely ended up in personal and archival correspondence collections. Figure 3.1 shows the number of extant letters for each year of Grotius's life, including both the outgoing and incoming correspondence.

The language of the letters is another noteworthy aspect of Grotius's development as a correspondent. As Figure 3.1 shows, the number of letters written in Latin compared to other languages sharply decreased. To make this trend visible, I have plotted the number of Latin letters as a fraction of the total letters for every year in Figure 3.1. Here we clearly see that Grotius wrote exclusively in Latin in the first few years of his career. Yet upon taking public office he began to write in Dutch to some friends, in French to colleagues and functionaries in France, and in German with colleagues and functionaries in the German-speaking parts of Europe. Nevertheless, throughout his life Grotius continued to correspond in Latin with government emissaries, scholarly colleagues, and his son. Since this chapter solely focuses on the Latin letters it will exclude roughly half of the total correspondence (see Figure 3.1 and Table 3.1 for a detailed overview by year). The excluded letters are of a more diplomatic and intimate nature, while scholarly and official letters were often written in Latin. The selection excludes, however, the letters to his brothers-in-law Johan and Nicolaas van Reichersberch and to his own wife, Maria van Reigersberch (1589–1653), as well as Willem van Oldenbarnevelt (1590–1639), the son of the famous state pensionary Johan van Oldenbarnevelt (1547–1619).

There are several reasons to focus solely on the Latin letters. First, it is the most frequently used language in Grotius's correspondence. Second, it allows for comparisons with Chapter 1's analysis of the language used in collections of *vitae*. Thirdly, and probably most importantly, the digital text analysis only works in one language at a time. To include the other languages—French, Dutch, German—would require a considerable investment of time with very little return given that the focus of this research is on the scholarly aspect of his correspondence.

This chapter will focus on three aspects of Grotius's scholarly correspondence and examine how these mechanisms helped establish a collective memory and identity among scholars more generally. The first section will focus on how Grotius tried to establish himself in the learned world, especially in his first epistolary exchanges with Isaac Casaubon and Justus Lipsius. The second section will unravel the citation and reference patterns in

**Table 3.1. Letters written to and from Hugo Grotius. In the columns: the year, the total amount of letters, the number of Latin letters and the percentage of Latin letters per year**

| Year | Total | Latin | % Latin | Year | Total | Latin | % Latin |
|---|---|---|---|---|---|---|---|
| **1597** | 4 | 4 | 100% | **1622** | 96 | 57 | 59% |
| **1598** | 2 | 2 | 100% | **1623** | 56 | 44 | 79% |
| **1599** | 7 | 7 | 100% | **1624** | 71 | 32 | 45% |
| **1600** | 6 | 6 | 100% | **1625** | 108 | 57 | 53% |
| **1601** | 8 | 8 | 100% | **1626** | 71 | 39 | 55% |
| **1602** | 9 | 9 | 100% | **1627** | 93 | 40 | 43% |
| **1603** | 10 | 10 | 100% | **1628** | 149 | 63 | 42% |
| **1604** | 12 | 12 | 100% | **1629** | 108 | 62 | 57% |
| **1605** | 13 | 13 | 100% | **1630** | 105 | 80 | 76% |
| **1606** | 21 | 21 | 100% | **1631** | 151 | 77 | 51% |
| **1607** | 16 | 16 | 100% | **1632** | 86 | 53 | 62% |
| **1608** | 46 | 45 | 98% | **1633** | 97 | 54 | 56% |
| **1609** | 29 | 27 | 93% | **1634** | 57 | 50 | 88% |
| **1610** | 18 | 17 | 94% | **1635** | 456 | 282 | 62% |
| **1611** | 17 | 12 | 71% | **1636** | 491 | 261 | 53% |
| **1612** | 32 | 31 | 97% | **1637** | 492 | 299 | 61% |
| **1613** | 61 | 55 | 90% | **1638** | 506 | 343 | 68% |
| **1614** | 78 | 74 | 95% | **1639** | 543 | 345 | 64% |
| **1615** | 53 | 46 | 87% | **1640** | 541 | 291 | 54% |
| **1616** | 53 | 49 | 92% | **1641** | 540 | 262 | 49% |
| **1617** | 61 | 56 | 92% | **1642** | 489 | 222 | 45% |
| **1618** | 31 | 24 | 77% | **1643** | 603 | 264 | 44% |
| **1619** | 9 | 3 | 33% | **1644** | 604 | 338 | 56% |
| **1620** | 15 | 13 | 87% | **1645** | 233 | 137 | 59% |
| **1621** | 102 | 66 | 65% | ***Total*** | *7459* | *4378* | *59%* |

the correspondence. Who was cited in the correspondence of Grotius, and in what context? The citation patterns in the correspondence will be analysed as a form of memory and identity formation. The third section addresses the virtues and values that were extolled in the Grotius correspondence. What kind of virtues and values did Grotius and his correspondents foster, and how? The digital textual analysis of the virtues will inform us about what kind of characteristics were important for Grotius's imagination of the learned world.

Grotius's correspondence offers a good case study, not only because the letters are digitised, but also because Grotius was an ideal member of the Republic of Letters. He was not an average early modern scholar, but rather a polymath who stood out in the breadth of his interests and political engagement. Grotius studied and wrote about classical Greek and Latin philology, biblical criticism, and law, but also about the political aspects of religion. Before we continue with the analysis of identity formation in Grotius's correspondence, we will briefly introduce his life.[24]

Hugo Grotius was born in a wealthy patrician family in Delft in 1583.[25] His father, Jan Huygenszoon de Groot (1554–1640) was a well-educated and by all accounts successful merchant who instructed his son from an early age in literature and other branches of learning. Jan de Groot was also well connected to contemporary scholars in the Dutch Republic, such as Simon Stevin (c. 1548–1620) and Justus Lipsius (1547–1606). As a youngster, Hugo went to the elite Latin school in Delft and matriculated at Leiden University in 1594 at the age of eleven. At the young university, he was taught Greek by Bonaventura Vulcanius (1534–1614), lodged in the house of theologian Franciscus Junius (1545–1602) and was acquainted with philology by one of his favourite professors, Joseph Justus Scaliger (1540–1609).[26] After approximately four years at Leiden University, Grotius ventured on a diplomatic journey to France, most probably in the company of statesman Johan van Oldenbarnevelt, who was Advocate of Holland at the time.[27] On this tour around France, he visited many statesmen, ambassadors, nobility, and scholars. In May 1598, at the age of fifteen, he even received his doctorate in Roman and canon law from the University of Orléans.[28] When he returned to Delft in the same month, this formal and informal education prepared him for both scholarship and statesmanship.

In the years that followed, Grotius established himself in the European world of learning. He sent his first scholarly work, the *Syntagma arateorum*, to Justus Lipsius in 1600; collected data for the French polyhistor

24 This section takes Henk Nellen's 2007 biography as its main source of biographical information: see Henk Nellen, *Hugo de Groot: een leven in strijd om vrede, 1583–1645* (Amsterdam: Balans, 2007); the English translation was published as *Hugo Grotius: A Lifelong Struggle for Peace in Church and State, 1583–1645*, trans. J. C. Grayson (Leiden; Boston: Brill, 2015). This chapter will generally draw on the English translation, with references to more detailed information about the events described in the footnotes. See also this more recent concise biography: Henk Nellen, *Geen vredestichter is zonder tegensprekers: Hugo de Groot, geleerde, staatsman, verguisd verzoener* (Amsterdam: Athenaeum, 2021). See also Randall Lesaffer and Janne E. Nijman, eds., *The Cambridge Companion to Hugo Grotius* (Cambridge: Cambridge University Press, 2021).

25 Nellen, *Grotius*, pp. 24–33. The business of Jan de Groot, Hugo's father, was flourishing during Hugo's youth, but went bankrupt around 1600. Subsequently, Jan was charged with cases of financial malversations and renounced his Delft citizenship in 1601, see ibid., pp. 84–86.

26 Ibid., pp. 35–40.

27 Ibid., pp. 45–46.

28 Ibid., pp. 49–50.

Jacques-Auguste de Thou's *Historiae sui temporis* (discussed in Chapter 2.4); and worked closely with scholarly friends in Leiden—professor of poetics Daniel Heinsius, Johannes Meursius, extraordinary professor of eloquence Dominicus Baudius (1561–1613), and Simon Stevin. On matters of law and state Grotius worked together with people such as Oldenbarnevelt and Johannes Wtenbogaert (1577–1644), leader of the Remonstrant movement in the Republic.[29] In 1609, Grotius published an influential book on the legal status of the high seas titled *Mare liberum* (The free sea), one of the fruits of his legal work.[30] Up until 1613, when Grotius was appointed Pensionary of Rotterdam, he had a magnificent career in statesmanship and politics.

Grotius pleaded for religious tolerance in the context of the religious pluralism that characterised the Dutch Republic. He became involved in the theological controversy between the Remonstrants (also called Arminians), who supported Leiden professor Jacobus Arminius (1560–1609), and the Counter-Remonstrants (also called Gomarists), who supported the Calvinist theologian Franciscus Gomarus (1563–1641). After Arminius's death, his Leiden chair was filled by Conrad Vorstius (1569–1622), who had Socinian leanings, causing major upheaval. Grotius defended the Leiden authorities' right and power to appoint a professor, even against the wishes of religious authorities. Grotius summed up this position in his pamphlet *Ordinum pietas* (The piety of the states) and continued to push an agenda of religious tolerance for years on behalf of the States General together with Oldenbarnevelt.[31] Eventually this led to a conflict with Maurice of Nassau (1567–1625), *stadtholder* for most of the provinces of the Dutch Republic. In 1619, Oldenbarnevelt was sentenced to death and Grotius to life imprisonment for their political and religious activities.

Grotius eventually escaped and fled to Paris two years later, where he stayed for the rest of his life. He would continue his scholarly and political work in Paris and published *De iure belli ac pacis* (On the law of war and peace) on the legal status of war.[32] In Paris, Grotius became the ambassador of Sweden (1634–1645) and continued to publish about, and argue in favour of, the unity of Christians and a tolerant Christianity. On his way back from Sweden he died in a shipwreck on 28 August 1645.[33] His body was

29 Ibid., pp. 56–82. For the relationship between Daniël Heinsius and Grotius, see J. P. Heering, 'Kunstbroeders en rivalen: Heinsius en Grotius', in *Voorbeeldige vriendschap: Vrienden en vriendinnen in theologie en cultuur*, ed. H. J. Heering, 47–56 (Groningen: Styx, 1993).

30 Hugo Grotius, *Mare liberum* [...] (Leiden: Lodewijk Elzevir, 1609).

31 Hugo Grotius, *Ordinum Hollandiae ac Westfrisiae pietas* [...] (Leiden: Ioannes Patius, 1613).

32 Hugo Grotius, *De iure belli ac pacis libri tres* [...] (Paris: Nicolaus Buon, 1625).

33 Nellen, *Grotius*, pp. 725–734.

moved to his birthplace of Delft and he was buried in the Nieuwe Kerk on 3 October 1645.[34] After his death, Grotius was praised as a great statesman and lawyer.[35] The celebrated artist and sculptor Rombout Verhulst designed and made a clay model for a tomb monument in the church in Delft, but it was never executed.[36]

## 3.1 Hierarchy, Collective Goals, and Conduct

The early modern letter demanded an advanced competence in style and rhetoric, since different social contexts asked for specific rhetorical strategies.[37] One would employ a different tone and language in an intimate letter to a friend, compared to a missive or even a letter to one's parents. Moreover, the stage of the correspondence also mattered. The rhetoric in the first letters that initiated a correspondence exchange were often wholly different to the letters between two correspondents after a sustained epistolary exchange over the course of years. In this section, I want to show how early modern scholars used a specific scholarly rhetoric that aimed at initiating, but especially at maintaining and continuing an epistolary and scholarly connection.[38] More specifically, this section will focus on the role of hierarchy as well as the centrality of common ideals and collective identity in the exchange between Grotius and fellow scholars.

According to Anne Goldgar in her book *Impolite Learning*, from the end of the seventeenth century onwards the most important common concern in the learned world was the conduct of its members. Moreover, the conduct of scholars was primarily what bound them as a community. Based on her close

34 Grotius was buried next to Maurice, Prince of Orange, on the left side of the choir; see Nellen, *Grotius*, pp. 734–735, especially n. 58. The tomb memorial was erected a century later in 1781 by members of the Grotius family.

35 See, e.g., Jacob Ploos van Amstel, *Een tempel voor Hugo de Groot: een rede van Jacob Ploos van Amstel uit 1774*, ed. Jan Bloemendal and Hans Krist Ploos van Amstel (Amersfoort: Florivallis, 2010). For Grotius's posthumous influence and controversies surrounding his work, see Nellen, *Grotius*, pp. 737–758.

36 M. van Notten, *Rombout Verhulst beeldhouwer 1624–1698, een overzicht zijner werken* (The Hague: Nijhoff, 1907), p. 80; cited in Frits Scholten, *Sumptuous Memories: Studies in Seventeenth-Century Dutch Tomb Sculpture* (Zwolle: Waanders Publishers, 2003), p. 59, n. 194.

37 For the functions of different Neo-Latin paratext, see Karl A. E. Enenkel, *Die Stiftung von Autorschaft in der neulateinischen Literatur (ca. 1350–ca. 1650): Zur autorisierenden und wissensvermittelnden Funktion von Widmungen, Vorworttexten, Autorporträts und Dedikationsbildern* (Leiden; Boston: Brill, 2015).

38 On the topic of epistolary continuity, see Schneider, *Culture of Epistolarity*, pp. 56–68.

reading of an extensive amount of early modern correspondence, Goldgar argued that learned men 'identified much more strongly with the concept of a community to be defended by behavioural rules' than with ideology, religion, or any other philosophical framework.[39] In epistolary interactions, scholars would find a common sense of community bound by manners and a code of conduct. This has also been called the moral economy of knowledge, since one's membership of the learned community depended on one's adherence to certain codes of conduct, norms, and values.[40] Correspondence is an interesting source for the study of these prevalent norms and values, since most letters tend to stress the relationship between the sender and addressee and contain many value-laden compliments and remarks.

The correspondence of a scholar thus reflects what I would call a dance of obligation. The first letter can often be interpreted as an invitation to this dance. A reciprocal to-and-fro of obligations and gratitude then follows. This sense of obligation stems from the increasing culture of civility from the late sixteenth century onwards. In that period, the word 'civility' referred to correct social behaviour. Manuals on behaviour such as Erasmus's *De civilitate morum puerilium* (On civility in children), which was very popular and appeared in print all over Europe, helped propagate ideals of civility.[41] Courtesy manuals often stressed the transactional and dynamic character of courtesy, where small courtesies such as favours and invitations helped to create and reinforce social relationships.[42] When one performed courtesy for someone, this courtesy could either be accepted or rejected. If one accepted, one was obliged. Good manners dictated that one returned the favour to sustain the reciprocal and transactional relationship. Manners

39 Anne Goldgar, *Impolite Learning: Conduct and Community in the Republic of Letters, 1680–1750* (New Haven, CT; London: Yale University Press, 1995), p. 8.

40 For use of the term 'moral economy of knowledge', see, e.g., Steven Shapin, *A Social History of Truth: Civility and Science in Seventeenth-Century England* (Chicago; London: University of Chicago Press, 1994); Lorraine Daston, 'The Moral Economy of Science', in *Constructing Knowledge in the History of Science*, ed. Arnold Thackray, 3–24 (Chicago: University of Chicago Press, 1995); Marian Füssel, '"On the Means of Becoming Famous in the Learned World": Practices in Scholarly Constitution of Status and the Emergence of a Moral Economy of Knowledge in the Eighteenth Century', in *Scholars in Action: The Practice of Knowledge and the Figure of the Savant in the 18th Century*, vol. 1, ed. André Holenstein et al., 123–144 (Leiden; Boston: Brill, 2013).

41 Desiderius Erasmus, *De civilitate morum puerilium* [...] (Basel: Froben, 1530). The book was extensively published in the first few years in, for example, Salzburg (1530), Leipzig (1530, 1531, 1532, 1534), Cologne (1531, 1532), London (1532), Augsburg (1532, 1537), Strassbourg (1534), and Vienna (1536). For the modern edition see 'De pueris statim ac liberaliter instituendis', in ASD vol. 1, no. 2, pp. 1–78. See also Anna Bryson, *From Courtesy to Civility: Changing Codes of Conduct in Early Modern England* (Oxford; New York: Oxford University Press, 1998), pp. 46–50.

42 Ibid., pp. 122–124.

and courtesies were thus an exchange currency, which served as social debt and credit.[43]

In Grotius's letters, we see a similar exchange of social debt and credit in the 'learned exchange' (*commercium literarium*). The correspondence shows that the community of scholars was bound by a sense of hierarchy as well as common goals. Throughout his life, Grotius changed the tone and style of his letters depending on his hierarchical position in the learned world as well as the goal of the letter. As we will see (and as might be expected), in the early years some letters had the distinct goal of establishing a correspondence (with varying success), while later letters focused more on maintaining a relationship and a correspondence.

Grotius was considered a child prodigy. In his early years he was already in contact with senior learned men. This did not mean that he exchanged letters with them on equal footing. In the few letters that remain from his early years, most are attempts to initiate a learned exchange with the prominent scholars of his time, such as Justus Lipsius,[44] Jacques Auguste de Thou (1553–1617),[45] Isaac Casaubon (1559–1614),[46] and Bonaventura Vulcanius.[47] In the initiation phase of a correspondence with senior learned men, young Hugo would introduce himself and stress the unequal relationship with the addressee.

To take one example: Grotius put himself in a subordinate position in his first letter to Justus Lipsius, whom he probably only met as a young boy.[48] The unequal relationship is obvious from the pompous salutation: 'Hugo Grotius greets Justus Lipsius, the prince of letters'.[49] The letter is an attempt to meet Justus Lipsius, a scholar Grotius admired. As a gift to Lipsius, Grotius sent his *Syntagma Arateorum* (1600).[50] It was an edition of the astronomical treatise from Germanicus (24 BC–19 AD), based on the *Phaenomena* of Aratus of Soloi in Cilicia (c. 275 BC).[51] The presentation of a gift is an attempt to ensure a sense of obligation and exchange. If Lipsius accepted the letter and the gift, a reciprocal relationship could start. To further accentuate his low rank

43 Ibid., pp. 124–125.

44 Hugo Grotius to Justus Lipsius, 14 September 1600, *BW* I, p. 14, letter no. 17.

45 Hugo Grotius to Jacques Auguste de Thou, 1 April 1599, *BW* I, p. 7, letter no. 8.

46 Hugo Grotius to Isaac Casaubon, 18 February 1602, *BW* I, p. 22, letter no. 29.

47 Hugo Grotius to Bonaventura Vulcanius, 1600, *BW* I, p. 12, letter no. 14.

48 Lipsius left Leiden when Hugo Grotius was seven years old.

49 'I. Lipsio literarum principi H. Grotius salutem.' Hugo Grotius to Justus Lipsius, 14 September 1600, *BW* I, p. 14, letter no. 17; Justus Lipsius, *Iusti Lipsi epistolae*, ed. Aloïs Gerlo et al., 13 vols. (Brussel; Leuven; Paris: Peeters, 1978–), vol. XIII, no. 2862, pp. 226–227. Hereafter referred to as *ILE*.

50 Hugo Grotius, *Batavi Syntagma Arateorum* [...] (Leiden: Officina Plantiniana, apud Christophorum Raphelengium, 1600).

51 Nellen, *Grotius*, pp. 56–57; also see *ILE* XIII, p. 227, n. 3.

in the social hierarchy, Grotius showers Lipsius with compliments on his recently published panegyric on the Archduke Ferdinand and Archduchess Isabelle, edited together with Pliny the Younger's famous *Panegyricus* on the emperor Trajan. Grotius exemplifies Lipsius's greatness, stressing that Lipsius's laudatory address compares favourably with Pliny's *Panegyricus*:

> We have seen your *Panegyricus*[52] on the Archdukes, as well as the other Panegyric by Pliny, which is slightly less than yours. Herein, you are considered by everyone to have annotated things so varied and worthy of you that you arouse my desire for you (incredible for you; intolerable for me). I am only a peer for you because the aura of Scaliger supports me.

> Vidimus Panegyricum tuum ad Duces et alterum Plinii paulo minus quam tuum, in quo tam varia et te digna annotasse ab omnibus iudicaris, ut incredibile vobis, intolerabile nobis suscites desiderium tui, cui pares non essemus, nisi nos reficeret aura Scaligerana.[53]

The letter stresses the importance of high-quality textual scholarship. To finish this passage, Grotius refers to his teacher Scaliger and the high standard of scholarship he represented. Scaliger, after all, was Grotius's great inspiration and was cited extensively in his *Syntagma arateorum*.[54] Grotius emphasises that Lipsius clearly works at the highest scholarly level, that of Scaliger, which may be compared with Pliny's eloquence.

To increase his chance of establishing a lasting correspondence with Lipsius, Grotius presents himself as a scion of the patrician Grotius family. Young scholars such as Hugo could not yet draw on their own network or achievements and they would often refer to their illustrious teachers (in this case Scaliger) and the achievements and standing of their family. When one was low in the hierarchy, it helped to associate oneself with mentors and family who were higher in the social hierarchy. In this letter, Grotius refers to his father and his deep respect for Lipsius:

> We are all extremely devoted to you, especially however my father Jan de Groot, who always cared for you and valued you greatly, and now greets you in a most obliging manner.

52 Justus Lipsius, *Dissertatiuncula apud Principes* [...] (Antwerp: Plantiniana apud Ioannem Moretum, 1600).

53 Grotius to Lipsius, 14 September 1600, *BW* I, p. 14.

54 Nellen, *Grotius*, p. 56, n. 13.

> Sumus omnes tui quam studiosissimi, praesertim vero pater meus I. Grotius, qui te semper coluit, maximique fecit, et nunc quam officiosissime salutat.[55]

Grotius presents himself as kind and humble, like a civilian approaching a king at an audience. He invokes such a social setting explicitly by calling Lipsius a 'prince of letters' and by offering Lipsius a gift. The gift is a work to show his abilities, while stressing how much he owes to Lipsius intellectually from reading his work. Additionally, Grotius is quick to point out that it is not just he who can grow because of Lipsius, but also his father, who seems to have been an unofficial student of Lipsius and is thankful for his wisdom and works. That Hugo's father greeted Lipsius in 'a most obliging manner' ('officiosissime salutat') underscores the indebtedness of both Hugo and his father to Lipsius. The invocation of family is important in two ways. First, Hugo can position himself within a lineage and history of learning that extends to the life of his father. Secondly, Grotius can climb the social ladder by referring to his father, whom Lipsius apparently knew more intimately. Grotius tries to inscribe himself into an existing scholarly community by building on the identity of the Grotius family as a scholarly family.

Lipsius replied roughly a month later. He had received the letter and book and extended his thanks in a carefully constructed style that is both honourable and colloquial.[56] He considered Hugo to be a worthy successor of the name of Grotius, which brought forth his uncle, professor Cornelis Grotius (1544–1610), as well as Hugo's father, Jan de Groot (1554–1640). Lipsius addresses Grotius as a superior and teacher:

> Could it not be that when I think of you, I think of your father? He is an old student—he wouldn't deny it—and friend of mine. And you when you at your young age—so to say—have accomplished with hard work and innate brilliance what few of more advanced age accomplish. I congratulate you, my dear young man, I congratulate you; and with my mind and with my pen I encourage you to continue with straight steps on this road of glory. If only, per my innermost wish, it would be allowed to come back to see you, your father, uncle, and old friends there after things have calmed down.

55 Grotius to Lipsius, 14 September 1600, *BW* I, p. 14.

56 On Justus Lipsius's style in letters, see Justus Lipsius, *Principles of Letter-Writing: A Bilingual text of Justi Lipsi Epistolica institutio*, ed. and trans. R. V. Young and M. Thomas Hester (Carbondale: South Illinois University Press, 1996).

> An non sic esset, cum patrem tuum, cum te cogito: quorum ille vetus mihi et discipulus — non abnuet — et amicus: tu in ista aetatula — sic loquendum est — industria et ingenio effecisti, quod pauci in perfectiore illo aevo? Gratulor mi adolescens, gratulor: et animo ac stilo te excito, ut recto pede laudis hanc viam curras. Atque utinam, intimum meum desiderium, tranquillatis rebus fas te, patrem, patruum, et veteres ibi amicos revisere![57]

In this letter, Lipsius positions himself as a teacher vis-à-vis Hugo, since he clearly assesses the initial letter as a sort of test. In other words, Lipsius embraces his elevated hierarchical position in the learned world. Hugo has passed the test and Lipsius congratulates him on his spirit and style and his Aratea edition. In his reply, Lipsius points out that Grotius's father was a student of his as well. Moreover, he stresses a related set of virtues that will ensure Grotius's path to glory: hard work (*industria*) and innate brilliance (*ingenium*). Lipsius ends the letter with an encouraging message: 'as that famous author [Euripides] says, "God reveals a road to what we had not expected". Let Him preserve and increase your modesty, my Grotius, your virtue, and your fame.'[58]

The second and last exchange between Lipsius and Grotius is in the winter of 1601–1602. Again, Grotius greets Lipsius as the prince of letters ('Lipsius literarum princeps') and offers another book, his drama *Adamus exul* (Adam in exile).[59] This letter can be perceived as a second attempt to engage in a learned exchange with Lipsius, since Grotius repeats the rhetorical strategy of the first letter. Grotius presents his book as follows: 'I send you, most bright Lipsius, the *Tragoedia Sacra* [i.e., *Adamus exul*]' and expects Lipsius's response. 'I know you would cry out [this is] "a work full of dangerous risk"'.[60] This exclamation ('periculosae plenum opus aleae') from Horace's *Odes* acknowledges that Grotius knows Lipsius well, and stresses their mutual appreciation of Horace's work. Grotius stresses the importance of tolerance in the face of hostile critics, something Lipsius had endured more than once. Hugo continues to conjure an image of critics

57 Justus Lipsius to Hugo Grotius, 16 November 1600, *BW* I, letter no. 18, p. 14; *ILE* XIII, no. 2895, p. 291.

58 '[U]t ille ait, τῶν ἀδοκήτων πόρον εὗρε θεός. Is te, mi Groti, tibi modestiam, virtutem, famam servet augeatque.' Ibid.

59 Hugo Grotius to Justus Lipsius, 1 November 1601, *BW* I, no. 25, p. 20. Hugo Grotius, *Sacra in quibus Adamus exul tragoedia aliorumque eiusdem generis carminum cumulus consecrate Franci principi* (The Hague: Albertus Henrici, 1601).

60 'Inclamabis scio *Periculosae plenum opus aleae*'. Grotius to Lipsius, 1 November 1601.

who condemn others and argues that it is essential that Lipsius reads and comments on his work. He writes: 'It is necessary that someone who is about to write foresees that some people loath, others condemn and the majority tear apart with insults whatever they read.'[61] Yet Grotius also presents himself as someone who is ready to face the criticism and to stay neutral: 'And look, I openly confess that I have not privileged either of the parties.'[62]

Lipsius responded four months later with a solemn letter.[63] He had received Grotius's *Adamus exul*, and 'read part of it and on that basis approved of the entire work.'[64] Lipsius compliments Grotius on his work, but warns him of the path he has taken towards writing stylistically pleasing pieces and steers him to continue his pursuit of the study of wisdom.

> Intellectual philosophy is a great part of these studies, which you have successfully presented a sample of in your *Adamus* [*exul*]. Continue to love knowledge without the [hot] air and learning without novelty. He who strides from the public road and looks for small paths, seeks errors and twists. Ignore that, thus God wants you honourable here [on earth] and blessed elsewhere [in the afterlife]. Leuven, February 3, 1602. I greet your father and uncle from the heart.

> Intellectualis philosophia magna studii huius pars est, quam feliciter in Adamo tuo libasti. Perge, hanc ama, et sine vento scientiam, et sine novitate doctrinam. Publica via qui excidit, et semitas quaerit, errores et flexus quaerit. Sperne, ita Deus te hic honestum, alibi beatum velit. Lovanii III Nonas Februar. ∞ IↃCII. Patrem et Patruum ex animo saluto.[65]

Lipsius distances himself from the young Grotius in this letter and implies he has wandered off the well-trodden, safe, and straight path. While in the previous letter Lipsius offered Grotius encouragement to stay on the path of glory, modesty, and virtue, here he foreshadows the many errors and deviations Grotius will find on the erroneous paths he has chosen to tread. Lipsius fell out of favour in Calvinist circles in Leiden when he returned to Leuven and re-entered the Catholic Church after having lived and lectured

61 'Scripturum praevidere necesse est, nauseantes nonnullos, damnantes alios, convitiis proscindentes, quicquid legunt, plurimos.' Ibid.

62 'Et tamen, ecce, profiteor palam nihil dedisse alterutri partium.' Ibid.

63 Justus Lipsius to Hugo Grotius, 3 February 1602, *BW* I, no. 28, p. 21; not yet in *ILE*, would appear in part XV.

64 'Adamum tuum vidi, partem legi, et ex ea omnia probavi.' Ibid.

65 Lipsius to Grotius, 3 February 1602.

at the Protestant Leiden University between 1578 and 1591. Lipsius warned Grotius not to be led astray and mingle in theological discussions.[66] Grotius was willing to set aside the confessional differences between them and focus rather on the scholarly aspect of their possible friendship. In the eyes of Grotius, Lipsius still held a high position in the hierarchy of scholarship, regardless of his confessional attitudes. Yet the last letter shows that Lipsius closed the door on Grotius, probably because he veered too far away from the proper confessional path. Lipsius's apparent confessional inconstancy caused him to fall out of favour amidst the Scaliger alumni in Protestant Leiden. After merely four letters, the correspondence between Grotius and Lipsius came to a grinding halt. Their letters show that if the central ideal—in this instance, the confession and 'right' path to virtue—was no longer shared, the foundation for membership of a community also fell through.

Grotius was more successful in applying a similar strategy in his first letter to Isaac Casaubon in February 1602, just two weeks after Lipsius had written his last reply. Casaubon, unlike Lipsius, was a dedicated Huguenot scholar and had more contacts with Protestant scholars throughout Europe.[67] In Grotius's introductory letter to Casaubon, he first mentions a letter from Casaubon to Meursius from more than six months earlier, in which Casaubon had asked Meursius to greet Grotius, Scriverius, and Heinsius.[68] Grotius uses a rhetorical question to flatter and bond with Casaubon:

> And whose esteem, I ask, could be more important for me after Scaliger's, whose esteem by its own merit is always the greatest, than of the man whom Scaliger himself preferred above him? For the world acknowledges the two of you [i.e., Scaliger and Casaubon] as the *duumviri*[69] to restore the learned cause and as princes of both languages [Latin and Greek].

> Et cuius, quaeso, gratia mihi pluris sit post Scaligeri, quae suo merito semper maxima, quam eius, quem sibi praefert ipse Scaliger? Vos enim

66 On the two-sided reputation of Lipsius as an excellent scholar and inconstant religious man, see Jeanine De Landtsheer, '*Pius Lipsius* or *Lipsius Proteus*?', in *Between Scylla and Charybdis: Learned Letter Writers Navigating the Reefs of Religious and Political Controversy in Early Modern Europe*, ed. Jeanine De Landtsheer and Henk Nellen, 303–349 (Leiden; Boston: Brill, 2011).
67 See Dirk van Miert, 'The Limits of Transconfessional Contact in the Republic of Letters around 1600: Scaliger, Casaubon, and their Catholic Correspondents', in *Between Scylla and Charybdis*, ed. De Landtsheer and Nellen, pp. 367–408.
68 Isaac Casaubon to Johannes Meursius, 7 Id. Aug. (of Mart.), 1601, in Johannes Meursius, *Opera*, vol. 11, ed. Giovanni Lami (Florence: apud Tartinium et Franchium, 1772), p. 89.
69 Literally 'two men', referring to a pair of magistrates in Roman antiquity.

> agnoscit Orbis restituendae Rei literariae Duumviros, et Principes utriusque linguae.[70]

This type of deferential letter signifies the difference in his status and place when compared to Casaubon in the hierarchy of the learned world. Grotius adorns Casaubon with compliments and considers him to be a high official in the quest to restore learning itself ('restituendae Rei literariae duumviri').

The political image of calling an illustrious man a *duumvir* or a 'prince of both languages' ('princeps utriusque linguae') had profound implications. Taken together, Scaliger, Lipsius, and Casaubon were considered the *triumviri* of the learned world, since Grotius referred in other letters to the triumvirate of Scaliger, Lipsius and Casaubon.[71] These three scholars were credited with restoring the learned world after its downfall in the wake of religious upheaval, and especially restoring those classical texts that were considered highly valuable. In the historical context of the Roman Republic, the *duumviri* were tasked to coordinate the government amidst a time of many civil wars and upheavals. In this letter to Casaubon, it is noteworthy that Grotius characterises both Casaubon and Scaliger as *duumviri*, but leaves out Lipsius, probably because the relationship between Grotius and Lipsius had stalled after their last exchange. Additionally, the rift between Catholic Leuven and the Protestant north caused the alumni of Scaliger, such as Grotius and Casaubon, to turn against the Catholic Lipsius. In the eyes of Grotius, Scaliger and Casaubon were now the *duumviri* who were tasked with saving scholarship after the religious and political distress and upheaval of the late sixteenth century. The French Wars of Religion created a fissure between Catholics and Protestants around Europe, which

70 Hugo Grotius to Isaac Casaubon, 18 February 1602, *BW* I, p. 22, letter no. 29.

71 Charles Nisard, *Le Triumvirat littérarie au XVIe siècle: Juste Lipse, Joseph Scaliger et Isaac Casaubon* (Paris: Amyot, [1852]). Nisard borrowed the phrase from Antoine Teissier; see *Les éloges des hommes scavans*, vol. 2 (Utrecht: François Halma, 1696), p. 446: 'Lui-même [i.e., Lipsius] étoit un Prince parmi les doctes de son temps, et Scaliger, Casaubon, et lui étoient les Triumvirs, comme on les nommoit, de la République des Lettres'. Teissier provided no source. In 1605, the German poet Friedrich Taubmann dedicated his edition of Plautus to 'Optimis maximis reipublicae litterariae triumviris Iosepho Scaligero, Iusto Lipsio, Isaaco Casaubono'. Taubmann to Scaliger, Lipsius and Casaubon, 22 July 1605, in Joseph Justus Scaliger, *The Correspondence of Joseph Justus Scaliger*, vol. 6, ed. Paul Botley and Dirk van Miert (Geneva: Droz, 2018), p. 98, ll. 1–2; and likewise, p. 98, l. 13, and pp. 99–100, ll. 55, 74, and 82. Scaliger responded by warning Taubmann to be aware of the 'nebulo' ('scoundrel') Kaspar Schoppe, who took it on himself 'to bark with the most impudent writings against the Triumvirs to whom you sent your Plautus-edition'. Scaliger to Taubmann, 19 June 1606, in ibid., p. 444, ll. 11–12: 'triumviros illos quibus Plautum tuum misisti impudentissimis scriptis allatrare instituit [sc. Schoppius]'. These references were found and shared by Dirk van Miert.

had taken its toll on the ideals of tolerance and the social cohesion of the European community of learned men.[72] Grotius thus invoked this context and explicitly referred to the collective quest to restore this community and learning as a collective goal and ideal that bound Protestant learned men.

Casaubon accepted Grotius's letter and continued in the considerate and respectful tone with a resort to the public goal of learning, echoing Lipsius's reference to the path of glory that learning can provide.

> Also, before it was clear to me whether I was known to you even by name, I loved you out of general respect for learning and wished you all the very best. Now what do you think [I thought] after receiving your letter? From that letter of yours I learned that you also place some value in my trifles. I know what makes you tick, and I am not so foolish or puffed as not to understand how much I owe to the great Scaliger on that account.

> Etiam antequam constaret mihi, an vel de nomine tibi innotuissem, publico literarum respectu te amavi, omniaque tibi fausta optavi: quid censes nunc post acceptam epistolam tuam? Ex illa [epistola] cognovi, te quoque nugis nostris pretium aliquod ponere. Scio quid te potissimum moveat, neque ita desipio aut τετύφωμαι, ut quantum magno Scaligero etiam isto nomine debeam, non intelligam.[73]

Casaubon explicitly accepts the correspondence with Grotius and embraces his higher position in the learned hierarchy with the question, 'Now what do you think after I have accepted your letter?' Casaubon extends compliments and claims that 'everyone admired and fell silent at this novel prodigy of your innate brilliance'.[74] Grotius was clearly seen as a prodigy at just eighteen years old when Casaubon wrote this letter in 1602. Scaliger, their common object of devotion, features prominently in their correspondence as the epitome of learning and a shared example.

In this exchange, three aspects—hierarchy, common goal, and reference to lineage—come to the fore. Grotius was clearly aware of his low hierarchical position, particularly in relation to Casaubon, whom he dubbed a *duumvir* and prince of both languages. The imagined community thus relied on a sense of hierarchy and Grotius wanted to inscribe himself into

72 Mack P. Holt, *The French Wars of Religion, 1562–1629*, 2nd ed. (Cambridge; New York: Cambridge University Press, 2005).

73 Isaac Casaubon to Hugo Grotius, 10 July 1602, *BW* I, letter no. 32, p. 24.

74 'Mirari omnes et stupere ad hoc novum ingenii tui portentum.' Ibid.

this scholarly community. The common goal that Casaubon and Grotius shared was the restoration and continuation of the learned world. The community of which they considered themselves part aimed to restore the riches of learning. Finally, Grotius and Casaubon worshipped a mutual idol and example: Joseph Justus Scaliger. In their letters, the two men bond over their shared mentor and the excellent scholarship he represented.

Between 1602 and 1611 there are only six extant letters between the two men. Despite the low frequency of correspondence, Grotius developed a scholarly friendship with Casaubon. In these years, Grotius and Casaubon exchanged ideas about books, scholarship, politics, and religion.[75] From 1611, the tone and intensity of the correspondence and the nature of their relationship changed. Although Casaubon was twenty-five years older than Hugo, their friendship developed into an equal scholarly friendship in which they esteemed each other highly.

Casaubon moved to England in October 1610, after the assassination of King Henri IV of France (1553–1610). Under the protection of King James I (1557–1625), Casaubon received a pension and would dedicate his time to scholarship, much of which consisted of Protestant polemical work. In the following years, Grotius wrote letters to Casaubon to convince James I of the necessity of the pacification of the religious factions in the Dutch Republic.[76] In April 1613, Grotius travelled to England as part of a delegation of the States General.[77] On this trip, he met Casaubon for the first and only time. Casaubon had organised an audience with James I and introduced Grotius to theologians in England.[78] From this point onwards, the tone, style, and intimacy of the letters between Casaubon and Grotius considerably changed. In his first letter from London from 1611, Casaubon wrote to Hugo:

> Great man, ever since the will and miraculous providence of God brought me to this kingdom [of England], I have often wanted to establish an exchange of letters [*commercium literarum*] with you. For I love you since long with sincere affection and for as much as I owe, I mightily venerate your virtue, that you pair with equal piety.

75 G. H. M. Posthumus Meyjes, 'Twee vrienden, Isaac Casaubon en Hugo de Groot', in *Voorbeeldige vriendschap: Vrienden en vriendinnen in theologie en cultuur*, ed. H. J. Heering, 37–46 (Groningen: Styx, 1993).

76 Nellen, *Grotius*, pp. 139–140.

77 Ibid., pp. 149–154.

78 Ibid., pp. 153–162.

> Vir magne, ex quo me Dei voluntas et admiranda providentia in hoc regnum [Angelorum] traduxit, volui saepe commercium literarum tecum instaurare. nam et amo te dudum sincero affectu, et tuam virtutem pari pietate iunctam summopere, pro eo ac debeo, veneror.[79]

The tables have turned. Casaubon states that he is indebted to Grotius, rather than the other way around. Additionally, he adorns Grotius with compliments. It is now Casaubon who invites him to engage in a frequent exchange of letters: 'I have often wanted to establish an exchange of letters with you'.[80] In the letter Casaubon talks about the English king and the political situation, as well as his recent booklet against the Jesuit Fronton du Duc (1558–1624), *Epistola ad Frontonem Ducaeum* (Letter against Fronton du Duc), to criticise the revision of Greek authors by the Jesuits under the leadership of Fronton du Duc.

Grotius responded roughly a month later; in the letter he acknowledges the excellent scholarship of Casaubon:

> Casaubon, most learned of men, I had diligently read your *Letter against Fronton du Duc* even before I received the copy you sent to me, and I had urged other famous men in our republic to attentively read the same book. I now pass over the brilliance of your fluid style, which is not, as that of so many authors today, concealed by a coloured outer surface, but is full of sanity, full of power.

> Librum ad Frontonem tuum, Casaubone virorum doctissime, iam ante quam eum te mittente accepi, legeram diligenter, aliisque claris in nostra hac republica viris eundem ut sedulo legerent auctor fueram. Omitto nunc limpidissimum illud flumen orationis, quae non, ut hodie multorum, summam cutem pigmentis quibusdam obducta, sed tota plena est sanitatis, plena roboris.[81]

The friendship between the two now consisted of reading each other's work, commenting on the recent developments in the learned world as well as the political and religious developments in England and the Dutch Republic. In this letter, Grotius affirms Casaubon's central position in the learned world by stating that he urges many bright men read his work, effectively

79 Isaac Casaubon to Hugo Grotius, 11 December 1611, *BW* I, letter no. 218, p. 191.

80 '[V]olui saepe commercium literarum tecum instaurare'. Ibid.

81 Hugo Grotius to Isaac Casaubon, 7 January 1612, *BW* I, letter no. 219, p. 191.

trying to enlarge the circle of people that should read Casaubon's work. At this point in their friendship, Grotius is in a position to commend the style and scholarship of Casaubon as well. Much like the comparison between Lipsius's edition of Pliny and other editions of lower quality, Grotius stresses the excellence of Casaubon's work, the 'brightest flow of eloquence', its sanity and strength. This elevation of scholarly work helped to foster a sense of a community to which the virtue of scholarly excellence is central. Grotius emphasises that Casaubon is not merely excellent but also has a similar view on religious issues.

In this small sample of the correspondence, young Hugo tries to become part of a learned community and establish a scholarly exchange. Through careful references and salutations, the writer stresses the embedded hierarchy of the community in terms reminiscent of Cicero's *De re publica* and ancient Rome. Grotius addresses Lipsius as the 'prince of the learned' (*princeps literatorum*) and Casaubon as a *duumvir* of the learned world. Such appearances showed the structure of the community and the collective sense of action. In most of the letters, the scholars express a public or shared goal. That Lipsius and Grotius disagreed on their common goal (and what the 'right' path was), and the subsequent silence between the two scholars, shows the importance of such a common goal. In contrast, the correspondence and friendship between Casaubon and Grotius is an example of the ideal scholarly friendship: a horizontal friendship focused on a discussion of scholarly work, politics, and religion, all in service of the common good.

## 3.2 Citation Analysis: Collective Memory and Historical Awareness

The letters of Grotius and his correspondents are rich in citations and references to the Church Fathers, classical writers, and reformers from the sixteenth century. The mentioning of names in correspondence served a range of different purposes. It could be a reference to scholarly work ('see Augustine, *Confessions*'), an interaction ('I talked with Wytenbogaert yesterday'), but also the idea or *exemplum* of a person ('the aura of Scaliger'). This section maps the different references to people in the correspondence of Grotius. To enhance the explanatory and analytical value of the names mentioned in the corpus, it is important to clarify and classify different types and functionalities of referrals and referents, as well as their consequences for the construction of a shared sense of history and identity.

In this section, we move from the close-reading perspective of the previous section to a macro perspective, much like the approach taken with the *vitae* in Chapter 1. Historians are increasingly paying attention to the networks and citation patterns of scholars and scientists. Rather than looking at each letter on its own, they have started to look at correspondence as a dataset. Aggregating and visualising the contextual data of a correspondence (such as place and time of the sending of a letter) can foreground patterns in the exchange such as the geographical spread of correspondents or the chronological development of a correspondence.[82] For example, historians have used quantitative analysis of correspondences to identify shifts in the structure of collaboration networks of physicists working on general relativity in the 1950s.[83]

This study uses the method of citation analysis and named entity recognition. The former entails the study of the citations and mentions of people and works in a certain corpus of letters. If one maps all the citations in a corpus of journal articles, one can see how scholars and scientists are building and relying on another's work on a macro level.[84] This section takes a similar approach to the correspondence of Grotius and looks at the occurrences of proper names in the letters. Since it is difficult to untangle and disambiguate references to contemporary scholars, and because this chapter is concerned with the formation of identity and memory, the focus of the analysis is on the proper names of people who preceded Grotius and were not his contemporaries.[85]

Apart from referring to people via a literal reference to a book or passage, name-dropping in correspondence had a cultural function. To mention a name could invoke the virtues, thought, authority, and example a certain person represented. Such *exempla*—often considered as common predecessors of the community—were essential to foster a shared set of virtues and values in the community. Citations would reaffirm a collective sense of knowledge and memory that the scholars were building on. To mention

82 See, e.g., João Gama Oliveira and Albert-László Barabási, 'Human Dynamics: Darwin and Einstein Correspondence Patterns', *Nature* 437 (2005): 1251.

83 Roberto Lalli et al., 'The Dynamics of Collaboration Networks and the History of General Relativity, 1925–1970', *Scientometrics* 122, no. 2 (2020): 1129–1170.

84 Yves Gingras, 'Mapping the Structure of the Intellectual Field Using Citation and Co-Citation Analysis of Correspondence', *History of European Ideas* 36 (2010): 330–339. The co-citation analysis is a more specific form that looks at the names and works cited alongside one another in a particular text, such as a journal article or a letter.

85 The dividing line has been drawn in 1583, the birth year of Grotius. Proper names of people who lived before 1583 were included in the analysis.

a name such as Horace, Livy, or Erasmus could refer to a symbol of virtue and established knowledge, as Timothy Hampton argued:

> The reader who comes upon the name of a heroic ancient exemplar in a text has come upon a single sign which contains folded within it the entire history of the hero's deeds, the whole string of great moments which made the name a marked sign in the first place. Expressed differently, we might say that the name is a noun with a verb phrase (the various great deeds) condensed inside it.[86]

The appearance of a name in the correspondence of Grotius should thus be seen as a complex reference which needs to be carefully unpacked. It can be a direct citation of a person or a work, but often also refers to a set of deeds and ideals that a historical person represented. The reference to a historical figure also affirmed the social status of the correspondent, since it signalled to another correspondent that he was erudite and cultured. A closer look at the letters of Grotius helps to differentiate between the different functions the appearance of a name could have.

This analysis focuses on three discernible categories of people in Grotius's correspondence: Church Fathers and early Christian apologists; Graeco-Roman pagan writers from antiquity; and sixteenth-century theologians and reformers. Of course, the correspondence mentions people outside these three categories, for example scholastic philosophers, family members, nobility, or colleagues. The three categories were chosen after meticulous analysis of the mentioned names and partly enable us to answer our question about the role of memory in shaping communal identities and the mechanisms of creating a sense of community.

### 3.2.1 Church Fathers

The Church Fathers emerge in the correspondence as historical sources, religious authorities, but also as examples of how to navigate religious discord. The citations of Church Fathers and early Christian apologists are abundant, especially in the letters written by Grotius himself. As can be seen in Figure 3.2, the letters often refer to the Greek Church Fathers Justin Martyr (c. 100–c. 165) and John Chrysostom (c. 347–c. 407), as well as the Latin Church Fathers Augustine of Hippo (354–430) and Jerome

86 Timothy Hampton, *Writing From History: The Rhetoric of Exemplarity in Renaissance Literature* (Ithaca, NY: Cornell University Press, 1990), p. 25.

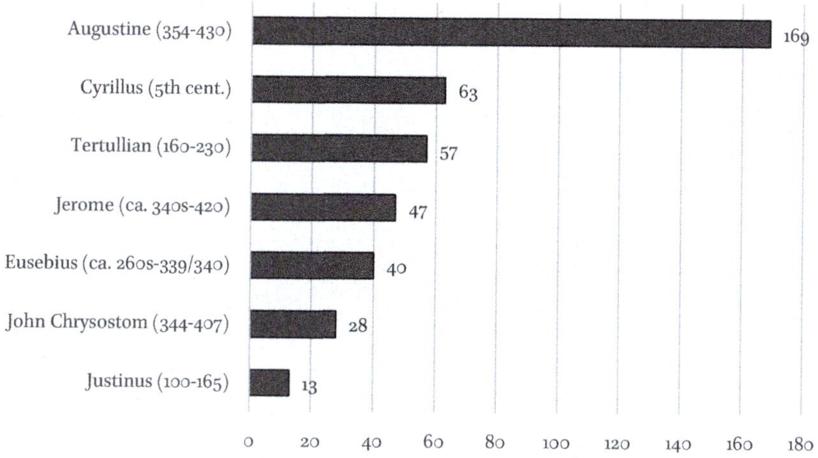


Figure 3.2. Number of letters in Grotius's correspondence which mention early Christian Church Fathers and apologists

of Stridon (c. 342/347–420). Additionally, the letters often refer to the church historian Eusebius of Caesarea (c. 260/265–c. 339/340) and the early Christian apologist Tertullian (c. 155–c. 220). Augustine stands out, since his name appears about three times more often than most of the other names.

Helped by Gerardus Vossius, Grotius pleaded for religious toleration in the Arminian–Gomarist church conflict during the Twelve Years' Truce (1609–1621). The conflict focused on the question of predestination, but quickly developed into wider political, social, and personal conflicts. In his work and letters, Grotius drew on patristic literature to argue for toleration. The early Christian church, around the third and fourth centuries, was often in turmoil and Christian thinkers and theologians did not agree on the interpretation and rites of the Christian faith. In Grotius's view, this did not break the unity of the early Church, despite the various theological and religious positions, and his work on the Church Fathers should be seen as a historical study to remedy this contemporary religious conflict.

Grotius interpreted the Church Fathers as historical and theological sources. If we take a closer look at the references to Augustine, half of them refer directly to his works in the context of Grotius's sustained study of the Church Fathers. Augustine was at the centre of Reformation religious debates, since Martin Luther (1483–1546), a former Augustinian monk, based his views on grace and predestination on the writings of Augustine. Grotius was particularly interested in Augustine's views on predestination and his role in the Synod of Carthage in 418, also called the Council of Africa.[87] Augustine was cited as a historical source in Grotius's correspondence between roughly 1610 and 1620. Such references often take the form of a reference to a short title and a chapter number, such as 'Augustini lib. 12 contra Faustum cap. 7' ('Book 12, Chapter 7 of Augustine against Faustus') or 'Augustinus ad Bonif. lib. I, c. 1' ('Augustine to Boniface book I, chapter 1') followed by a direct citation of the works of Augustine.[88] In these types of appearances, the focus is entirely on the work of Augustine, rather than the person or his deeds. Among these direct citations are also the more abstract references to Augustine's views without a quotation, which often contain a summary of 'Augustine's views' ('sententiae Augustini'). These are not precise references and do not refer to a certain passage or work by Augustine, but rather conflate Augustine and his work.

Grotius considered Augustine to be a representative of a period in the development of the early church. For example, Grotius would say 'the period after Augustine' or 'this is most certain before the age of Augustine'.[89] Rather than referring to specific time periods, Grotius often attributed time with a certain person: 'during the time of Justinian', 'Tertullian's age', and 'this is most certain before the age of Augustine'.[90] Part of Grotius's historical argument in his *Ordinum Pietas* (*Piety of the States*, 1613) was to show that 'after Augustine's time there was a diversity of scholarly opinions about predestination and the relationship between grace and free will'.[91] Grotius divided history into historical periods and associated these periods with these Christian writers. The citation of the Christian writers thus reveals Grotius's conceptualisation of the history of the early Christian church as a succession of Christian thinkers.

The other portion of the citations are embedded in the contexts and associations with Augustine. Augustine was often mentioned alongside the Greek Church Fathers, the *patres graeci*. In many letters, Grotius offered

87 See, e.g., Arnoud Visser, *Reading Augustine in the Reformation: The Flexibility of Intellectual Authority in Europe, 1500–1620* (New York: Oxford University Press, 2011), pp. 129–130.

88 Gerardus Vossius to Hugo Grotius, 16 September 1614, *BW* I, letter no. 372, p. 353; and 'ante Augustini aetatem certissimum est' in Hugo Grotius to Gideon van den Boetzelaer, early December 1615, *BW* I, no. 438, p. 441.

89 '[P]ostrema Augustino aetate' in Hugo Grotius to Isaac Casaubon, 13 May 1614, *BW* I, no. 337, p. 313; Hugo Grotius to Jean de Cordes, 30 October 1634, *BW* V, no. 1959, p. 288.

90 'Iustini tempore videtur'; 'Tertulliani aevo'; and 'ante Augustini aetatem certissimum est'; all in Grotius to De Cordes, 30 October 1634, *BW* V, no. 1959, p. 288.

91 'Post Augustini tempora de praedestinatione et gratiae cum libero arbitrio convenientia variatum inter scholasticos opinionibus fuit.' Hugo Grotius, *Ordinum Hollandiae ac Westfrisiae Pietas (1613)*, ed. and trans. E. Rabbie (Leiden; Boston: Brill, 1995), p. 56.

his correspondent a complete enumeration of all the *patres graeci* who held the same—or in some cases differing—theological views. Such use of Augustine can be seen, for example, in a letter from 1611 to Anthonius Wallaeus (1573–1639), a preacher and teacher in Middelburg.[92] In a discussion about the possibility of free will and the necessity of grace, Grotius pleaded that he followed the position of the Church Fathers, including Augustine.

Augustine's intellectual authority was flexible, because his large, disparate body of work could be used to support contrasting theological and historical arguments.[93] A clear example of an attempt by Grotius to put the position of Augustine in a historical and theological perspective can be found in an elaborate letter written to the States' ambassador in Paris, Gideon van den Boetzelaer (1569–1634), in which Grotius reacted to a treatise by Pierre Du Moulin (1568–1658), a Huguenot minister in France. Grotius offers an interpretation of the relationship between church and state in the ancient church.[94] He points to the importance of the rites of the ancient church above the recently invented rites in Protestant churches. The Church Fathers were a model and their writings the source of the practice of rites in Protestant churches. Grotius argued that there is no reason to favour one Church Father over another (as Du Moulin seemed to do with Augustine): 'Why should the authority of Augustine be greater than of all the Church Fathers, who lived in older and more pure times, within four hundred years after the birth of Christ?'[95] It is clear that Grotius presented the Church Fathers as historical figures on an equal footing, rather than as spiritual or religious inspirational figures. Yet at the same time, Grotius cited Augustine considerably more often than any other Church Father. Moreover, he referred to Augustine not only as a historical source, but also as an *exemplum*—a moral example of good conduct. In the same letter to Boetzelaer, Grotius

92 Hugo Grotius to Anthonius Wallaeus, 11 November 1611, *BW* I, no. 215, pp. 185–187. For background on this exchange, see Nellen, *Grotius*, pp. 140–141.

93 Arnoud Visser argued that 'the sheer size of Augustine's oeuvre also facilitated an extremely varied reception. Far from being a monolithic influence, Augustine could be adduced to illuminate almost any aspect of Christian life and hence also support strikingly different, even contrasting theological arguments.' Visser, *Reading Augustine*, p. 4.

94 Hugo Grotius to Gideon van den Boetzelaer, early December 1615, *BW* I, no. 438, pp. 431–448; translated into English as Hugo Grotius, *Of the Government and Rites of the Ancient Church, Conciliation of Grace and Free-will, Certainty and Assurance of Salvation, Government of the Highest Powers in Church-affairs. In a letter To the States Embassador* (London: Printed for the translator, 1675).

95 '[C]ur maior esse debeat Augustini auctoritas, quam omnium Patrum, qui aevo superiore ac puriore, hoc est, intra quadringentos a nato Christo annos vixerunt'. Grotius to Van den Boetzelaer, *BW* I, p. 432.

draws a parallel with Augustine's thought and his own wishes to unite Christians throughout Europe:

> Holy Scripture demonstrates plainly that we fulfil our vocation and election firmly and certainly when we are led by faith, virtue, knowledge, continence, patience, piety, love, and charity. Hence, consenting with the ancient authors, Augustine proclaimed: as much as charity grows, fear is diminished.
>
> quod clare demonstrat sacra Scriptura, cum nos iubet per fidem, virtutem, cognitionem, continentiam, patientiam, pietatem, amorem et charitatem, vocationem atque electionem nostram firmam sive certam efficere. Unde Augustinus veteribus consentiens praedicat, quantum crescit charitas, tantum minui timorem.[96]

Grotius used the authority of Augustine, who was considered the most important Church Father on both sides of the religious divide, to find common ground with Calvinists. Since Augustine was also regularly cited for polemical reasons, Grotius reinvented Augustine as a universal proclaimer of moderation, peace, and tolerance. In the letter to Van den Boetzelaer, Grotius transforms Augustine into a symbol and rhetorical tool to remind Calvinists of the importance of his irenicist ideals. It is therefore noteworthy that Grotius repeats the adage from Augustine: 'when charity grows, fear decreases'.[97] For Grotius, the values of unity, truth, and love for your fellow man had a strong appeal and he appropriated Augustine to reflect and propagate these virtues. Augustine became an *exemplum* in the reconciliation of conflicting views on the interpretation of scripture and with the sign of Augustine Grotius tried to appeal to fellow theologians and reformers, such as Du Moulin. Augustine was a historical example for Grotius of how to navigate towards concord in a Europe full of religious discord.

The citations of and references to Augustine show that Grotius considered him both a historical source and a non-controversial moral example who united Catholic thinkers as well as writers from various Reformed confessions. Therefore, the invocation of Augustine in a letter was often an attempt to refer to theological common ground. For example, in the letter to Gideon

96 Ibid., p. 440.

97 'Quantum enim illa [caritas] crescit, ille [timor] decrescit; et quantum illa [caritas] fit interior, timor pellitur foras. Maior caritas, minor timor; minor caritas, maior timor.' Augustine, *In epistolam Ioannis ad Parthos*, tractatus IX.

van den Boetzelaer, Grotius expresses his admiration for Augustine and presents him as a shared example. The moral and theological rectitude of Augustine and the gravitas of his words were meant to serve as a means of uniting the various confessions. Grotius appropriated Augustine's words to convey his irenicist message: when tolerance and charity grows, fear is diminished.

### 3.2.2 Graeco-Roman Writers

Early modern scholarly correspondence is generally filled with references to Graeco-Roman writers. Figure 3.3 shows that Roman authors had a central role in the Grotius correspondence: the historian Tacitus (56–117), the philosopher Seneca (4 BC–65), the orator Cicero (106 BC–43 BC), and the poet Martial (40–104) were cited frequently. From Greek antiquity, Plato (424 BC–348 BC) and Aristotle (384 BC–322 BC) are the most cited philosophers, with a comparable number of citations of the poets Euripides (480 BC–406 BC) and Homer (800 BC–750 BC). Together, the citations show the strong presence of Graeco-Roman culture in the letters of Grotius and his correspondents and the central position of this culture in their historical consciousness.

Tacitus was referred to most often, mostly because Grotius was working on his *Annales et Historiae de rebus Belgicis* (Annals and histories of the Low Countries' War, 1657), a history in the style of Tacitus's *Historiae*.[98] Tacitus was a source of stylistic inspiration and admiration for Grotius, as well as a historical source for the history of the Low Countries. An example of this dual characterisation of Tacitus can be seen in a letter to Nicolas-Claude Fabri de Peiresc (1580–1637). Grotius refers to Tacitus as a possible source for the history of the Low Countries as well as a 'great writer' ('Tacitus magnus scriptor'). The latest edition of Tacitus from 'our Lipsius' ('Lipsius noster'), Grotius continues, has shown 'we people of Holland are particularly indebted to him [Tacitus] for all of our knowledge of our antiquity.'[99] Tacitus was thus considered an excellent writer as well as a critical source

98 Jan Waszink, 'Shifting Tacitisms: Style and Composition in Grotius's *Annales*', *Grotiana* 29, no. 1 (2009): 85–132; Waszink, 'Lipsius and Grotius: Tacitism', *History of European Ideas* 39, no. 2 (2014): 151–168.

99 'Tamen cum nuper Tacitum, magnum, quod inter omnes convenit, scriptorem, sed cui Batavi totam vetustatis nostrae notitiam peculiariter debemus, percurrerem, natae sunt mihi sub manum emendationes aliquot, seu conjecturae potius dicendae sunt, sed quarum pars magna decrevit, ex quo ultimam Lipsii nostri editionem inspexi.' Hugo Grotius to Nicolas-Claude Fabri de Peiresc, 13 January 1628, *BW* III, no. 1212, p. 222.

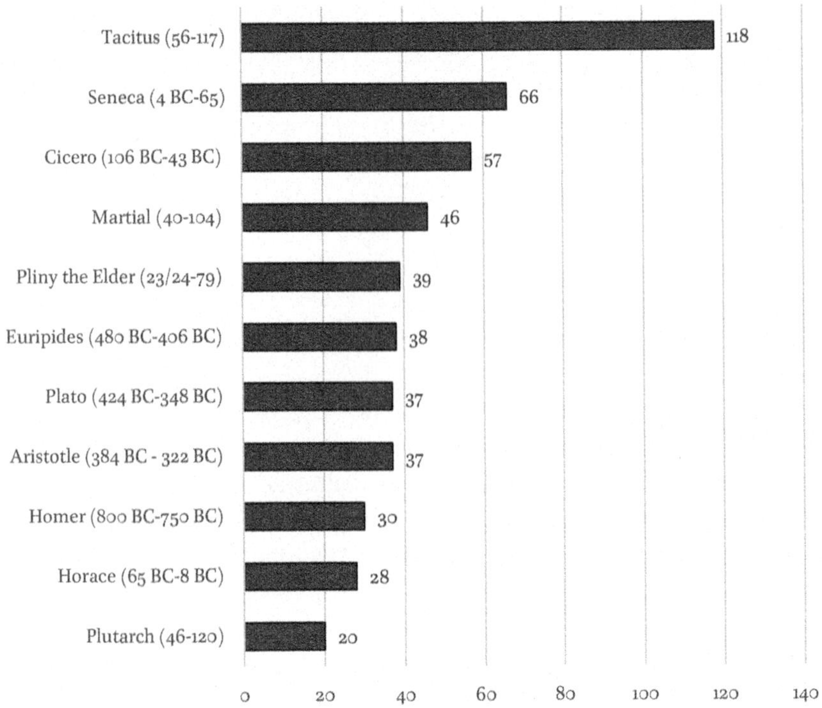


Figure 3.3. Number of letters in Grotius's correspondence which mention writers and thinkers from Roman and Greek antiquity

of historical knowledge of the supposed forefathers of the Low Countries, the Batavians. In their letters, Grotius and his correspondents shared their admiration for Tacitus and the style of history he represented. Additionally, Tacitus's *Germania* and *Historiae* helped build a patriotic identity for the Low Countries.[100] According to Tacitus, the Batavians were the bravest of the German tribes. This resonated with Grotius's ideal of the Low Countries as a state independent of Roman Catholic influence. Grotius established this continuity between the Batavians in Tacitus and the resistant Dutch

100 Two citations from Tacitus are the core of this so-called Batavian myth: *Germania*, 29, and *Historiae*, bk. ch. IV, l. 14. For the development of the so-called Batavian myth, see I. Schöffer, 'The Batavian Myth During the Sixteenth and Seventeenth Centuries', in *Some Political Mythologies: Papers Delivered to the Fifth Anglo-Dutch Historical Conference*, ed. J. S. Bromley and E. H. Kossmann, 78–101 (The Hague: Martinus Nijhoff, 1975); H. Kampinga, *De opvattingen over onze oudere vaderlandsche geschiedenis bij de Hollandsche historici der XVIe en XVIIe eeuw* (Utrecht: Hes, 1980), esp. pp. 56–86; Isidore Leonard Leeb, *The Ideological Origins of the Batavian Revolution: History and Politics in the Dutch Republic, 1747–1800* (The Hague: Martinus Nijhoff, 1973).

Republic against the Spanish Kingdom in *De antiquitate reipublicae Batavicae* (The antiquity of the Batavian Republic, 1610).[101]

Most of the references to Tacitus are a combination of philological interest in his *Historiae* as well as an allusion to the scholarly virtue of style that Tacitus represented. Two letters between Johan Isaaksz Pontanus (1571–1639) and Hugo Grotius show how they often mentioned Tacitus. Pontanus was a historical writer who composed a history of Amsterdam (*Historia urbis et rerum Amstelodamensium*, 1611), the Kingdom of Denmark (*Rerum Danicarum historia*, 1631) and the county of Guelders (*Historia Geldrica*, 1639). Like Grotius, Pontanus admired Tacitean history and he asks for Grotius's opinion of his work in the first letter.[102] The letter approaches Tacitus as a source for philological scholarship: Pontanus proposes 'rebelles' should be replaced with 'imbelles' and offers a 'truer reading' ('verior lectio') of various passages.[103] Grotius responded to all the philological points Pontanus raised.[104] In this exchange, Tacitus's *Germania* was used for philological dissection, but without any reference to the person of Tacitus. The references to Tacitus are either direct citations from his work that consider it a trustworthy historical source or allude to the scholarly example Tacitus set as a great writer and historian.

Ovid and Cicero are cited extensively in the correspondence as masters of poetry and oratory, as becomes clear in a letter to Daniel Heinsius. First, Grotius praises a speech Heinsius wrote: 'Therefore, when I compare your former speeches with the one you now brought forth, I think that your eloquence has developed from the subtlety of Seneca to the elegance of Pliny, and has now been perfected into the eloquence of Cicero.'[105] Elsewhere in this letter Grotius asks Heinsius, 'For what else ought we to expect from an Ovidian poet than a Ciceronian orator?'[106] The comparison of Heinsius's speech to the work of Seneca, Pliny and Cicero is one of the highest forms of praise and reveals that Grotius considered Heinsius's work as a successful

101 Hugo Grotius, *Liber de Antiquitate Reipublicae Batavicae* (Leiden: Officina Plantiniana Raphelengii, 1610); see also the modern English–Latin edition: Hugo Grotius, *The Antiquity of the Batavian Republic: With the Notes by Petrus Scriverius*, ed. and trans. Jan Waszink et al. (Assen: Van Gorcum, 2000).

102 Johan Isaaksz Pontanus to Hugo Grotius, 1 December 1615, *BW* I, no. 439, pp. 448–450.

103 Ibid., p. 449.

104 Hugo Grotius to Johan Isaaksz Pontanus, 20 January 1616, *BW* I, no. 444, pp. 459–460.

105 'Itaque cum priores tuas orationes cum his postremis confero, ita iudico, eloquentiam tuam a subtilitate Senecae ad Plinii elegantiam venisse, nunc in perfectissima Ciceronis facundia consummatam.' Hugo Grotius to Daniel Heinsius, 18 February 1609, *BW* I, letter no. 159, p. 139.

106 'Quid enim aliud exspectare debebamus ab Ovidiano poeta quam Ciceronianum oratorem?' Ibid.

continuation of the tradition of these classical authors. Moreover, the association with these classical authors, who were considered the pinnacle of poetry (Ovid) and prose (Cicero), also celebrated the literary friendship between Heinsius and Grotius.

The proverbial wisdom of both Cicero and Seneca was popular in the correspondence, too. Cicero and Seneca were presented as sources of wisdom on how to act. From earlier letters by Grotius, we know that he took Cicero and Seneca as examples, as he wrote to his brother, Willem de Groot: 'I go through the meaning of the Greek poets diligently, after the example of Cicero and Seneca.'[107] In later letters, proverbs and metaphors from Graeco-Roman authors are used to convey advice or observations. An example of the authority of these classical authors can be found in a letter to Grotius from Johan Witten, councillor to the Duke of Mecklenburg.[108] Witten quotes from Seneca and Cicero in a letter to Grotius offering guidance on how to achieve peace. First, Witten refers to Seneca's *De Ira*: 'Never does someone want to hit an enemy so hard, says Seneca, that one leaves the hand in the wound and one cannot withdraw himself from the strike.'[109] This, Witten argues, is not in the interest of achieving a peaceful pact. Witten further backs up the point that hostility solves nothing with a quotation from Cicero: 'And Cicero aptly noted in his *Letters to Friends*, book 10, epistle 6, that this rock caused many shipwrecks.'[110] Witten referred to Seneca and Cicero as sources of wisdom and proper moral conduct. Since both Witten and Grotius shared an education grounded in these classical authors, such references to common wisdom forged and reaffirmed the bond between the two men. In this way, the proverbs of authors such as Seneca and Cicero functioned as a method of proclaiming a shared sense of identity based on the moral examples of these authors. The references to classical examples were signs that refer to the shared examples in their learned community.

In yet another letter, Grotius assumes the role of the wise advisor and uses a well-known Horatian adage to add weight to his words. Grotius expresses

107 'Ego in vertendis Graecorum Poëtarum sententiis naviter pergo, Ciceronis et Senecae exemplo.' Hugo Grotius to Willem de Groot, 10 January 1621, *BW* II, letter no. 612, p. 40.

108 See Nellen, *Grotius*, pp. 473–474.

109 'Numquam velit quisquam, inquit Seneca, tam graviter hostem ferire, ut relinquat manum in vulnere et se ab ictu revocare non possit.' J. Witten to Hugo Grotius, 26 July 1633, *BW* V, letter no. 1861, p. 154. Reference to Seneca, *De ira*, II, 35, 1: 'Numquid velit quisquam tam graviter hostem ferire ut relinquat manum in vulnere et se ab ictu revocare non possit?'

110 'Quodque bene notat Cicero ad famil. lib. 10, epist. 6. atque ad quem scopulum multi naufragium fecerunt.' J. Written to Hugo Grotius, 26 July 1633, *BW* V, letter no. 1861, p. 154. Reference to Cicero, *Epistulae ad Familiares* X, 6: 'postremo ut pacem esse iudices non in armis positis, sed in abiecto armorum et servitutis metu.'

his surprise to Georg Müller, secretary of the Swedish court, that the Swedish field marshal Johan Banér (1596–1641) does not hold the chroniclers of his deeds in high regard.[111] Grotius reminds Müller that unrecorded deeds are lost to posterity:

> But anyhow, if we listen to Horace, who was a soldier and courtier:
> There is little to distinguish hidden valour from buried cowardice.
>
> Atqui, si Horatium audimus, qui et miles fuit et aulicus:
> Paulum sepultae distat inertiae
> Caelata virtus.[112]

The quote from Horace originated from an Ode in which he wrote to Marcus Lollius (c. 55 BC–2 AD) about the immortality one can achieve through the laudation by a *vates sacer*, a divine poet, in this case Horace himself.[113] Without the poet, the virtue and deeds of great men would remain unrecorded and forgotten by posterity. By invoking this specific adage, Grotius speaks to Müller as Horace spoke to Lollius. For Grotius, Horace is exemplary for thinking about virtue and the quote in the letter to Müller adds historical weight to their epistolary exchange. Horace was a common historical exemplar who represented the virtue of wisdom and giving good advice.

The citation from the work of lofty figures of antiquity, be it Horace, Cicero, or Ovid, was a commonplace for learned men in the seventeenth century. Grotius adorned his letters with wisdom from antiquity to strengthen his message with credibility and wisdom from ancient authors. Additionally, citations of writers from antiquity refer to sources and the quest to find the best edition or textual interpretation. These citations allowed Grotius and his correspondents to show their shared knowledge of, and admiration for, ancient authors. The authority of authors such as Homer, Horace, or Cicero was uncontested; they were the basis of a shared historical consciousness.

111 'Novam D. Bannerii victoriam, quam postrema literarum Nobilitatis vestrae pars affirmat, nuntiavi ministris regiis, qui sive pacem cupiunt, sive, quod magis reor, bellum, certe nostra prospera lucro suo deputant. Miror ita contemni a D. Bannerio scripturarios, ut ne suae quidem gloriae divulgatores secum habeat.' Hugo Grotius to G. Müller, 16 April 1639, *BW* X, letter no. 4071, p. 262.

112 'Atqui, si Horatium audimus, qui et miles fuit et aulicus: Paulum sepultae distat inertiae // Caelata virtus.' Grotius to Müller, 16 April 1639, ibid. Reference to Horace, *Odes* IV, 9, 29–30: 'paulum sepultae distat inertiae // celata virtus.' Translation based on Horace, *Odes and Epodes*, ed. and trans. Niall Rudd (Cambridge, MA; London: Harvard University Press, 2004), pp. 245–247.

113 Paul Winsor Sage, 'Vatic Admonition in Horace *Odes* 4.9', *The American Journal of Philology* 115, no. 4 (1994): 565–586 (esp. pp. 575–576).

Therefore, the extensive citation of pagan, classical authors reflected a shared identity and memory based on the tenet that learned men were working in the tradition of the ancients. In this tradition, Ovid and Horace were the masters of poetry and Cicero the master of oratory. With these common *exempla*, Grotius and his correspondents could foster common ground and bond over uncontroversial knowledge and wisdom. Furthermore, their use of proverbs and adages from classical authors helped to forge a bond between two correspondents. The authority of these proverbs was unquestioned; the inclusion of these in the letters affirmed the bond of two correspondents. The citations of, for example, Horace's *Odes* as well as adages from Cicero confirm that this is shared wisdom that one could build upon.

### 3.2.3 Reformers and Religious Thinkers

Sixteenth-century reformers, religious thinkers, and theologians are the last group of proper names that were frequently referred to in Grotius's correspondence. Figure 3.4 shows that a select group of these writers play a significant role: Jean Calvin (1509–1564), Martin Luther (1483–1546), Martin Bucer (1491–1551), and Philipp Melanchthon (1497–1560) are cited frequently.

The two central reformers, Calvin and Luther, figure prominently in letters that discuss their teachings of predestination. Predestination refers to the idea that all events and outcomes in life are decided beforehand (and thus 'predestined') by God. The pressing issue was the tension between God's omniscience and omnipotence on the one hand, and the possibility for an individual to obtain salvation by living a virtuous Christian life on the other hand (thereby exercising one's free will). Luther affirmed the Augustinian teaching of predestination which entailed that one's potential for salvation was predestined.[114] Calvin's teaching of predestination is often dubbed 'double' predestination, since he contended that God's predestination applied both to salvation and damnation: one was destined either for salvation or damnation and there was nothing one could do about it, since it was subject to God's will and omnipotence.[115] The discussion about predestination

114 See Visser, *Reading Augustine*, pp. 24–27, 36–37. For an overview of Augustine's views on predestination, see James Wetzel, 'Predestination, Pelagianism and Foreknowledge', in *The Cambridge Companion to Augustine*, ed. Eleonore Stump and Norman Kretzmann, 49–58 (Cambridge: Cambridge University Press, 2001).

115 See Matthew Levering, *Predestination: Biblical and Theological Paths* (Oxford; New York: Oxford University Press, 2011), ch. 4, 'The Reformation and Early Modern Period: Causal Chains', pp. 98–134, esp. pp. 101–110.

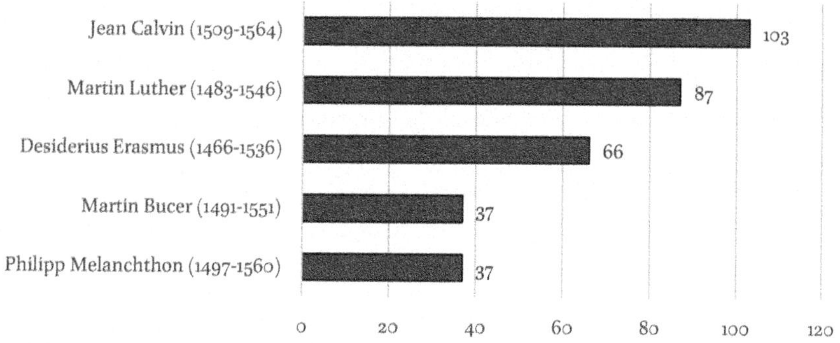


Figure 3.4. Number of letters in Grotius's correspondence which mention writers and thinkers from the sixteenth century

erupted in the Dutch Republic and especially at Leiden University, where the Leiden professors Arminius and Gomarus hotly debated its status.[116] Grotius was directly involved in the debate as he defended Arminius's conception of predestination to many of his correspondents. The references to Calvin, Luther, and Melanchthon in Grotius's correspondence were often made in discussions about their theology. More specifically, most of these references were made in discussions about their teachings of predestination in the context of the Arminian conflict in Leiden. Grotius tried to convince many fellow learned men of the aptness of Arminius's interpretation of the issues of predestination and the role of free will.[117]

Grotius had a nuanced sense of history and the aforementioned reformation thinkers are viewed through this perspective, especially in the correspondence with Gerardus Vossius.[118] In the 1640s, Grotius and Vossius extensively discussed the history of the diverse reformations throughout Europe, and the confessionalisation of the Dutch Republic in particular.[119] A good example of a historical and theological discussion that amply cites religious reformers is a letter from Gerardus Vossius to Grotius from 1642.[120] Vossius commended Grotius on his annotations to the New Testament:

116 Keith D. Stanglin, *Arminius on the Assurance of Salvation: The Context, Roots, and Shape of the Leiden Debate, 1603–1609* (Leiden; Boston: Brill, 2007).

117 Camilla Boisen, 'Predestination', in Lesaffer and Nijman, *Cambridge Companion to Hugo Grotius*, esp. pp. 224–227.

118 Jan Waszink, 'Historical Writings', in ibid., pp. 315–338.

119 For an introduction to Gerardus Vossius's view on history, see Gerardus Vossius, *Geschiedenis als wetenschap*, ed. with introduction and annotations by Cor Rademaker (Baarn: Ambo, 1990).

120 Gerardus Vossius to Hugo Grotius, 17 June 1642, *BW* XIII, letter no. 5752, pp. 264–269.

> On the things you have annotated in the New Testament; the more eager I deal with them, the more I admire them and became utterly speechless. Nothing similar or so great has appeared thus far. Of course, our enemies prefer to be silent about them [the annotations], rather than denounce them.
>
> Quae ad Novum Testamentum adnotasti, quanto studiosius verso tanto magis exosculor ac stupeo prorsus. Nihil hactenus tale vel tantum prodiit. Sane inimici de iis tacere quam vituperare illa malunt.[121]

Before explaining his criticism, Vossius points out that they are colleagues in the interpretation of the New Testament, unlike the silent 'enemies' ('inimici'). Vossius invokes the idea of 'enemies' here, which strengthened their sense of a community of course. He praises Grotius's scholarship by stressing its quality and innovativeness. The reformers are cited extensively in a passage of the letter where Vossius tries to distinguish the teachings of Calvin from those of Martin Bucer. Vossius critically assesses how both Calvin and Bucer interpreted the Scripture:

> The only question that then arises is whether Bucerism is more adverse to them than Calvinism. It would be easy for anyone to settle this controversy if the Scriptures could not be misunderstood by Calvin and those who follow him. I think, therefore, it is worthwhile to see how the doctors of the old church understood these. If one would do this, Bucer would win without a doubt.
>
> Solum igitur quaestio oritur an Bucerismus ei[s] magis adversetur Calvinismo. Facile esset cuivis hanc decidere controversiam nisi Scripturae etiam possent male intelligi a Calvino et eum sequentibus. Puto igitur operae esse videre quomodo eas intellexerint veteris ecclesiae doctores. Quod si fiat sine dubio vicerit Bucerus.[122]

For both Vossius and Grotius, the interpretations of reformers such as Calvin and Bucer should be critically assessed compared with the Church Fathers. Vossius mentions that 'Bucer would win without a doubt' ('sine dubio vicerit Bucerit'), creating the sense of a competition with two camps.

More importantly, Grotius considered a lineage of those who wanted to unite Christians throughout Europe. When Israel Jaski showed him the

121 Vossius to Grotius, 17 June 1642, p. 265.
122 Vossius to Grotius, 12 June 1642, p. 267.

irenic work of the French theologian Théophile Brachet de La Milletière (1596–1665), Grotius pointed out that Brachet stood in a long tradition of irenicist thinkers.

> Théophile Brachet is not the first to try to bring together Christians who are spread into so many parts. Erasmus, Cassander, Melanchthon, Witzel, Casaubon, and many others whose teachings have been published, ran the same course. These men did not allow themselves to be deterred from the great and troublesome work by the difficulties that were too manifest for them than to be overlooked. Yet they entrusted the result of a happy outcome to the omnipotent God, of whom it is to know the times and ages. They have done what suited pious men who love the truth along with Christian peace.

> Recolligere christianos in tot membra sparsos non primus conatur D. Milleterius. Cucurrerunt idem stadium Erasmus, Cassander, Melanchthon, Wicelius, Casaubonus multique alii, quorum et syllabi sunt editi. Neque isti viri a magno et odioso labore passi sunt se absterreri difficultatibus sane manifestioribus quam ut eas non viderent. Sed successus regimen Deo omnipotenti permittentes, cuius est nosse tempora et articulos temporum,[123] fecerunt, quod pios ac veritatis simul et pacis christianae amantes decebat.[124]

Grotius emphasised a shared history of irenicism. He pointed to irenic thinkers who preceded the work of Brachet and 'ran the same course' ('cucurrent idem stadium'). The lineage started with Erasmus, who was followed by George Cassander (1513–1566), Philip Melanchthon (1497–1560), Georg Witzel (1501–1573), and finally Grotius's close friend Isaac Casaubon (see Section 3.1).[125] By extension, Grotius himself was part of this history and considered Brachet within this tradition. In this letter, Grotius refers to the sayings of Jesus to stress the universal quality of the Christian teachings: 'It is not for you to know times or epochs which the Father has fixed by His own authority'.[126] This enumeration of names reveals Grotius's historical

123 Reference to Daniel 2:21: 'Et ipse mutat tempora, et articulos temporum, constituit reges et admovet reges: dat sapientiam sapientibus, et scientiam iis qui scientiam cognoscunt.'

124 Hugo Grotius to I. Jasky, 23 February 1641, *BW* XII, no. 5074, p. 125.

125 See also Dirk van Miert, 'The Curious Case of Isaac Casaubon's Monstrous Bladder: The Networked Construction of Learned Memory within the Seventeenth-Century Reformed World of Learning', in *Memory and Identity in the Learned World: Community Formation in the Early Modern World of Learning and Science*, ed. Koen Scholten et al., 307–341 (Leiden; Boston: Brill, 2022).

126 'Dixit autem eius: Non est vestrum nosse tempora vel momenta quae Pater posuit in sua potestate.' *Actus Apostolorum* 1:7.

narrative of attempts to bring together Christians. Grotius mentions Erasmus and Melanchthon relatively often in his correspondence; however, references to Cassander and Witzel are rare. Nevertheless, these passages suggest that Grotius considered himself to be part of an irenic, learned community with a long tradition.

In an early letter to Casaubon, Grotius melancholically lamented that the world had largely abandoned the kind of scholarly qualities exemplified by Erasmus and Melanchthon: 'Now the prudent simplicity of Erasmus and the sweetest soul of Melanchthon do not have many imitators'.[127] The extensive citation of humanists and religious scholars from the sixteenth century reflects the contemporary debates about the ideas of these reformers. Moreover, the references show that Grotius and some of his correspondents considered religious thinkers such as Erasmus and Melanchthon to be scholarly examples and especially Erasmus as the founding father of irenic thought. Throughout the correspondence, Grotius associated himself with Erasmus. After all, Erasmus was a famous scholar from Rotterdam and Grotius himself became the pensionary of Rotterdam in 1613. When he moved from The Hague to Rotterdam to become the city's Pensionary, he wrote to Casaubon, 'I moved from the hometown of [Janus] Secundus to the city of Erasmus'.[128] Grotius propagated the idea of Erasmus as the pride of Rotterdam and went to great lengths to erect a bronze statue of Erasmus in the city's central square.[129] Moreover, as the public intellectual of Rotterdam, Grotius fashioned himself as the successor of Erasmus, actively propagating this persona, as we will see below.

Other correspondents also associated Grotius with Erasmus, albeit indirectly. The correspondent G. Möller wrote to Grotius in 1633 and stressed the shared example of Erasmus. During Möller's trip to Basel, he paid tribute to Erasmus and stressed the fame he gave to the city of Basel:

> We also arrived in Basel in good fortune, as we wanted. If the place would not commend itself enough, the bones and ashes of the great Erasmus alone would make it famous.

127 'Nunc vero prudens Erasmi simplicitas, et dulcissima Melanthonis anima, adeo imitatores non habent, ut publica non unarum partium rabies in hos crudelissima incurrat.' Hugo Grotius to Isaac Casaubon, 9 May 1612, *BW* I, letter no. 236, pp. 206–207.

128 '[E]x Secundi patria in Erasmi civitatem migro'. Hugo Grotius to Isaac Casaubon, 22 June 1613, *BW* I, letter no. 272, pp. 248–251.

129 Lucy Schlüter, *Standbeelden van Erasmus in Rotterdam: 1549–2008* (Rotterdam: Stichting Erasmushuis Rotterdam, 2008); Jochen Becker, *Hendrick de Keyser: Standbeeld van Desiderius Erasmus in Rotterdam* (Bloemendaal: H. J. W. Becht, 1993).

> Venimus autem feliciter Basileam, ac voluimus. Locus si non seipsum satis commendaret, solius magni Erasmi ossa et cineres celebrem facerent.[130]

Throughout the letter, it becomes clear that Möller is an admirer of Grotius. He calls him a hero, whose 'unique humanity' and 'extensive erudition' deserve to be admired.[131] The reference to Erasmus in the passage above should thus be seen as flattery. The allusion to Erasmus and the fame he received in Basel was yet another way to express his admiration for Grotius. Möller praises Erasmus to show he belonged to the same community that admired Erasmus. The collective praise of Erasmus helped create a bond between Grotius and Möller.

The correspondence venerates Erasmus as an intellectual and moral example. Throughout the correspondence, we find allusions to the scholarly qualities of Erasmus. For example, at the end of a letter Casaubon from 1614 Grotius ensured that Rotterdam is the first place one should visit in Holland:

> But I remind you over and over that the first port for you is Rotterdam. For no other city in Holland ought to look on Casaubon before Rotterdam, which gave birth to Erasmus, so very similar to Casaubon.

> Sed illud etiam atque etiam moneo ut primus tibi portus sit Rotterodam. neque enim debet alia in Batavia urbs Casaubonum videre ante illam, quae simillimum Casaubono Erasmum genuit.[132]

In this letter Grotius compliments Casaubon on his scholarly qualities by comparing him to Erasmus. At the same time, Grotius connected Erasmus to Rotterdam and presented him as a scholarly exemplar. The memory of Erasmus is kept alive by the correspondents of Grotius, too. For example, in a letter to Grotius, Gerardus Vossius remembers Erasmus's birthday. At the end of a long letter, Vossius signs it with a short remembrance:

130 G. Möller to Hugo Grotius, 4 July 1633, *BW* V, letter no. 1857, pp. 148–149.

131 The full first sentence of the letter reads: 'Quamvis, vir excellentissime clarissime, audax et impudens videri possim, qui ad tantum te heroem scribere non vereor, me tamen facile excusat, ut puto, partim singularis vestra humanitas, quam omnes praeter diffusissimam eruditionem tuam merito mirantur, ut quae nostro hoc seculo vix in quoquam iunctim reperiuntur, partim etiam singularis tua in me benevolentia id exigit, qua me saepissime affectum memini, quum me accedentem benevole toties admisisti.' G. Möller to Hugo Grotius, 4 July 1633, *BW* V, letter no. 1857, pp. 148–149.

132 Hugo Grotius to Isaac Casaubon, 4 May 1614, *BW* I, letter no. 334, pp. 309–310.

> Amsterdam, 1641, 28 October, that is on the very birthday of Erasmus, due to the memory of this hero, now the page comes to an end, I would almost begin a new one. For when I think of him and compare our times, he would have had pity on the church.
>
> Amstelodami, MDCXLI, XXVIII Octobris, h[oc] e[st] ipso natali Erasmi, cuius herois memoria pagina hac exeunte pene ordirer novam. Nam eum quando cogito et tempora nostra confero, ecclesiae miserescit.[133]

This recollection of the birthday of Erasmus shows that Vossius and Grotius considered him worthy of remembrance and placed directly him in the context of their scholarly and religious affairs. Their remembrance of Erasmus's birthday is reminiscent of how the birth days of saints were commonly remembered in the early modern period.[134] Throughout the correspondence, Grotius and his correspondents remembered Erasmus in an intimate fashion as part of their scholarly community.

Grotius stressed to Johann Friedrich Gronovius (1611–1678), professor at Leiden University, that both of them were among the admirers of Erasmus now, but that his virtues remained largely ignored while he lived:

> Because if such a great man [as Casaubon] had no shortage of detractors, and I mean not within the hoard of Jesuits, but among us Dutchmen, then it is small wonder if the same critique is aimed at me, who follows in Casaubon's famous footsteps. Erasmus himself now has admirers of his virtues there [in Holland]; but certainly, while he lived, to nobody he was known less than to his own people. Therefore, he also spent the most important part of his life anywhere rather than there [The Low Countries].
>
> Quod si illo viro tanto non defuere obtrectatores, non iam dico in Ignatiano grege, sed apud Batavos nostros, quid mirum est, si nos tam praeclara vestigia secutos paria sequuntur? Ipse etiam Erasmus habet nunc ibi suarum virtutum admiratores; at certe, dum vixit, nullis minus quam suis fuit cognitus. Quare et ubivis potius locorum quam ibi potissimam aetatis suae partem voluit agere.[135]

133 Gerardus Vossius to Hugo Grotius, 28 October 1641, *BW* XII, letter no. 5443, pp. 600–601.

134 For the remembrance of saints before and after the Reformation, see Freddy C. Domínguez, 'Relics and Saints: Commemoration and Memoralization of the Holy Dead', in *A Companion to Death, Burial, and Remembrance in Late Medieval and Early Modern Europe, c. 1300–1700*, ed. Philip Booth and Elizabeth Tingle, 393–417 (Leiden; Boston: Brill, 2021).

135 Hugo Grotius to J. F. Gronovius, 7 May 1639, *BW* X, letter no. 4096, p. 308.

Grotius compared Casaubon and Erasmus directly (and by extension himself) as irenicist thinkers who went against the grain. Grotius worried that the voices of detractors would drown out the potential for glory for irenicist thinkers, such as Erasmus, Casaubon, and Grotius himself. The memory of these irenicist thinkers should be safeguarded and cherished among the learned community in the Low Countries.

Erasmus was an exemplar for Grotius that represented both erudition and irenicism. The extensive citation of Erasmus in the Grotius correspondence rarely refers to his written works, such as Erasmus's *Adages*, but often refers to the moral example set by Erasmus, as can be seen in the examples above. Grotius associated himself with Erasmus as the pensionary of Rotterdam, and his correspondents made allusions to Erasmus to compliment Grotius.[136] Grotius thus became an 'Erasmus reborn' ('Erasmus redivivus').[137]

All in all, Grotius had a distinct view of the learned world and its collective history. First, the heroes Grotius looked up to were Lipsius, Scaliger, and Casaubon and many contemporaries considered Grotius their successor. For example, a letter from Johannes Isaac Pontanus to Grotius mentions the triumvirate of Lipsius, Scaliger, and Casaubon, and also characterises Grotius as their worthy successor:

> Except that we may have to cry out with Domitius Afer, 'The system of the triumvirate has perished'. Our recent heroes have gone—Lipsius, Scaliger, Casaubon—whose erudition prevails, like their highest authority does for everyone. But they have passed and have gone to a common place. And therefore, above others who in small numbers rise up to the glory of our ancestors, I have you amongst my princes, my Grotius, to whom I see I must take refuge.
>
> nisi ut cum Domitio Afro tandem exclamandum habeamus 'Triumv[iri] hoc artificium periit'.[138] Fuerunt illi nuper heroes nostri, Lipsius, Scaliger, Casaubonus; quorum ut eruditio, ita summa quoque apud omnes autoritas plurimum pollebat; sed fuere et in communem abiere locum. Itaque prae

136 See also Nellen, *Grotius*, pp. 186–187.

137 Jan Paul Heering, *Hugo Grotius as Apologist for the Christian Religion: A Study of His Work* De Veritate Religionis Christianae *(1640)* (Leiden; Boston: Brill, 2004), p. xix.

138 Plinius, *Epist.* II, 14, 11, reads: 'Tum intermissa causa "Centumviri," inquit, "hoc artificium periit."'

reliquis, qui in paucis maiorum nunc gloriae succrescunt,[139] inter principes te habeo, mi Groti, ad quem confugiendum video.[140]

This letter showcases the way in which a sense of history was conveyed and cherished among learned men in the correspondence of Grotius. First, Pontanus put himself in a humble position in relation to Grotius, whom he considered to be a prince of learning. Secondly, he conjured an image of a community with a history and 'our heroes': the triumvirate of Lipsius, Scaliger, and Casaubon. These men were, according Pontanus, their 'ancestors' and as a prince, Grotius lived up to the glory of these ancestors. Thirdly, Pontanus constructed a link with classical antiquity by applying a reference to the Roman orator Gnaeus Domitius Afer († 59 AD) to their current historical situation. Finally, Pontanus extolled the virtue of erudition as the main quality of those in their historical community. All of these mechanisms helped to create a sense of a learned community with a history, tradition, and distinct scholarly virtues.

## 3.3 Scholarly Virtues and Values

In the final part of this chapter, we will look at the specific virtues that were extolled by Grotius and his correspondents. Much like the analysis in Chapter 1, this section draws on a word count of an entire corpus of letters. The digital text analysis has been carried out with AntConc with the same methodology that we applied to the *vitae* (see Chapter 1.1). Distant reading with digital text analysis allows us to distil the virtues and values that were commonly used in Grotius's correspondence.

Virtues play a central role in the works and correspondence of Grotius. The role of ethics and virtues is central to his works of law and theology, and even his plays.[141] The word *virtus* itself appears often in the correspondence (635 times) and indicates the significance of morality and virtuous behaviour in Grotius's thinking. The list of virtues that appear in his correspondence can be seen in Figure 3.5. The results have also been numerically presented in Table 3.2. In this section, we will take a closer look at virtues cited most

139 A reference to the saying 'Et se gloriae seniorum succrevisse et ad suam gloriam consurgentes', in Livy, *Ab urbe condita*, bk. 10, ch. 13, ll. 17.

140 Johan Isaaksz Pontanus to Hugo Grotius, 1 December 1615, *BW* I, no. 439, p. 450.

141 Mark Somos, 'Virtue', in Lesaffer and Nijman, *Cambridge Companion to Hugo Grotius*, pp. 91–117.

often: friendship (*amicitia*), trust and faith (*fides*), erudition (*eruditio*), and innate brilliance (*ingenium*).

**Table 3.2. Virtues and the count, relative frequency and chance virtues appear in a random letter**

| Virtue | Count | Relative frequency | Chance in letter |
|---|---|---|---|
| Amicitia/Amicus | 2842 | 0.16733% | 1.88841% |
| Fides | 1529 | 0.09003% | 1.01597% |
| Eruditio | 736 | 0.04333% | 0.48905% |
| Amicitia | 734 | 0.04322% | 0.48772% |
| Virtus | 635 | 0.03739% | 0.42194% |
| Ingenium | 617 | 0.03633% | 0.40997% |
| Prudentia | 544 | 0.03203% | 0.36147% |
| Pietas | 507 | 0.02985% | 0.33688% |
| Diligens/tia | 446 | 0.02626% | 0.29635% |
| Studiosus | 327 | 0.01925% | 0.21728% |
| Sapientia | 249 | 0.01466% | 0.16545% |
| Integritas | 228 | 0.01342% | 0.15150% |
| Humanitas | 216 | 0.01272% | 0.14352% |
| Doctrina | 205 | 0.01207% | 0.13622% |
| Doctus | 172 | 0.01013% | 0.11429% |
| Probus | 146 | 0.00860% | 0.09701% |
| Industria | 125 | 0.00736% | 0.08306% |
| Augustus | 111 | 0.00654% | 0.07376% |
| Modestus | 99 | 0.00583% | 0.06578% |
| Tolerans | 16 | 0.00094% | 0.01063% |

### 3.3.1 Friendship—*amicitia*

The virtue of *amicitia* appears twice in Figure 3.5 and Table 3.2. The highest ranking 'amicitia/amicus' includes words relating to friendship such as 'friend' ('amicus') in all declensions.[142] The result for the virtue of *amicitia* does not include the references to *amicus*. *Amicitia* in itself (referred to solely as an ideal or virtue in the correspondence) ranks fourth.[143] Friendship as a virtue and ideal was prominent in scholarly exchange throughout the early modern period and played a central role in Grotius's correspondence as well.

142 This includes: 'amicitia', 'amicitiae', 'amicitiam', 'amicitiarum', 'amicitiis', 'amicitias', 'amicus', 'amica', 'amicum', 'amici', 'amicae', 'amico', 'amice', 'amicorum', 'amicarum', 'amicis', 'amicos', 'amicas', 'amicabilis', 'amicabile', 'amicabili', 'amicabilem', 'amicabiles', 'amicabilia', 'amicabilium', 'amicabilibus'.

143 This is only the reference to the word *amicitia* and its declensions in singular and plural.

Friendship was often thought of as a bond of trust between two like-minded people and was therefore part of many official and personal relationships in the early modern period.[144] The ideal of *amicitia* was a mix of the ideal of friendship based on Cicero's *De amicitia* (On friendship) and chivalric ideals of friendship as 'wedded brothers'.[145] Early modern friendship was a social bond and came with duties. For example, two people who were in a friendship were expected to fulfil each other's requests, as well as honour and defend the other's reputation.[146] Learned men also committed themselves to friendship in early modern learned exchanges. 'Friends hold all things in common' was the first adage in Erasmus's *Adagia*, and this became a commonplace among writers and scholars throughout the early modern period.[147] The idea that a partnership based on shared intellectual pursuits was more fundamental and lasting than any other bond was prevalent among learned men.

Grotius was fond of the many and close friendships he maintained throughout his life. He complimented his correspondents, for example Gerardus Vossius, on 'the fruits of our friendship' ('de fructibus amicitiae nostrae').[148] The idea of *amicitia* is often followed by a possessive pronoun, as in *amicitia tua* ('your friendship') and *amicitia nostra* ('our friendship'). Friendship was a social bond that one could engage in. For example, Hugo Grotius recommended a friendship to Gerardus Vossius in 1622. This type of friendship was a social bond among scholars. In this letter, Grotius talked about Claude Saumaise (1588–1653), a well-known French classical scholar, and urged Vossius to initiate a friendship with him:

> I am confident that this [friendship] request will not be denied to you. The man himself [Salmasius] is most worthy of your friendship: he is learned beyond example and unusually modest at that.
>
> Id tibi postulanti negatum non iri confido. Vir ipse est amicitia tua dignissimus: doctus supra exemplum, rarae ad hoc modestiae.[149]

144 Daniel T. Lochman, Maritere López, and Lorna Hutson, eds., *Discourses and Representations of Friendship in Early Modern Europe, 1500–1700* (Farnham; Burlington, VT: Ashgate, 2011).

145 Reginal Hyatte, *The Arts of Friendship: The Idealization of Friendship in Medieval and Early Renaissance Literature* (Leiden: Brill, 1994); Alan Bray, *The Friend* (Chicago; London: University of Chicago Press, 2003); Daniel T. Lochman, and Maritere López, 'Introduction: The Emergence of Discourses: Early Modern Friendship', in *Discourses and Representations of Friendship*, ed. Lochman et al., pp. 1–26, esp. pp. 3–9.

146 Hyatte, *Arts of Friendship*, p. 27.

147 Eden, *Friends Hold All Things in Common*.

148 Hugo Grotius to Gerardus Vossius, 6 April 1630, *BW* IV, letter no. 1490, pp. 178–181 (p. 179).

149 Hugo Grotius to Gerardus Vossius, 1 September 1622, *BW* II, letter no. 784, pp. 242–243.

This exchange reveals the centrality of friendships in the learned world of Grotius. Friendship is a proposition that one made that can be denied or accepted. Friendship can be offered to someone, and when it is accepted, two people are then linked in this friendship.

This sense of indebtedness is clear in a 1637 letter Grotius wrote to Peter Holing (1607–1652?), a close colleague of Grotius in Paris between 1628 and 1632. Grotius wanted to respond to two letters from Holing his reply was twofold.[150] In the second letter, Grotius explains that the obligation of their friendship compelled him to respond to the second letter: 'I replied profusely to both your letters as I was indebted to our friendship and to the dignity of the argument which your letters contained.'[151] Friendship is thus an important ideal in Grotius's conception of the learned world. If one accepted a friendship with another learned man, one had to fulfil the requests made by this friend. Another letter from Grotius to Georg Michael Lingelsheim (1556–1636) is an example of the use of *amicitia*.[152] Grotius greeted Lingelsheim in this letter as 'the best of men and friends' ('Pridem est, virorum et amicorum optime [...]'), already referring to their bond as a friendship between virtuous men. Grotius wanted advice on a manuscript he had been working on and appealed to their longstanding friendship. He 'broke the silence' in their correspondence with this edition and hoped he could rely on the 'intimacy of our old friendship'.[153] In the letter to Lingelsheim, Grotius appeals to the friendship and asks for a favour. Both the letter to Holing and to Lingelsheim shows that friendship tied learned men together in a social bond of obligation and that the virtue of friendship was explicitly extolled. From the correspondence of Grotius *amicitia* appears as a bond between two like-minded scholars. It is a kind of vow to aid one another towards a common goal: the advancement of the learned cause. The trope of *amicitia nostra* ('our friendship') reveals that Grotius and his correspondents stressed their social indebtedness to each other thus strengthening their sense of a wider community tied together by individual friendships.

150 Hugo Grotius to Peter Holing, 21 April 1637, *BW* VIII, letter no. 3038, pp. 234–236. The other response letter is Hugo Grotius to Peter Holing, 1 August 1635, *BW* VI, letter no. 2207, pp. 119–122.

151 'Respondi utrisque literis tuis ut debebam liberaliter pro amicitia nostra proque dignitate argumenti, quod tuae literae continebant.' Grotius to Holing, 21 April 1637, *BW* VIII, p. 234.

152 Hugo Grotius to G. M. Lingelsheim, 10 May 1623, *BW* II, letter no. 832, p. 291.

153 The full sentence reads: 'Abrupit tamen silentium Stobaeana mea editio, cuius copiam nulli quam tibi prius facere debui pro veteris amicitiae nostrae necessitudine.' Ibid.

### 3.3.2 Trust and Faith—*fides*

Trust (*fides*) is a central concept for Grotius, both in his work and his personal communication with fellow learned men. In the correspondence, Grotius signed many letters with 'Yours, in the best trust, H. Grotius' ('Tuus fide optima H. Grotius').[154] As part of the *amicitia* between learned men, trust was a prerequisite for a positive exchange between scholarly equals.

In its legal and political context, trust (*fides*) was important because it was a prerequisite for Grotius's programme of pacifism and unification. Hans Blom has argued that 'trust is the quintessence of the larger social order that reaches beyond the small area of friendships and family ties.'[155] Examples of such a larger social order are diplomacy, learned exchanges, and religious discussions. Two parties could only enter negotiation and agreements in good faith (*bona fide*) if there was trust. In his 1602 manuscript *De fide et perfidia* (On trust and treachery), Grotius explained the foundational role of trust for justice.[156] Trust was not merely a requirement for equal exchanges, but also for war: 'faith must be kept with the enemy in every way (as Cicero maintains), and even (so Ambrose specifies) when one is dealing with a treacherous enemy.'[157] According to Grotius, trust was a root principle in law, both in theory and practice. For example, during the Thirty Years' War, the concept of trust was a binding principle in the discussions about the confessional disputes.[158] Grotius applied this principle of *fides* not only in his legal, religious, and political work, but also in his diplomatic and learned exchanges.

In the correspondence, trust often appears together with the word 'diligent' in Grotius's correspondence, often forming the conjunction 'fide ac diligentia' in the ablative case, which then refers to faith and diligence as a source or the way something was done. Trust and diligence

154 See, e.g., Hugo Grotius to Petrus Bertius, 30 November 1613, *BW* I, letter no. 306, p. 283.

155 Hans Blom, 'Hugo Grotius on Trust, Its Causes and Effects', in *Trust and Happiness in the History of European Political Thought*, ed. László Kontler and Mark Somos, 76–98 (Leiden; Boston: Brill, 2018), p. 97. See also Peter Schröder, 'Trust (*Fides*)', in Lesaffer and Nijman, *Cambridge Companion to Hugo Grotius*, pp. 118–137.

156 Arthur Eyffinger, '*De fide et perfidia / On Trust and Treachery*, Latin text constituted from the Ms., with English translation and introduction by Arthur Eyffinger', *Grotiana* 36, no. 1 (2015): 79–117, especially its introduction, 'On Good Faith and Bad Faith: Introductory Note', pp. 79–105.

157 Hugo Grotius, *Commentary on the Law of Prize and Booty*, ed. with introduction by Martine Julia van Ittersum (Indianapolis: Liberty Fund, 2006), ch. 8, p. 172, with a reference to Cicero, *De Officiis* III, 108; cited in Eyffinger, '*De fide et perfidia*', p. 94.

158 Ralf-Peter Fuchs, 'Trust as a Concept of Religious Plurality During the Thirty Years' War', in *Trust and Happiness*, ed. Kontler and Somos, pp. 302–319.

are conceptualised coupled virtues throughout the correspondence by Grotius and his correspondents. In a letter to his brother, Willem de Groot, Grotius compliments him on carrying out some business for him: 'Given the faith and diligence with which you take care of my business, I ought to prioritise all my business according to your wishes.'[159] Faith and diligence are the norms of good work. This conjunction of virtues is used by some of Grotius's correspondents as well, for example Johannes Wtenbogaert, who wrote to Grotius in 1613 that he was in Leiden where Grotius carried out his business 'with faith and diligence'.[160] In later diplomatic letters, Grotius used the term again to refer to two central virtues that he deemed essential for carrying out a task ('negotium curare'). In a letter to Frederick, Landgrave of Hesse-Eschwege (1617–1655), Grotius stresses these virtues to demonstrate his loyalty and commitment:

> I, not able to do as much as I want to, but only what was in my capacity for the wishes of Sir de Siraumont, have shown that I am very prepared to do and help with everything that pertains to the glory and usefulness of your Highness, and I will continue to do this with the utmost trustworthiness and diligence, as someone who desires to attend the same issues, most excellent and illustrious prince.
> Most prepared and obliged to serving you, your Highness,
> H. Grotius.
>
> Ego qui non quantum volo tantum possum, quantum meae erat facultatis ad postulata domini Siraumondi ostendi paratissimum me omnia ea facere atque iuvare quae ad gloriam atque utilitatem Celsitudinis vestrae pertinent, idque facere porro pergam summa cum fide ac diligentia, ut qui rebus ipsis apparere cupiam, excellentissime et illustrissime princeps,
> Celsitudini vestrae servire paratissimus obligatissimusque,
> H. Grotius.[161]

The same Frederick ended up in a conflict with George II, Landgrave of Hesse-Darmstadt, about the possession of the city and University of Marburg. Grotius presented himself as an advisor and diplomat, who would

159 'Ea fide ac diligentia mea negotia curas, ut desideriis tuis omnia mea negotia debeam praevertere.' Hugo Grotius to Willem de Groot, 1 August 1625, *BW* II, letter no. 997, pp. 463–467.
160 '[F]ide ac diligentia curavi'. Johannes Wtenbogaert to Hugo Grotius, 13 September 1613, *BW* I, letter no. 282, pp. 260–261.
161 Hugo Grotius to Frederick of Hesse-Eschwege, 13 February 1644, *BW* XV, letter no. 6702, p. 123.

carry out the tasks given to him with 'the utmost faith and diligence'.[162] The conjunction of the virtues faith and diligence thus referred to the central characteristics of how a virtuous man should act justly and with perseverance.

The concept of faith and trust (*fides*) in the correspondence of Grotius is thus a central virtue not just of learned men, but of anyone who acted in a morally just way. Faith and diligence were central concepts in communicating observations in the early modern period, and scholars stressed these characteristics to showcase their trustworthiness and integrity.[163] During religious discussions, acting in good faith and trust (*bona fide*) was an essential requirement for having a discussion at all. The concept of trust as something that is required for just warfare played a major role in Grotius's legal work, especially in his *De iure belli ac pacis* (On the law of war and peace, 1625). In their correspondence, Grotius and his correspondents stressed the linked virtues of faith and diligence to refer to the commitment they had undertaken to carry out certain tasks and diplomatic missions.

### 3.3.3 Innate Brilliance and Erudition—*ingenium* and *eruditio*

The *vitae* expressed the virtues of innate brilliance and erudition as a process in the life of a learned man (as we already explained in Chapter 1.2). First, one was gifted with intelligence and innate brilliance. Then, through education and scholarship one could polish (*erudire*) these qualities into erudition. This conception of the lifetime of a learned man is repeated in the correspondence: innate brilliance can be cultivated (or worshipped) by other learned men. The fruits of this innate brilliance then showcase someone's erudition. By repeatedly stressing these virtues to one another, the learned men in Grotius's correspondence show how a shared identity helped form a sense of a community. By extolling another man's erudition, the letter writer signals to the other learned man that they were part of the learned world. Correspondents often complimented each other on their innate brilliance and the erudition they displayed in their letters and works.

*Ingenium* (innate brilliance) was often used as a compliment in the last few sentences of a letter, both by Grotius and his learned correspondents. This virtue was also a way to express the scholarly aspect of an epistolary exchange. For example, Grotius signed many of his early letters (1606–1610)

162 '[S]umma cum fide ac diligentia'. Ibid.

163 Dirk van Miert, 'Concluding Observations on Communicating Observations', in *Communicating Observations*, ed. Van Miert, pp. 223–247 (pp. 229–230).

to Daniel Heinsius with 'H. Grotius, admirer of your innate brilliance'.[164] This underscores Grotius's admiration for Heinsius as well as his subordinate role in the scholarly hierarchy at this point in his scholarly career. Grotius repeatedly signed off with by complimenting Heinsius's scholarly brilliance, which helped foster a scholarly bond between the two men. In another letter to Heinsius, Grotius signed off with 'H. Grotius, cultivator of your innate brilliance and incomparable erudition'.[165] Echoing the 'admirer of your innate brilliance' trope, he signed many letters 'H. Grotius, cultivator of your innate brilliance'.[166] Innate brilliance was a compliment that was extended to scholars that Grotius looked up to.

The works of scholars were dubbed 'monuments of innate brilliance' and in the correspondence, Grotius spelled out what a learned man could do. Benjamin Aubery du Maurier (1566–1636), French ambassador to the States General of the Dutch Republic, lost his wife and Grotius offered consolation in the following way:

> Whatever you do is remarkable and amounts to an example. Some time it might be stated: Bear the death of your wife as Du Maurier carried it. […] Reconsider the virtues of that man, teach them to your children, and propagate to posterity what you are capable of through the famous monuments of his brilliance.
>
> Quicquid agis, in conspicuo est, et in exemplum abit. Fac dicere aliquando liceat: Ita fer uxoris mortem ut Maurerius tulit. […] Virtutes illius animo recole, liberis trade, et quod potes illustribus ingenii monimentis ad posteritatem propaga.[167]

Here, Grotius sums up the shared goals for learned men: to cultivate the virtues, to teach and exchange one's knowledge, and bring forth monuments of innate brilliance and erudition for later generations. Grotius offers

164 'Tui ingenii admirator H. Grotius'. Hugo Grotius to Daniel Heinsius, September 1606, *BW* I, letter no. 81; Hugo Grotius to Daniel Heinsius, 11 November 1606, *BW* I, letter no. 87; Hugo Grotius to Daniel Heinsius, 26 February 1608, *BW* I, letter no 112; Hugo Grotius to Daniel Heinsius, 31 May 1608, *BW* I, letter no. 135; Hugo Grotius to Daniel Heinsius, 5 July 1608, letter no. 138; Hugo Grotius to Daniel Heinsius, July 1608, *BW* I, letter no 140.

165 'Cultor ingenii, atque eruditionis tuae incomparibilis Hugo Groti [*sic*]'. Hugo Grotius to Daniel Heinsius, c. 1 July 1603, *BW* I, letter no. 43.

166 'Tui ingenii cultor H. Grotius'. Hugo Grotius to Daniel Heinsius, *BW*, letters 95, 99, 106, 115, 121.

167 Hugo Grotius to Benjamin Aubéry du Marier, 27 February 1621, *BW* II, letter no. 618, pp. 47–56.

guidance to Du Maurier on the death of his wife and consoles him that he will be an example for others. He thus advises him (albeit using the imperative tense) to do what learned men are meant to do: cultivate, teach, and leave works for posterity.

Just as faith and diligence (*fides ac diligentia*) appear together in the correspondence, innate brilliance and erudition are often juxtaposed as a compliment. In one of his earliest letters, Grotius bestows Daniel Heinsius with an overly lofty praise. Grotius compares Heinsius with the sun, the 'great eye of the world', and says that 'he is not merely a prince of the learned youth', and finally: 'you share with us the glory of your most brilliant innate brilliance and learning'.[168] His copious sentences reflect the ideal of the learned man: a prince of learning in pursuit of glory, achieved through innate brilliance and learning that can be shared with everybody. Grotius received such compliments later in his life as well: for example, the scholar Lucas Holstenius (1596–1661) writes to Grotius that he 'eagerly admires the greatness of [his] innate brilliance and erudition'.[169] The trope that one is a cultivator, admirer, or benefactor of 'innate brilliance and erudition' abounds in the correspondence of Grotius. It is rich in references to *ingenium* and *eruditio*: Grotius calls his brother's work 'evidence of innate brilliance and erudition',[170] a correspondent calls himself 'a great follower and admirer of your innate brilliance',[171] and Grotius states that a book of Franciscus Junius (1591–1677) 'has a lively image of your innate brilliance and erudition'.[172]

If we look at all the virtues together, we can draw the conclusion that learned men expressed different virtues in the medium of the letter compared to those in the *vitae* in Chapter 1. Compared to the *vitae* (see Chapter 1, Figures 1.1–1.3), the virtues of *fides*, *amicitia*, and *eruditio* all stand out. Although the medium of the letter is more intimate, these scholarly virtues

168 Hugo Grotius to Daniel Heinsius, 1 March 1603, *BW* I, letter no. 38, pp. 30–31. The full sentence reads: 'Sed quemadmodum magnus ille Mundi oculus cum in caetera astra per se lucis egentia, tum in ipsam quoque siderum plebem iubar atque aureum illud lumen transfundit caelesti largitione, quod tamen solius suum est, ita tu, cui iniuriam fecerit, qui iuventutis tantum literariae principem dixerit, gloriam splendidissimam ingenii atque eruditionis tuae nobiscum communicas, qui in eodem quidem spatio tecum currimus, sed ignobili studio, atque aeterna nocte damnamur, nisi quantum prohibent radii sideris tui.'

169 Lucas Holstein to Hugo Grotius, 5 November 1627, *BW* III, letter no 1193, pp. 193–194.

170 '[T]estimonium ingenii atque eruditionis'. Hugo Grotius to Willem de Groot, 7 May 1636, *BW* VII, letter no. 2583, pp. 133–134.

171 '[E]tiam a cliente tuo tuique ingenii admiratore maximo'. Matthias Bernegger to Hugo Grotius, 13 February 1637, *BW* VIII, letter no. 2960, p. 97.

172 '[I]pse vivam habet ingenii eruditionisque tuae picturam'. Hugo Grotius to Franciscus Junius, 31 May 1638, *BW* IX, letter no. 3606, pp. 339–340.

were also repeated throughout the correspondence. However, the virtue of *amicitia* is much more prevalent in the correspondence than in the *vitae* because this virtue was used to strengthen and emphasise the relation between two correspondents. First, the ideal of friendship, *amicitia*, was extolled as a bond between two equals. Secondly, the repetition of the faith (*fides*) in one another helped form a bond of trust. Friendship and faith are relational virtues that were functional in the strengthening of individual relations. The scholarly letter was thus a medium that was particularly well suited to extolling relational virtues and building a more intimate imagined community of friendship and trust.

## 3.4 Conclusion

The correspondence of Hugo Grotius has proven to be a rich source for the study of the formation of scholarly identity. Grotius forged new alliances by citing and remembering authors, scholars, and authorities that were commonly studied in his circle of friends, and he reinforced this shared intellectual ground by praising scholarly virtues such as *amicitia* and *eruditio*.

Any correspondence was a unique, intimate medium to express and imagine a learned community. To build a relationship with fellow learned men through letters, one had to adhere to the hierarchy of the community as well as the tacit code of conduct which prescribed an approach to a learned man who was at the top of the hierarchy, as evidenced by the exchange between Lipsius and Grotius in his younger years. Together with his correspondents, Grotius had a distinct view of the history of the learned community in which he worked, as evidenced by the ample citation of writers from antiquity, religious writers, and reformers.

From the Grotius correspondence emerges a learned world that is built upon intimate, personal connections between individuals. The community was not seen as a group of people who worked together, but rather as an amalgamation and entanglement of individual friendships. Grotius stressed these friendships by referring to them with possessive nouns: 'our friendship' (*nostra amicitia*) and 'your friendship' (*tua amicitia*). Grotius and his correspondents thereby conveyed the sense of confidential and faithful bonds between correspondents: friendships held together by trust and faith. The sense of a community was created in the correspondence by an implicit hierarchy, a common cultural corpus, canonical figures, and a shared scholarly goal. By means of these mechanisms, correspondence became a medium that helped to imagine a community.

When Grotius responded to Gerardus Vossius's request to befriend his son, Isaac Vossius, he asked Vossius:

> But do you think that your son ought to be commended to me, first as Vossius the son of Vossius, then on account of his innate brilliance, his erudition, his modesty, on account of which he ought to be dear to anyone, even to a previously unknown person?
>
> Filiumne vero etiam tuum mihi commendandum censes, primum Vossium Vossii filium, deinde eius ingenii, eius eruditionis, eius modestiae, ob quae cuivis etiam ante ignoto carus esse debeat?[173]

Grotius's question shows the order of things in assessing a potential friendship: first one's name, then one's scholarly virtues, and finally one's modesty. These are the foundational virtues that solidify a friendship. Of course, Grotius affirmed Gerardus Vossius's request and in doing so he praised the scholarly friendship between Isaac Vossius and Johann Friedrich Gronovius:

> I have seen with great pleasure with what friendship he [Isaac] and [Johann Friedrich] Gronovius—a most delightful pair—have embraced each other, who now share their itineraries with each other, Gronovius from Italy to Holland, your [son] from Holland to Italy.
>
> Summa mea cum voluptate vidi, qua amicitia se mutuo complectantur ille et Gronovius, par lectissimum, qui nunc itinera inter se commutant, hic ex Italia in Batavos, tuus e Batavis in Italiam.[174]

Both Isaac Vossius and Johann Friedrich Gronovius travelled extensively through Europe as scholars. This letter hints at the comradery such voyages, with their shared itineraries, could bring. The scholarly journeys many young scholars undertook represented a foundational element of the bond that connected scholars in Europe. In the next chapter we will explore how such scholarly voyages throughout Europe allowed scholars to experience, represent, and form learned communities.

173 Hugo Grotius to Gerardus Vossius, 23 November 1641, *BW* XII, letter no. 5479, pp. 640–641.
174 Ibid.

# 4. The *Peregrinatio Literaria*: Experiencing, Representing, and Forming Learned Communities

**Abstract:** This chapter investigates the role of scholarly travel (*peregrinatio literaria*) in sustaining the Republic of Letters. It explores how learned figures like Isaac Vossius, Nicolaas Heinsius, and Joannes Kool travelled across Europe, visiting key intellectual centres and interacting with notable scholars. These journeys helped scholars create and maintain a transnational network, reinforcing the idea of a pan-European community of knowledge. The chapter highlights how these scholarly travels were distinct from typical Grand Tours, focusing more on scholarly exchange and research. Such travels allowed scholars to engage with broader European intellectual traditions, thus contributing to the Republic of Letters' collective identity.

**Keywords:** Scholarly travel, *peregrinatio*, knowledge exchange, scholarly identity, early modern academic mobility

> Yet, there is this caution here: namely that it [the travelling] should not be merely with pleasure, but with profit. Anyone can ramble, roam, or wander, but few research and learn; that is, truly travel.
>
> Sed ista hic cautio: ut id fiat non cum voluptate solum, sed cum fructu. Vagari, lustrare, discurrere quivis potest: pauci indagare, discere; id est, vere peregrinari.[1]

1 This passage was originally part of a letter, see Justus Lipsius to Philippe de Lannoy, 3 April 1578, in Justus Lipsius et al., *Epistolarum centuriae duae* [...] (Leiden: Ex officina Plantiana, apud Franciscum Raphelengium, 1590), epist. XXII., vol. I, pp. 33–40; also in *ILE* I, pp. 59–61. The Latin letter was later published as Justus Lipsius, *Epistola de peregrinatione Italica* (Leiden: Franciscus Haegerus, 1592). The letter also appeared in a travel manual by Thomas Erpenius, see

Scholten, Koen. *Rethinking the Republic of Letters: Memory and Identity in Early Modern Learned Communities*. Taylor & Francis Group, 2025
DOI 10.5117/9789048559855_CH04

The well-known humanist Justus Lipsius promoted the humanist ideals of travelling for young ambitious men in his *Epistola de peregrinatione Italica* (Letter on travelling in Italy), a popular letter that was republished many times. It was also translated into Dutch in the seventeenth century to serve as a manual for future travellers. A virtuous and successful Grand Tour, in the eyes of the humanist, thus meant balancing both pleasure (*voluptas*) and usefulness (*utilitas*). The young, studious scholar did not ramble or wander, but spent his travelling days doing research and learning.

A wide array of people travelled around Europe in the early modern period: diplomats, ambassadors, merchants, and students, but also pilgrims, colonists, and pirates.[2] Additionally, both young and experienced artists were known to be fond of travelling. For example, Netherlandish artists travelled through Europe to learn new methods and acquire fame as court artists.[3] The traditional early modern journey, passing through France, Italy, Germany, and the Low Countries, is often dubbed the Grand Tour.[4] Many young Dutch students and scholars travelled across the continent to study at

Justus Lipsius, 'De Ratione cum fructu peregrinandi, & praesertim in Italia', in Thomas Erpenius, *De peregrinatione Gallica Utiliter instituenda Tractatus* [...], pp. 105–128 (Leiden: Franciscus Hegerus, 1631). The letter was translated into Dutch in a travel manual, see Justus Lipsius, 'Een groot oordeel van dien grooten en uytsteeckenden Justus Lipsius, over het reysen', in Nicolaes van Ravesteyn, *Wegh-Wyser* [...] (Amsterdam: Nicolaes van Ravesteyn, 1647). The Dutch passage reads: 'Doch men lette hier, dat dat geschiede niet alleen met vermaeck, maer oock met vrucht: een yeder kan wel swieren en snuffelen; weynige ondersoecken en leeren; dat is, recht reisen.'

2 For a typology of early modern travellers, see William H. Sherman, 'Stirrings and Searchings (1500–1720)', in *The Cambridge Companion to Travel Writing*, ed. Peter Hulme and Tim Youngs, 17–36 (Cambridge: Cambridge University Press, 2002).

3 Frits Scholten et al., eds., *Art and Migration: Netherlandish Artists on the Move, 1400–1750* (Leiden; Boston: Brill, 2014).

4 For a brief overview of literature on the Grand Tour, see: Ludwig Schudt, *Italienreisen im 17. und 18. Jahrhundert* (Vienna; Munich: Schroll-Verlag, 1959); Anna Frank-van Westrienen, *De Groote Tour: Tekening van de educatiereis der Nederlanders in de zeventiende eeuw* (Amsterdam: Noord-Hollandsche Uitgeversmaatschappij, 1983); Jeremy Black, *The British Abroad: The Grand Tour in the Eighteenth Century* (Stroud: Alan Sutton, 1992); Jás Elsner and Joan-Pau Rubiés, eds., *Voyages and Visions: Towards a Cultural History of Travel* (London: Reaktion Books, 1999); Jeremy Black, *Italy and the Grand Tour* (New Haven, CT; London: Yale University Press, 2003); Black, *France and the Grand Tour* (London: Palgrave Macmillan, 2003); Michael G. Brennan, ed., *The Origins of the Grand Tour: The Travels of Robert Montagu, Lord Mandeville (1649–1654), William Hammond (1655–1658), Banaster Maynard (1660–1663)* (London: Hakluyt Society, 2004); Rosemary H. Sweet, *Cities and the Grand Tour: The British in Italy, c. 1690–1820* (Cambridge; New York: Cambridge University Press, 2012); Gerrit Verhoeven, *Anders reizen? Evoluties in vroegmoderne reiservaringen van Hollandse en Brabantse elites (1600–1750)* (Hilversum: Uitgeverij Verloren, 2009), translated into English as *Europe Within Reach: Netherlandish Travellers on the Grand Tour and Beyond, 1585–1750*, trans. Diana Webb (Leiden; Boston: Brill, 2015); Rosemary Sweet et al., eds., *Beyond the Grand Tour: Northern Metropolises and Early Modern Travel Behaviour* (London;

a foreign university, learn a language, build up a network, obtain a doctoral degree, and become acquainted with local customs and forms of civility. For noblemen and scholars alike, travelling was a formative experience, as well as a credential necessary to further one's career in statesmanship or academia. Although this tradition of travelling via universities (*peregrinatio academica*) and the educational journey to Italy (*iter italicum*) did change in form and destinations throughout the early modern period, this conviction that travelling was an important step towards adulthood remained constant.[5]

Travelling was an essential part of becoming a renowned scholar in the learned world. The scholarly journeys of three young, ambitious humanist scholars take centre stage in this chapter: Isaac Vossius (1618–1689), Nicolaas Heinsius (1620–1680), and finally Joannes Kool (1672–c. 1708). As argued by historians Jás Elsner and Joan-Pau Rubiés, a humanist education was essential for a systematic empirical approach to travelling. It offered both philological tools and classical literary models, as well as a positive view of travels as a means of education.[6] This positive attitude towards travelling can also be observed in the passage of Lipsius's letter above. The balance between enjoyment and serious research was a characteristic of a successful Grand Tour, and was even more important for travelling scholars. Throughout the seventeenth century, most travellers increasingly moved away from studying and obtaining doctoral degrees, and rather travelled to explore foreign cultures, signalling the beginning of so-called 'Grand Tourism'.[7] Vossius, Heinsius, and Kool, however, did take Lipsius's advice to heart: while they also enjoyed the beauty and splendour of foreign lands, they spent most of their time researching manuscripts and acquiring books.

This chapter will assess their scholarly travels in the light of community formation.[8] The idea that travelling was essential for the formation and maintaining of the scholarly community throughout the early modern

New York: Routledge, 2017); Alan Moss, 'Gemaakt op reis: Identiteitsvorming in verslagen van de Nederlandse educatiereis (1648–1713)' (PhD diss., Radboud University, 2022).

5 Ad Tervoort, *The iter italicum and the Northern Netherlands. Dutch Students at Italian Universities and their Role in the Netherlands' Society (1426–1575)* (Leiden; Boston: Brill, 2005); W. T. M. Frijhoff, *La societé neerlandaise et ses gradués, 1575–1814: Une recherche sérielle sur le statut des intellectuels à partir des registres universitaires* (Amsterdam: APA, 1981); Hans Bots and Willem Frijhoff, 'Academiereis of educatiereis? Noordbrabantse studenten in het buitenland, 1550–1650', *Batavia Academica* 1 (1983): 13–30.

6 Elsner and Rubiés, *Voyages and Visions*, p. 46.

7 See, e.g., Verhoeven, *Europe Within Reach*.

8 In the modern period, travelling helped create a collective sense of Europe after the Second World War; see Richard Ivan Jacobs, *Backpack Ambassadors: How Youth Travel Integrated Europe* (Chicago: University of Chicago Press, 2017).

period is a trope in historical literature.[9] For example, David Lux and Harold Cook argued that 'travel, more than any other activity, established the weak ties by which knowledge could be exchanged.'[10] Lux and Cook stressed that knowledge could only travel if there was trust, and travelling allowed scholars to meet each other and build a relationship of trust for decades. More generally, they argued that 'the international exchange of learning therefore depended upon an exchange of personal visits.'[11] This view of the centrality of travelling in the learned communities is the starting point of this chapter. Yet few historians have examined the mechanisms of community formation that ultimately formed pan-European learned and scientific communities. This chapter will look at a particular learned community: Dutch philologists (textual scholars) who travelled to Italy. Heinsius, Vossius, and Kool were all part of a Dutch tradition of textual scholarship, inspired by Joseph Justus Scaliger (1540–1609), Daniel Heinsius (1580–1655), and later Jacob Gronovius (1645–1716). This chapter shows how travelling helped to form collective scholarly identities and what constituted such a scholarly identity. Scholars could meet fellow scholars who shared the same values and subscribed to a collective goal of learning during what I propose to call a *peregrinatio literaria*. It was this particular type of learned journey throughout Europe that enabled the formation of a pan-European imagined community.

This chapter is divided into two distinct parts. The first part explores the characteristics of *scholarly* travel, as opposed to the more common Grand Tour and *peregrinatio academica*. The travels of Isaac Vossius and Nicolaas Heinsius will be taken as examples of scholarly travel. Both scholars ascended the social and scholarly ladder throughout their careers, in which travel played an important part. I will argue that the *peregrinatio literaria* is a type of journey that is markedly different from the more common Grand Tour on the one hand, and the *peregrinatio academica* on the other, and that it was an essential aspect of the Republic of Letters. The second part

9 For the role of travelling in the learned world, see Kasper Risbjerg Eskildsen, 'Exploring the Republic of Letters: German Travellers in the Dutch Underground, 1690–1720', in *Scientists and Scholars in the Field: Studies in the History of Fieldwork and Expeditions*, ed. Kristian H. Nielsen et al., 101–122 (Aarhus: Aarhus University Press, 2012). For the role of travelling in seventeenth-century natural history, see Paula Findlen, *Possessing Nature: Museums, Collecting, and Scientific Culture in Early Modern Italy* (Berkeley: University of California Press, 1998), ch. 4, 'Pilgrimages of Science', pp. 155–193.

10 David S. Lux and Harold J. Cook, 'Closed Circles or Open Networks?: Communicating at a Distance During the Scientific Revolution', *History of Science* 36 (1998): 179–211 (p. 183).

11 Ibid., p. 202.

of this chapter will look more closely at the personal visits that Lux and Cook only hinted at. The unique travelogue of the aspiring scholar Joannes Kool, a student of professors Johann Georg Graevius (1632–1703) and Jacob Gronovius (1645–1716), is an intricate and comprehensive account of his travel to Italy. On the way, he met many fellow scholars and his travelogue shows us how these personal visits worked and how they helped form and maintain a sense of community. It showcases the mechanisms of scholarly identity formation on a *peregrinatio literaria* from the perspective of Kool himself, as well as that of the French and Italian scholars who identified and accepted him as a part of their local learned communities. Furthermore, his detailed accounts of visits to scholars abroad illustrate the vital part travelling played in representing and forming learned communities. In short, this chapter will argue that these scholarly travels were fundamental for the creation and sustenance of the scholarly community of classical scholars, which connected scholars from the Low Countries, France, and the Italian peninsula. Travelling was the intergenerational glue that helped scholars such as Vossius, Heinsius, and Kool to experience, represent, and form the pan-European learned community of philologists.

## 4.1 The Typology of a *peregrinatio literaria*

The learned community of textual scholars had a positive attitude towards travelling. As mentioned at the end of the previous chapter, Hugo Grotius praised the comradery between Isaac Vossius and Johann Friedrich Gronovius, since they 'now share their itineraries with each other'.[12] The travels of these men were part of a series of scholarly travels made by Dutch textual scholars to France and Italy, upon which I will elaborate later in the chapter. This section will first outline Vossius and Heinsius's journeys, which will be followed by a discussion of the historiography and typology of the Grand Tour, to consequently argue that their journeys are representative of a scholarly journey which I will term the *peregrinatio literaria*.

Isaac Vossius travelled to England, France, and Italy between 1641 and 1644.[13] These journeys have been well documented due to the many letters Isaac sent to his parents and a monograph written by F. F. Blok.[14] The voyages

12 Hugo Grotius to Gerardus Vossius, 23 November 1641, *BW* XII, letter no. 5479, pp. 640–641.

13 Y. H. Rogge, 'De reis van Isaac Vossius (1641–1645)', *Oud Holland* 18, no. 1 (1900): 3–20.

14 Transcriptions of these letters were published in F. F. Blok and C. S. M. Rademaker, 'Isaac Vossius' Grand Tour, 1641–1644: The Correspondence between Isaac and his Parents', *Lias* 33, no. 2

of Isaac Vossius and Nicolaas Heinsius were supported by their fathers, the scholars Gerardus Vossius (1557–1649) and Daniel Heinsius. Isaac Vossius first travelled to England, where he visited London, Oxford, and Cambridge.[15] Afterwards, he travelled to Paris, where he met and stayed with Hugo Grotius.[16] Here, Isaac also met Johann Friedrich Gronovius, who was taking a break from his Grand Tour in Paris in October 1641.[17] Between February and May 1642, Isaac was in Florence, and mainly spent his time working in the Bibliotheca Laurentiana Medicea.[18] He was welcomed in the Florentine learned circles, especially by Giovanni Battista Doni (1594–1647) and Carlo Dati (1619–1676).[19] After a month-long stay in Napels, Isaac travelled to Rome, where he was hospitably received by Lucas Holstenius (1596–1661), whom Isaac could give a letter of recommendation on behalf of his father Gerardus Joannes Vossius. In Rome, Vossius worked days on end in the Vatican Library and the Barberiniana.[20] After a few months in Rome, Isaac travelled back to Florence, and subsequently to Padua, Venice, and Milan. On his return journey from Italy, Vossius obtained his doctorate in law at the University of Orléans in 1643.[21] In July 1643, Vossius returned to Paris and worked as a secretary to Hugo Grotius.[22] More than a year later in August 1644, Isaac started his journey home, arriving in The Hague in October to meet his family. From 1649 onwards, he worked in the service of Queen Christina of Sweden (1626–1689) as her tutor, librarian, book collector, and scholar.[23]

Nicolaas Heinsius embarked on several voyages between 1641 and 1653. His first journey was to England in 1641, where he worked in the Bodleian Library in Oxford. In the summers of 1644 and 1645, Nicolaas visited the

(2006): 150–216; Blok and Rademaker, 'Isaac Vossius' Grand Tour, 1641–1644: The Correspondence between Isaac and his Parents. Part II: Isaac Vossius in Italy (with 3 ills.)', *Lias* 35, no. 1–2 (2008): 209–279; and Blok and Rademaker, 'Isaac Vossius' Grand Tour, 1641–1644: The Correspondence between Isaac and his Parents. Part III: Isaac in Paris', *Lias* 36, no. 2 (2009): 295–385; F. F. Blok, *Isaac Vossius en zijn kring: Zijn leven tot zijn afscheid van koningin Christina van Zweden, 1618–1655* (Groningen: Egbert Forsten, 1999); translated into English as F. F. Blok, *Isaac Vossius and His Circle: His Life Until His Farewell to Queen Christina of Sweden, 1618–1655*, trans. Cis van Heertum (Groningen: Egbert Forsten, 2000). The Dutch edition is cited here as Blok, *Vossius*.

15 Blok, *Vossius*, pp. 75–95.

16 Ibid., pp. 97–123.

17 Ibid., pp. 111–113.

18 Ibid., pp. 126–137.

19 Ibid., pp. 137–140.

20 Ibid., pp. 145–156.

21 Rogge, 'De reis van Isaac Vossius', p. 18.

22 Blok, *Vossius*, pp. 179–199.

23 On Vossius's acceptance of the position, see ibid., pp. 271– 281; and more generally on Vossius's time in service of the Swedish court, see ibid., pp. 253–503.

health resort Spa to recover from illness.[24] In September 1645, he embarked on a boat from Rotterdam to Paris, where he stayed for six months. Helped by many letters of recommendation, he enjoyed the company of a Parisian circle of learned men.[25] Between 1646 and 1648, he stayed in the Italian cities of Florence, Rome, Bologna, Venice, Milan, and Genoa. After this journey around Italy, he was invited to the Swedish court, to advise Queen Christina on philological matters and to collect books. At the end of 1649, he travelled to the court to meet Christina, René Descartes (1596–1650), and his old friend Isaac Vossius. Together with Vossius, Heinsius worked for Christina and travelled back and forth between Sweden and the Low Countries.[26] Nicolaas embarked on a second voyage to Italy at the end of 1651, to buy books and manuscripts in the Italian libraries for Queen Christina's collection. In the spring of 1653, he returned to his parental home in Leiden. The detailed historical work of A. H. Kan and F. F. Blok will serve as the main source for the itineraries of Nicolaas Heinsius.[27]

This section will introduce the concept of a scholarly journey—a *peregrinatio literaria*—to describe the voyages of scholars more accurately. Vossius and Heinsius's journeys are examples of such scholarly journeys, that differentiate themselves from the Grand Tour, the *peregrinatio academica* and the earlier *iter italicum*. As early as the fifteenth century, Dutch students would travel to Italy to obtain a university degree since there was a lack of universities in the north.[28] The journey to Italy was popular and the Italian universities were well attended by foreign students, with foreign attendance ranging between 10 per cent in Pisa and 50 per cent in Padua, at the extreme end. Most of these students came from the higher echelons of the German nations.[29] Historian Ad Tervoort concluded that the travelling helped bring Italian learned culture to the northern parts of Europe: 'In a wider sense students who went to Italy constituted a literate

24 F. F. Blok, 'Nicolaas Heinsius in dienst van koningin Christina van Zweden' (PhD diss., Leiden University, 1949), p. 21; hereafter cited as Blok, 'Heinsius'.

25 A. H. Kan, 'Nicolaas Heinsius in Italië (1646–1648)', *Onze Eeuw* 14, no. 3 (1914): 361–390 (esp. p. 372).

26 Blok, 'Heinsius', pp. 59–88.

27 A. H. Kan, 'Nicolaas Heinsius in Italië'; Kan, 'Nicolaas Heinsius in Italië (1646–1648) (II slot)', *Onze Eeuw* 14, no. 4 (1914): 57–74; Blok, 'Heinsius'; F. F. Blok, *Nicolaas Heinsius in Napels (april–juli 1647)* (Amsterdam; Oxford; New York: Noord-Hollandsche Uitgevers Maatschappij, 1984).

28 Jan den Tex, 'Nederlandse studenten in de rechten te Padua 1545–1700', *Mededelingen van het Nederlands Historisch Instituut te Rome*, 3rd series, 10 (1959): 45–165; H. de Ridder-Symoens, 'Brabanders aan de rechtsuniversiteit van Orléans, 1444–1556: Een socio-professionele studie', *Bijdragen tot de Geschiedenis* 61 (1978): 195–347; Frijhoff, *La société néerlandaise et ses gradués*.

29 Tervoort, *Iter italicum*, p. 13.

group of potentially successful and influential men on whom the culture of Renaissance Italy must have made an impression and who were likely to support or even propagate Renaissance culture'.[30] In the first half of the sixteenth century, the archetypical journey to Italy expanded to a more general *peregrinatio academica*, on which Northern European travellers would earn their degrees at French and German universities, moving away from the Italian universities of Bologna and Padua. For example, throughout the seventeenth century, the University of Orléans was one of the most popular destinations in Europe to obtain a law degree.[31]

The journey to Italy, the *iter italicum* and more generally the *peregrinatio academica*, developed into an educational journey—also called a *Bildungsreise*—towards the end of the sixteenth century and had its heyday in the long seventeenth century.[32] The goal of the journey was not to merely obtain a university degree, but also to learn about the local culture and language. A wide array of manuals and guides helped travellers to navigate the many parts and cultures of Europe to master the *ars apodemica*, the art of travel.[33] Anna Frank-van Westrienen discerned three different types of travellers in her landmark study of Netherlandish travel journals: diligent students, aspiring gentlemen from well-standing families, and Grand Tourists.[34] Not all travellers neatly fit in this typology of travellers, but in general most travellers can be identified by one or two of these categories of travellers. The main goal of the diligent student was, as Thomas Erpenius (1584–1624) wrote in his travel manual, 'to increase knowledge, which for the most part should be the sole target of travelling'.[35] Gerrit Verhoeven added considerable depth to the social aspects of travelling, in particular the composition of the travelling populace and the diverse travel goals they had. He distinguished between four general travel objectives: diplomacy, education, business, and

30 Ibid., p. 362.

31 C. M. Ridderikhoff, 'Orléans and the Dutch Revolt', in *The Great Emporium: The Low Countries as a Cultural Crossroads in the Renaissance and the Eighteenth Century*, ed. C. C. Barfoot and R. Todd, 59–82 (Amsterdam; Atlanta, GA: Rodopi, 1992).

32 Bots and Frijhoff, 'Academiereis of educatiereis?'.

33 Karl A. E. Enenkel and Jan L. de Jong, eds., *Artes Apodemicae and Early Modern Travel Culture, 1550–1700* (Leiden; Boston: Brill, 2019); Justin Stagl, *Apodemiken: eine räsonnierte Bibliographie der reisetheoretischen Literatur des 16., 17. und 18. Jahrhunderts* (Paderborn: Schöningh, 1983).

34 Frank-van Westrienen devoted a chapter to each of these types of travellers; see Frank-van Westrienen, *Groote Tour*, ch. 5, 'Student diligent', pp. 134–202; ch. 6, 'Heer in de leer', pp. 203–242; and ch. 7, 'Toerist in touw', pp. 243–312.

35 '[A]d augendum scientiae, qui unicus fere peregrinationis debet esse scopus.' Erpenius, *De peregrinatione Gallica utiliter instituenda Tractatus*, p. 18; partly cited in Frank-van Westrienen, *Groote Tour*, p. 136.

leisure.[36] The objective of education focused on *Bildung*, which included learning a language, acquainting oneself with local etiquette, and obtaining the odd academic degree. Verhoeven argued that 'from the seventeenth century onwards, the aim of a Grand Tour was to provide broad knowledge of painting, music, theatre, architecture and sculpture.'[37] The tour enabled students to study at a foreign university, become acquainted with foreign cultures and confessions, as well as learn and use foreign languages. Sons of noble families could learn courtly traditions throughout Europe and obtain a prestigious academic title abroad.[38] However, the secondary literature of the Grand Tour largely leaves out the journeys of scholars and scientists.

In recent years, historians have pointed to the scholarly aspects of educational travel. Dirk van Miert argued that the visit of Calvinist minister Abrahamus Trommius (1633–1719) to the grave of Erasmus in Basel could be interpreted as an identification with a scholarly identity and community.[39] In a similar vein, Paul Hulsenboom and Alan Moss pointed to the importance of sites and objects of memory in the construction of learned communities. Travellers engaged with a transnational history of learning, exemplified by the veneration of the pen of Lipsius by Jacob Gronovius, the laudatory poem Jacobus Kinschot wrote about the skull of Julius Caesar Scaliger (1458–1558), and the many references of esteem to the statue and epitaph of Erasmus.[40] Elsewhere, Alan Moss argued that literary and scholarly identities were formed in the visits to, and veneration of, sites of memory such as libraries, universities, as well as graves and memorials of famous scholars.[41] While scholars have touched upon these aspects of journeys that fostered a scholarly identity, this section will take a wide view and propose

36 Verhoeven, *Europe Within Reach*, pp. 52–81.

37 Ibid., p. 69.

38 The tour of noblemen is sometimes called a *Kavalierstour* in German historiography; see R. Babel, ed., *Grand Tour: Adeliges Reisen und europäische Kultur vom 14. bis zum 18. Jahrhundert* (Ostfildern: Jan Thorbecke Verlag, 2005); Th. Grosser, 'Reisen und soziale Eliten: Kavalierstour—Patrizierreise—bürgerliche Bildungsreise', in *Neue Impulse der Reiseforschung*, ed. Michael Maurer, 136–176 (Berlin: Akademie Verlag, 1999); Eva Bender, *Die Prinzenreise: Bildungsaufenthalt und Kavalierstour im höfischen Kontext gegen Ende des 17. Jahrhunderts* (Berlin: Lukas, 2011).

39 Dirk van Miert, 'Trommius's Travelogue: Learned Memories of Erasmus and Scaliger and Scholarly Identity in the Republic of Letters', *Early Modern Low Countries* 1, no. 1 (2017): 51–70.

40 Paul Hulsenboom and Alan Moss, 'Tracing the Sites of Learned Men: Places and Objects of Knowledge on the Dutch and Polish Grand Tour', in *Memory and Identity in the Learned World: Community Formation in the Early Modern World of Learning and Science*, ed. Koen Scholten et al., 257–306 (Leiden; Boston: Brill, 2022).

41 Moss, 'Gemaakt op reis', pp. 319–324.

a typology of the early modern scholarly journey, or what I have termed the *peregrinatio literaria*.

Three elements distinguished the *peregrinatio literaria* from the Grand Tour, the *peregrinatio academia*, and the earlier *iter italicum*. First, one of the primary goals of the scholarly journey was to meet fellow learned men throughout Europe. The contacts between Dutch and European scholars were maintained and renewed with such *rendezvous des savants*. With letters of recommendation from professors and renowned learned men, travellers could gain access to impressive and illustrious libraries in England, France, and Italy. The second goal of the travels of scholars was to collect and buy books for one's own collection, as well as for the collections of friends and colleagues. Finally, for young ambitious scholars the purpose of these travels was to carry out philological research. This meant working in obscure Italian libraries to copy and collate unique manuscripts. The journeys of Vossius and Heinsius are examples of such a *peregrinatio literaria*, characterised by the meeting of fellow scholars, the purchase of books, and the philological research in libraries.

The first, and arguably the most important, goal of a *peregrinatio literaria* after obtaining a university degree was to gain a reputation as a scholar in the European learned world. The names of Heinsius and Vossius had already become known in learned circles, because their fathers Daniel and Gerardus had extensive contacts with European scholars, especially in Paris and throughout Italy. Additionally, Heinsius could build on the fame of young scholars who had travelled before him: Isaac Vossius, Johann Friedrich Gronovius (1611–1671), and his son Jacob Gronovius (1645–1716). The many letters between J. F. Gronovius and Heinsius are evidence of the close friendship between the two scholars throughout their careers. Heinsius received letters about the latest developments in the learned world wherever he was.[42] As a token of their friendship, J. F. Gronovius introduced Heinsius to many of his scholarly friends, some of whom Gronovius had met on previous travels.[43]

Jacques (1591–1656) and Pierre Dupuy's (1582–1651) library—also called 'the cabinet'—was an important meeting place in Paris. During Nicolaas's first stay in 1645 the brothers Dupuy helped Heinsius with formal introductions

42 See the many letters in Pieter Burman, ed., *Sylloges epistolarum a viris illustribus scriptarum tomi quinque*, 5 vols. (Leiden: Samuel Luchtmans, 1727), vol. 3.

43 Blok, 'Heinsius', p. 103. For the scholarly journey of J. F. Gronovius, see Paul Dibon and Françoise Waquet, *Johannes Fredericus Gronovius, pèlerin de la république des lettres: recherches sur le voyage savant au xviie siècle* (Geneva: Librairie Droz, 1984).

to renowned French scholars. Heinsius met Jacques-Auguste de Thou (1609–1677), the son of the scholar and historian with the same name, in the library of the Dupuys. After their first meeting, De Thou allowed Heinsius to visit and work in the library of his father and the two maintained an epistolary exchange ever afterwards. When De Thou was ambassador of France in The Hague between 1657 and 1661, he and Heinsius often worked together to find and collect rare manuscripts. De Thou ordered specific manuscripts of Ovid to be sent to The Hague for Heinsius's critical edition of the classical poet.[44] Meetings with scholars such as the brothers Dupuy and De Thou were fundamental for his development as a European scholar of renown. Throughout his working life, Heinsius maintained an extensive correspondence with Jacques Dupuy and other French scholars.[45]

The friendship between Isaac Vossius and Nicolaas Heinsius was important to both their careers. They avoided discussions about their diverging Reformed and Remonstrant credentials, as well as the antagonism between their fathers in their letters, focusing instead on recently published books and the progress of their research.[46] The two men worked together for most of their scholarly lives and helped each other constantly, as their bond of friendship required. Because Isaac had travelled through Italy and France a couple of years earlier, Nicolaas could present himself as the next ambitious scholar from the Low Countries and rely on the contacts Isaac made during his voyage. In Rome, Lucas Holstenius (1596–1661) and Cassiano dal Pozzo (1588–1657) were exactly such valuable contacts. According to a letter Nicolaas wrote in Genova, he could reach highly placed Italian clergy with more ease than his own father:

> I am not such a despicable creature and so insensitive to contempt, that I will invite people to write who would rather be silent, since I have frequently been offended. If I wrote to a Cardinal or whatever ruler in Italy, I would have been certain that I would receive a reply.
>
> Ik ben niet zoo'n verachtelijk schepsel en zoo ongevoelig voor minachting, dat ik lui, die toch zullen blijven zwijgen, tegen hun zin zou willen uitnoodigen tot schrijven, terwijl ik al zoo dikwijls mijn hoofd heb gestooten.

44 Ibid., pp. 110–111.

45 Jacques Dupuy and Nicolaas Heinsius, *Correspondance de Jacques Dupuy et Nicolas Heinsius (1646–1656)*, ed. Hans Bots (The Hague: Nijhoff, 1971); Jean Chapelain, *Les lettres authentiques à Nicolas Heinsius (1649–1672): Une amitié érudite entre France et Hollande*, ed. with annotations and introduction by Bernard Bray (Paris: Honoré Champion Éditeur, 2005).

46 Blok, 'Heinsius', p. 77.

> Als ik aan een Kardinaal of welken vorst ook in Italië schreef, ben ik er zeker van, dat ik antwoord zou krijgen.[47]

Nicolaas was clearly convinced of his opportunities to gain access to powerful men and their libraries in Italy. His research in libraries and his meetings with learned men—a central aim of his journeys—are evidence of the scholarly aspect of his journeys.[48]

**Table 4.1. Travellers from the field of classical philology from the seventeenth-century Low Countries**

| Scholar | Destinations | Period | Sources |
|---|---|---|---|
| Joseph Justus Scaliger (1540–1609) | Rome, England | 1565–1566 | Bernays[49] |
| Justus Lipsius (1547–1606) | Vienna; Italy | 1565–1569; 1571 | Papy[50] |
| Gerardus Vossius (1577–1649) | Oxford, Cambridge | 1629 | Rademaker[51] |
| Erycius Puteanus (1574–1646) | Padua, Milan, Rome | 1603?–1606 | *NNBW*; Simar[52] |
| Thomas Erpenius (1584–1624) | France, Italy | 1608–1612 | *NNBW* |
| Johann F. Gronovius (1611–1671) | Oxford, Paris | 1637–1638 | Dibon and Waquet[53] |
| Isaac Vossius (1618–1689) | England, France, Italy | 1641–1645 | Blok, *Vossius* |
| Nicolaas Heinsius (1620–1681) | Oxford; France, Italy; Sweden; Italy | 1641; 1645–1648; 1649–1650; 1651–1653 | Blok, 'Heinsius' |
| Johann Georg Graevius (1632–1702) | Leipzig | 1649–1651 | *ADB* |

47 Nicolaas Heinsius to Daniel Heinsius, 13 June 1648; Dutch translation from the original Latin letter by Kan, who provided no transcription of the Latin source; see Kan, 'Nicolaas Heinsius in Italië (1646–1648) (II slot)', p. 73.

48 See also Blok, 'Heinsius', pp. 89–126.

49 Jacob Bernays, *Joseph Justus Scaliger* (Berlin: W. Hertz, 1855).

50 Jan Papy, 'Justus Lipsius, Rome en de Romereis: zoektocht naar een oude mythe?', *Kleio* 26, no. 3 (1997): 111–126.

51 C. S. M. Rademaker, *Life and Works of Gerardus Joannes Vossius (1577–1649)*, trans. H. P. Doezema (Assen: Van Gorcum, 1981), pp. 232–235.

52 Théophile Simar, *Etude sur Erycius Puteanus, 1574–1646, consideré spécialemente dans l'histoire de la philologie belge et dans son enseignement à l'Université de Louvain* (Louvain: Bureau de recueil, 1909).

53 Dibon and Waquet, *Johannes Fredericus Gronovius*.

| Scholar | Destinations | Period | Sources |
|---|---|---|---|
| Jacob Gronovius (1645–1716) | France, Spain, Italy; Pisa | 1671–1672; 1673–1675 | UBL, LTK 859; LTK 860 |
| Laurentius Gronovius (1648–1724) | Italy | 1679–1682; 1693–1695 | KB, KW 76 H 27; Hoogewerff; Moss[54] |
| Jacobus Tollius (1633/1640–1696) | German states, Italy | 1687–1692 | *ADB*; Van Vugt |
| Joannes Kool (1672–c. 1708) | France, Italy | 1698–1699 | BiASA, mss. 34 A 1–3 |

In a similar vein, Isaac Vossius met many scholars on his Grand Tour between April 1641 and October 1644.[55] In London, he was warmly welcomed by his uncle Franciscus Junius (1591–1677), who introduced him to Patrick Young (1584–1652) and Archbishop James Ussher (1581–1656), whom he helped to collate the letters of Ignatius of Antioch, one of the Apostolic Fathers.[56] Afterwards, he ventured to Paris, where he met and worked in the presence of Hugo Grotius for four months.[57] On his return journey, Isaac stayed almost a year at Grotius's home in 1644. During his stay in Paris, he met many prominent learned men such as Claude Sarrau (c. 1600–1651), Jean de Cordes (1570–1642), Jérôme Bignon (1589–1656), Jacques and Pierre Dupuy, Ismael Boulliau (1605–1694), Pierre Gassendi (1592–1655), and the Leiden professor Johann Friedrich Gronovius.[58] Isaac was welcomed into the Parisian homes and libraries of many of these learned men and spent much of his time on

54 G. J. Hoogewerff, 'Laurentius Theodorus Gronovius en zijn reizen naar Italië: I. De reiziger, zijn reisgenooten en zijn wedervaren', *Mededeelingen van het Nederlandsch Historisch Instituut te Rome*, 3rd series, 1 (1942): 35–56 (pp. 42–43); Hoogewerff, 'Laurentius Theodorus Gronovius en zijn reizen naar Italië: III. De briefwisseling met Antonio Magliabechi', ibid., 3 (1943): 160–167; Moss, 'Gemaakt op reis', *passim.*

55 For a short list of the learned men Isaac Vossius met on his Grand Tour, see Astrid Cecilia Balsem, 'Een biografie van de Bibliotheca Vossiana' (PhD diss. Leiden University, 2020), pp. 13–14.

56 Blok, *Vossius*, pp. 85–92. This work resulted in a copy of the manuscripts, see Isaac Vossius, 'Variae adnotationes de Ignatio, latine', 1642, Leiden, UBL, VGQ 72; cited in Balsem, 'Bibliotheca Vossiana', p. 20, n. 40. The resulting edition of the Ignatius letters is Isaac Vossius, ed., *Epistolae genuinae S. Ignatii martyris* […] (Amsterdam: Joan Blaeu, 1646).

57 Blok, *Vossius*, pp. 98–104. Gerardus Vossius thanked Grotius for receiving his son after his safe return: 'De reditu Isaaci mei magnopere gaudeo; hodie dies est quartus decimus, quod primum amplecti datum fuit. Ex sermonibus quos interea commutavimus, plurimum cepi voluptatis, nec vel, quantum profecerit, paenitet, vel impensae. Facile etiam video, quantum Excellentiae tuae debeam pro honore quem filio ac familiae meae habuisti. Nec beneficium tantum, cuius et ipse e sermonibus eius fructum percipio, unquam excidet animo cuiusquam nostri.' Gerardus Johannes Vossius to Hugo Grotius, 24 October 1644, *BW* XVI, letter no. 7111, pp. 88–89 (p. 88).

58 Blok, *Vossius*, pp. 106–113.

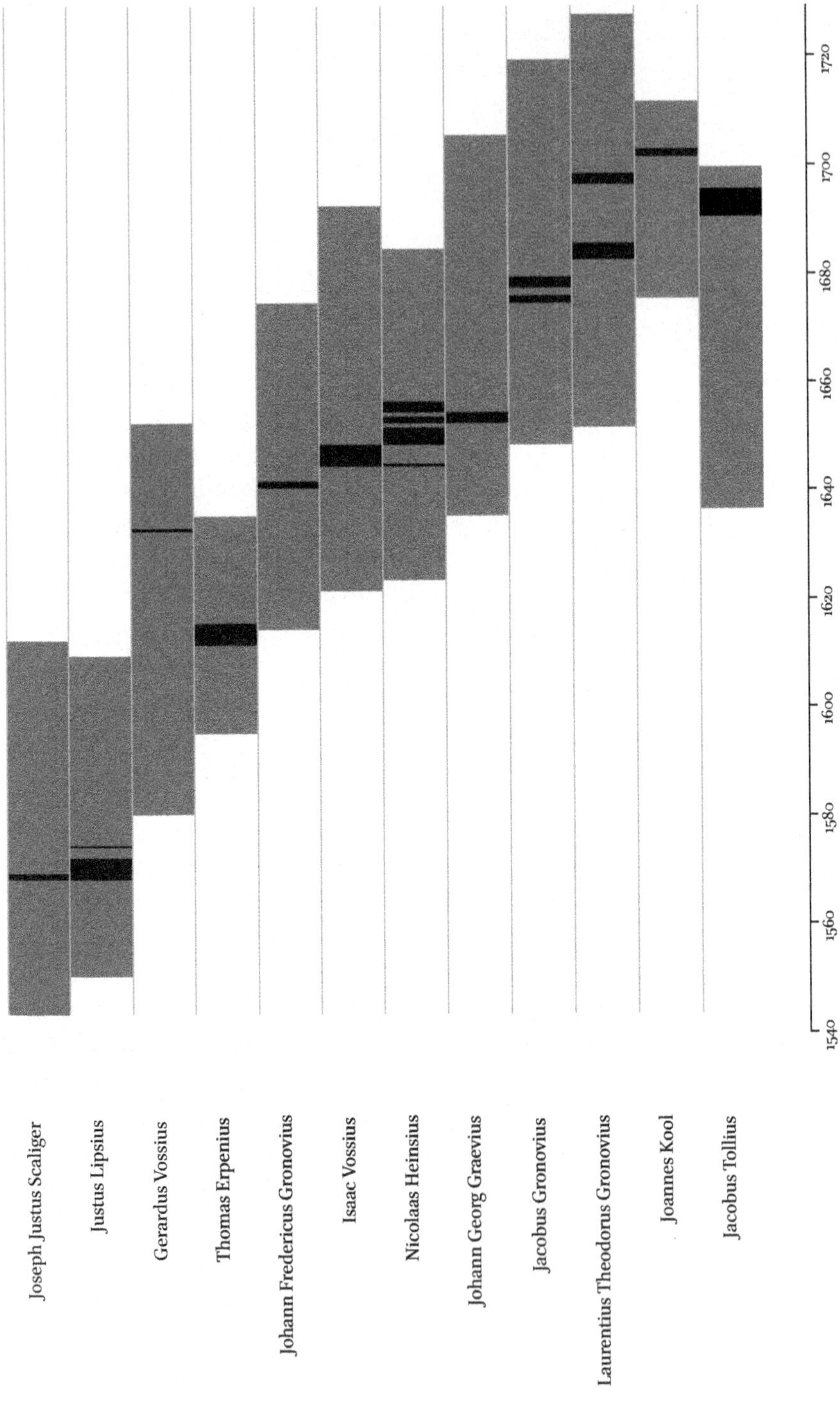


Figure 4.1. Travellers on a time scale to illustrate the repeated travels (in black) of 'Dutch' classical scholars

Figure 4.2. Anthony van Zijlvelt, *Eruditissimorum virorum effigies*, c. 1678–1691. Engraving and etching, 207mm × 157mm. Rijksmuseum: RP-P-OB-24.786. Depiction of the history of the field of biblical criticism and classical philology, with Justus Lipsius and Joseph Justus Scaliger as the two central teachers at the table. Isaac Casaubon and Johann Friedrich Gronovius also have a central place at the table

the copying and collation of rare manuscripts.[59] In Florence, Isaac was welcomed by the scholars Giovanni Battista Doni (1593–1647) and Carlo Roberto Dati (1619–1676). Isaac and Nicolaas held these two classical scholars

59 Ibid., pp. 97–123

in high regard, since they were among the few Italians who had profound knowledge of Greek and of antiquities.[60] In Rome, the rising scholars Lucas Holstenius and Gabriel Naudé (1600–1653) were crucial in providing access to the libraries as well as the local learned circles in Rome. In Paris, Isaac had received a letter of recommendation from Pierre and Jacques Dupuy addressed to Naudé to ensure he would indeed be received into the Roman learned circle.[61] Due to these connections, Isaac could work—albeit sporadically—in the Vatican Library and meet two well-known scholars, Leo Allatius (Allacci, 1586–1669) and Athanasius Kircher (1602–1680).[62]

Heinsius and Vossius were part of a stream of scholarly travellers to France and Italy that can be roughly divided into four generations (see Table 4.1 and Figure 4.1 for an overview of these travellers). First, Joseph Justus Scaliger and Justus Lipsius travelled extensively themselves and Lipsius professed the educational value of a scholarly travel. Both professors left their mark on the study of classical philology, especial at Leiden University (see Figure 4.2).[63] Many ambitious classical scholars who were inspired by Scaliger and Lipsius followed suit, such as Erycius Puteanus (1574–1646), Thomas Erpenius (1584–1624), as well as Johann Friedrich Gronovius.[64] Gerardus Vossius and Daniel Heinsius, both eminent historical and classical scholars around the 1620s and 1630s, inspired and encouraged their sons to continue this scholarly tradition. The second generation are the students and sons of these scholars, such as Isaac Vossius and Nicolaas Heinsius, who made a name for themselves with their extensive scholarship, innovative and extensive research on the various manuscripts, and their far-reaching networks in the learned world. The professors Johann Friedrich Gronovius and Johann Georg Graevius are also part of this generation, since they continued the work of Scaliger, Lipsius, and Gronovius Snr, were students of Daniel Heinsius, and perfected the philological knowledge of manuscripts. The next generation in the seventeenth century consisted of the sons of Gronovius, Jacob and Laurentius Theodorus. Throughout the century, the Netherlandish scholarly community profited from the good bonds between Dutch and Italian scholars. A decade later, Heinsius followed in Vossius's footsteps, and subsequently other scholars followed in their

60 Ibid., pp. 137–140.

61 Ibid., p. 118.

62 Ibid., pp. 151–154.

63 For the travel of Lipsius, see Papy, 'Justus Lipsius, Rome en de Romereis'.

64 J. Kleijntjes, 'Puteanus, Erycius', *NNBW*, vol. 6 (1924), pp. 1153–1154; Simar, *Etude sur Erycius Puteanus*; F. S. Knipscheer, 'Erpenius, Thomas', *NNBW*, vol. 8 (1930), pp. 495–496; Dibon and Waquet, *Johannes Fredericus Gronovius*.

footsteps in turn, including the sons of Johann Friedrich Gronovius. Jacob Gronovius studied in Oxford and Cambridge and travelled to Italy in the fall of 1673 to carry out research in the many libraries of Florence and Rome. A few years later, his brother Laurentius Theodorus Gronovius (1648–1724) travelled to Italy as well, first from 1680 to 1682, and once again from 1693 to 1695 as a steward and supervisor of three young Dutch travellers.[65] In the meantime, the Gronovius disciple and aspiring scholar Gisbert Cuper (1644–1712) established epistolary contact with Italian scholars as well, building on the strong network of his colleagues and mentors. The last scholar to travel as part of this tradition in the seventeenth century was Joannes Kool, who worked and identified as a disciple of Jacob Gronovius and Johann Georg Graevius.

The second characteristic of a *peregrinatio literaria* is the acquisition of books and manuscripts. Throughout the seventeenth century, there was a strong relationship between the publishing and bibliophile communities in Italy and the Low Countries. For example, the atlas maker and publisher Pieter Blaeu (1637–1706) made frequent visits to the Tuscan court to discuss the publication of town atlases and to explore a new book market.[66] The Heinsius family had good contacts with the Elsevier publishing family in Leiden. The father of Nicolaas, Daniel Heinsius, gave scholarly advice to the publisher Abraham Elsevier (1592–1652).[67] When Daniel Elsevier opened a shop in Amsterdam in 1638, Nicolaas Heinsius treaded into his father's footsteps and acted as an advisor and agent for the publishing house. The Elsevier publishing house developed into a favourite publisher for philologists such as J. F. Gronovius and Heinsius, because the books were neatly produced and the typography was flawless.[68] Heinsius published his

65 Hoogewerff, 'Laurentius Theodorus Gronovius I'; H. M. R. Leopold, 'Laurentius Theodorus Gronovius en zijn reizen naar Italië: II. Oudheidkundige waarnemingen en bevindingen', *Mededeelingen van het Nederlandsch Historisch Instituut te Rome*, 3rd series, 1 (1942): 58–87; Hoogewerff, 'Laurentius Theodorus Gronovius III'. The manuscripts of his travel journals: Laurentius Theodorus Gronovius, 'Itinerarium Laurentii Theodori Gronovii qui A.C. MDCLXXIX die III maii Haga Comitum profectus iter suscepit in Italiam et Lugdunum Batavorum rediit die XXI aprilis A.C. MDCLXXXIII', 1679–1682, The Hague, KB, 76 H 27 I; and Hoogewerff, 'Reize van Gronovius na en door Italië', 1693–1695, The Hague, KB, 76 H 27 II.

66 Henk Th. van Veen, 'A Tuscan Plan of Action for Joan Blaeu's Book of Italian Cities', *Lias* 18, no. 2 (1991): 221–227; Ingeborg van Vugt, 'The Structure and Dynamics of Scholarly Networks Between the Dutch Republic and the Grand Duchy of Tuscany in the 17th Century' (PhD diss., University of Amsterdam; Scuola Normale Superiore di Pisa, 2019), pp. 82–84.

67 Blok, 'Heinsius', pp. 92–93.

68 The success of the Elsevier editions also generated a discussion about the desirability of a high publication number for classical authors; see ibid., pp. 94–98.

*Poemata*, a book of his travel poetry, and the *Opera Ovidii* (Works of Ovid), his critical edition of Ovid's work, with Elsevier.[69]

As a student Nicolaas was known as a knowledgeable bibliophile by his friends in Leiden and they often asked his advice and recommendations on recently published books. For example, in 1637, Henrick Ploos van Amstel asked him about a recent edition of the letters of Pliny:

> I would like to know very much if the letter books of Pliny the Younger with the commentary of Johannes Maria Cataneus are valuable and if they can be acquired. Father maybe wants to buy them, so answer to this request, since we know that you know all the books.
>
> Zeer graag wil ik weten of de tien brievenboeken van Plinius Caecilius Secundus met commentaar van Johannes Maria Cataneus waardevol zijn en of ze te vinden zijn. Vader zou ze misschien wel kopen, geef daar dus antwoord op, want wij weten dat jij alle boeken kent.[70]

Within a few years, Nicolaas's passion for books was famed across Europe. Moreover, his role as a bibliophile grew more prominent in learned circles during his voyages to Paris and Italy. He played a central role in the seventeenth-century European book network. Letters and books formed part of a lively learned culture in which one could prove oneself by dedicating and gifting books. During this time, Heinsius would send many letters and books to the Florentine scholars Carlo Dati and Antonio Magliabechi (1633–1714).[71] In this way Heinsius was also up to date on the activity of Italian scholars and publishers.

During his journeys, Nicolaas bought many books for his own collection, as well as for the collections of his friends. He sent multiple chests filled with books to the Dutch Republic from the ports of Livorno, Genova, and Venice. The home front expected of travelling scholars to purchase books for

69 Nicolaas Heinsius, *Poematum nova editio, prioribus longe auctior*, 2 vols. (Amsterdam: Elsevier, 1666); Heinsius, *Opera P. Ovidii Nasonis editio nova* [...], 3 vols. (Amsterdam: Elsevier, 1664). For a detailed bibliography of Heinsius's manuscripts, see M. D. Reeve, 'Heinsius's Manuscripts of Ovid', *Rheinisches Museum für Philologie*, Neue Folge, 117, nos 1–2 (1974): 133–166; and Reeve, 'Heinsius's Manuscripts of Ovid: A Supplement', *Rheinisches Museum für Philologie*, Neue Folge, 119, no. 1 (1976): 65–78.

70 Henrick Ploos van Amstel to Nicolaas Heinsius, 18 February 1637, in Henrick Ploos van Amstel and Nicolaas Heinsius, *Brieven, 1635–1637*, ed. with introduction and annotations by Hans Kristian Ploos van Amstel, trans. Frank Tichelman (Woubrugge: Avalon Pers, 2007).

71 Ingeborg van Vugt, 'Using Multi-Layered Networks to Disclose Books in the Republic of Letters', *Journal of Historical Network Research* 1, no. 1 (2017): 25–51 (esp. pp. 36–41).

their family and friends. A scholar abroad would receive requests to search and purchase certain books that were hard to find in the northern parts of Europe. Isaac Vossius received a literary shopping list from his mother, Elisabeth Junius (c. 1585–c. 1659), during his journeys. Elisabeth asked Isaac to purchase special books for family, friends and neighbours in Paris, Rome, and Venice. For example, Isaac wrote to his parents from Venice that 'a large chest filled with colossal books will leave in a few days by boat'.[72] Additionally, Isaac and Nicolaas bought many books for their own collections. An uncle of Nicolaas expressed his amazement at the large sums of money Nicolaas invested in the purchase of books: 'I am extremely surprised that you spend so much money on coins and books, since the voyage itself is already expensive enough'.[73] As is well known, Isaac collected just as avidly as Nicolaas on his travels. During his Grand Tour and later travels throughout Europe, he bought and copied many manuscripts in English, French and Italian libraries.[74] His library was eventually bought by Leiden University and considerably enriched the library with rare manuscripts and books.[75]

In the latter part of their lives, Nicolaas and Isaac became literary advisors and agents for Queen Christina of Sweden.[76] They could both rely on the contacts they had already forged in the learned world. All the books that were published at the Elsevier publishing house in Leiden were immediately purchased for the Royal Library in Stockholm.[77] Heinsius's knowledge of books was well regarded among learned men in Leiden and at the Swedish court, although not everyone was pleased with his acquisitions. During his voyage through Italy, Italian scholars criticised Heinsius because he would try to trade rare manuscripts from local libraries for a printed edition. Nicolaas scouted many monastery libraries and other small libraries in search of unique manuscripts from antiquity. Octavius Ferrarius (1607–1682), professor at the University of Pisa, complained that 'certain people working for Queen Christina have loaded the spoils of libraries on boats'.[78] It became

72 'Interim expedivi arcam grandem et praegrandibus oneratam libris, quae propediem mari committetur'. Isaac Vossius to his father, 1 March 1643, in 'Isaac Vossius' Grand Tour [...] Part II', ed. Blok and Rademaker, p. 272.

73 'Ik verbaas mij er ten zeerste over, dat ge zooveel geld besteedt voor munten en boeken, daar de reis op zich zelf al kostbaar genoeg is'. Willem Goes to Nicolaas Heinsius, 30 May 1647; Dutch translation of the Latin original cited in Kan, 'Nicolaas Heinsius in Italië (1646–1648) (II slot)', p. 69.

74 Blok, *Vossius*, pp. 73–199.

75 Balsem, 'Bibliotheca Vossiana', pp. 103–118.

76 Blok, 'Heinsius', pp. 59–88; Blok, *Vossius*, pp. 267–285.

77 Isaac Vossius to Nicolaas Heinsius, 19 April 1651, in Burman, *Sylloges*, vol. 3, pp. 594–595.

78 '[N]eque ita pridem certi homines a Christina Augusta delegati bibliothecarum spoliis naves onerarunt'. Cited in Blok, 'Heinsius', p. 147.

clear to Italian scholars that Heinsius and Vossius traded in rare manuscripts at the expense of the collections of Italian libraries. This type of criticism appears to have been limited and did not turn Heinsius and Vossius into pariahs. Italian scholars and rulers regarded travelling bibliophiles and philologists highly because they confirmed the prestigious nature of the library collections in their native cities. An example of this high regard is Cosimo III's insistence on meeting the scholars from the North when he travelled to the Low Countries from 1667 to 1669: Heinsius himself, J. F. Gronovius, Graevius, and the well-known naturalist Jan Swammerdam (1637–1680).[79]

The third characteristic of a scholarly travel was studying, copying, and collating manuscripts in prominent libraries for days on end. The elder Heinsius, Vossius, and Gronovius stood in the Leiden philological tradition founded by Justus Lipsius (1547–1606) and Joseph Justus Scaliger (1540–1609).[80] This research was carried out first in the libraries, where Isaac and Nicolaas also met fellow learned men. One of these was the previously mentioned cabinet belonging to the brothers Dupuy, which housed the large library of Jacques-Auguste de Thou (1533–1617). His son and librarian Jacques-Auguste de Thou presided over the eleven thousand tomes. Nicolaas gained permission to study in his library and wrote to Vossius:

> I greeted the preseident [Jacques-Auguste] De Thou at the brothers Dupuy yesterday. He is so courteous to open his library, so that I can use it as much as I want and wander around in it.
>
> Praesidem Thuanum apud Puteanos fratres heri salutavi. Is bibliothecam suam liberaliter obtulit, ut tamquam mea uterer, quam et perlustrabo.[81]

79 G. J. Hoogewerff, *De twee reizen van Cosimo de' Medici, prins van Toscane, door de Nederlanden (1667–1669): Journalen en documenten* (Amsterdam: Johannes Müller, 1919), pp. l–liii. The travel journal of Cosimo III's travels has been translated into Dutch as Lodewijk Wagenaar, ed., *Een Toscaanse prins bezoekt Nederland: De twee reizen van Cosimo de' Medici, 1667–1669*, trans. Berti Eringa (Amsterdam: Uitgeverij Bas Lubberhuizen, 2014).

80 For the classical scholarship of Justus Lipsius and Joseph Justus Scaliger in Leiden, see Anthony Grafton, *Defenders of the Text: The Traditions of Scholarship in an Age of Science, 1450–1800* (Cambridge, MA; London: Harvard University Press, 1991); Grafton, 'Texts, Antiquities, and Their Scholarly Transformations in the Early Modern West', in *World Philology*, ed. Sheldon Pollock et al., 154–177 (Cambridge, MA; London: Harvard University Press, 2015), esp. pp. 157–160; Grafton, *Joseph Scaliger: A Study in the History of Classical Scholarship*, vol. 1: *Textual Criticism and Exegesis* (Oxford: Clarendon, 1983), pp. 118–133; Bernays, *Scaliger*, pp. 134–139. For the travel of Lipsius, see Papy, 'Justus Lipsius'.

81 Nicolaas Heinsius to Isaac Vossius, 12 August 1651; cited in Blok, 'Heinsius', p. 292, n. 5.

Together with his travel companion Lucas Langermann (1625–1686), a lawyer from Hamburg with an interest in manuscripts, Heinsius studied many manuscripts in the Parisian library. Langermann compared the variants of an old Prudentius manuscript, part of which appeared in a critical edition by Heinsius.[82] Before Heinsius travelled to Italy via Marseille, he diligently collated a Martial manuscript from the library of Jacques Dupuy and the *Ars Tactica* by the philosopher and military tactician Asclepiodotus (first century BC).[83]

Heinsius and Langermann arrived in the port of Livorno in July 1646. A few days later, they arrived in Florence. Heinsius received a hospitable reception by Paganino Gaudenzi (1595–1649), a professor and scholar in Pisa.[84] This warm welcome was due to their common friend, J. F. Gronovius, who had wholeheartedly recommended Heinsius to Gaudenzi. In a similar way, Heinsius also met the esteemed scholars Giovanni Battista Doni and Carlo Roberto Dati, who were friends of Gronovius and had previously spent time with Isaac Vosssius.[85] Gronovius had already befriended Gaudenzi, Doni, and Dati when he visited Florence in 1641. Subsequently, Gronovius recommended and introduced Isaac Vossius to these scholars when he arrived two years later. Vossius praised Dati's friendship in a 1647 eulogy, saying, 'no one is more loving of [Dati]' than himself, and telling Dati, 'like you had once received me in Florence with such friendship and kindness, you now address me with your letters now that I am absent'.[86] In turn, the connections and knowledge of Gronovius and Vossius greatly benefited Heinsius, since they helped him with letters of recommendation and introductions, as well as guiding him towards specific manuscripts and libraries, such as the Laurentiana and San Marco libraries.[87] As a token of friendship and comradery, Heinsius was appointed as member of two Florentine academies. During his second stay, he became a member of

82 Blok, 'Heinsius', p. 110.

83 Ibid., p. 103.

84 G. Brunelli, 'Gaudenzi, Paganino', *DBI*, vol. 52 (1999), pp. 676–678.

85 For Doni's position in the learned world, see Eric Bianchi, 'Bad Latin, Bad Manners: Giovanni Battista Doni, Marin Mersenne, and Literary Style in Seventeenth-Century Music Theory', *Music & Letters* 96, no. 2 (2015): 167–184 (esp. pp. 169–179); G. Formichetti, 'Doni, Giovanni Battista', *DBI*, vol. 41 (1992), pp. 167–170.

86 '[N]emo tui est amantior'; 'Cum enim tam amice et humaniter olim Florentiae exceperis, nunc etiam absentem litteris tuis compelles'. Isaac Vossius, 'Viro Cl. Carolo Dato Isaacus Vossius S.', in *Elogio di Carlo Roberto Dati recitato nella reale Accademia Fiorentina nell'adunanza del dì 30. di settembre 1790*, ed. Francesco Fontani, 66–67 (Florence: Gaetano Cambiagi Stampatore Granducale, 1790).

87 Henk Th. van Veen and Andrew P. McCormick, *Tuscany and the Low Countries: An Introduction to the Sources and an Inventory of Four Florentine Libraries* (Florence: Centro Di, 1984), pp. 28–30; Van Vugt, 'Structure and Dynamics', pp. 80–82.

the Accademia della Crusca and was elected president of the Accademia degli Apatisti.[88] This brought him into contact with a large portion of the intellectual life of the city, exponentially expanding the network his two predecessors had established. Thus, Italian and Dutch scholars gained good and close connections throughout the seventeenth century. These meetings on a scholarly voyage were often the start of lifelong epistolary exchanges. The Italian scholars took care of the access to libraries and introductions to the local learned world. In return, learned men such as Heinsius could give the Italian scholars access to the publishing houses in Leiden and Amsterdam to publish their work.

Gronovius and Isaac Vossius, however, met considerable resistance from the Catholic clergy who managed and controlled access to the libraries in Florence. To make matters worse, Gronovius was refused access to the manuscript collections of the libraries altogether.[89] To be able to work there, he had to bribe the librarian with a few *soldi* for every hour he worked in the library. If people entered the library, he had to hide his pen and paper, as if he was merely reading the manuscripts and not collating them.[90] Nor did Isaac Vossius easily gain access to the manuscripts at the Bibliotheca Laurentiana. He worked in the library every day in the spring of 1642, but could only work there a few hours per day, after the friars had left the library. During these precious hours, he worked on copies of Arrianus's *Tactica*, letters from Barnabas and Ignatius, and a text from Ptolemy.[91]

The religious tensions between Protestants and Catholics were the main cause for difficult access to Italian libraries for travelling scholars. They could only procure access by appealing to a local scholarly or political authority. Heinsius enjoyed preferential treatment due to his good connections in the learned world. In Florence he received an invitation for an audience with the Grand Duke of Tuscany, Ferdinand II, who afterwards granted him access to the Bibliotheca Laurentiana to work there as much as he wanted.[92] In the Laurentiana, Heinsius spent most of his time on the famous Virgil codex, the manuscripts of Claudianus and Curtius, and the letters of Pliny. During his second visit to the Laurentiana, he focused solely on the manuscripts of Ovid, collating as many as eighteen different manuscripts.[93] Langermann

88 Van Veen and McCormick, *Tuscany and the Low Countries*, p. 30.

89 Blok, 'Heinsius', pp. 155–156.

90 Johann Fredericus Gronovius to Nicolaas Heinsius, 6 December 1645, in Burman, *Sylloges*, vol. 3, p. 163.

91 Isaac Vossius to Gerardus Vossius, 7 March 1642, in 'Isaac Vossius' Grand Tour [...] Part II', ed. Blok and Rademaker, pp. 222–224.

92 Blok, 'Heinsius', 155–161.

93 Ibid., p. 181.

carried out his assignment from Isaac Vossius to copy Greek manuscripts on astrology by Maximus and Manetho.[94]

All in all, the travels of Heinsius and Vossius do not fit the categories of a *peregrinatio academica* nor of a Grand Tour. Rather, because the primary aim of their travels was to meet fellow scholars, acquire books, and work in libraries, the journeys can be characterised as examples of the *peregrinatio literaria*. This type of journey became a tradition in the Leiden community of textual scholarship, starting with Justus Lipsius and Joseph Justus Scaliger, followed by Johann Friedrich Gronovius, Isaac Vossius, and Nicolaas Heinsius, who in turn inspired the next generation, including Jacob Gronovius and Laurentius Gronovius, to undertake scholarly travel. Joannes Kool was the last to travel to Italy as part of a *peregrinatio literaria* in the seventeenth century. His travel journal offers an intricate account of the community and identity formation that could take place on a scholarly journey.

## 4.2 Joannes Kool in Italy, 1698–1699

Joannes Kool (1672–c. 1708) was a Dutch scholar who had more than one mission on his Grand Tour: his own and those of other learned men, such as Leiden University professor Jacob Gronovius (1645–1716). Kool studied at Utrecht University and obtained his doctoral degree in law from the same university in 1698. After graduation, he ventured to Rome as a learned man, well versed in antiquities and law, together with his friend and fellow doctor of law, Lucas van Voorst (1670–1738).[95] Kool's extensive travel journal carries the title *Journaal van een rijse gedaan door Italien &c in den Jaare Anno 1698* (Journal of a journey undertaken through Italy etc. in the year 1698; see Figure 4.3).[96]

94 Ibid., p. 156.

95 Lucas van Voorst, *Disputatio juridica inauguralis de hereditatis petitione*, doctoral diss., Utrecht University (Utrecht: François Halma, 1693). Lucas van Voorst became a magistrate in Utrecht in 1701, see Het Utrechts Archief (hereafter: HUA), TF2 Utrechtse Bestuurders 1618–heden, no. 1000: Lucas van Voorst.

96 Joannes Kool, 'Journaal van een Rijse gedaan door Italien &c in den Jaare anno 1698', 1698–1699, BiASA, mss. 34 A 1–3, 970 fols. in 3 vols. See Koen Scholten and Asker Pelgrom, 'Scholarly Identity and Memory on a Grand Tour: The Travels of Joannes Kool and His Travel Journal (1698–1699) to Italy', *Lias* 46, no. 1 (2019): 93–136 for an earlier presentation of this research. The manuscript was unearthed by Asker Pelgrom in the Biblioteca di Archeologia e Storia dell'Arte in Rome in 2017 and he reconstructed the provenance of the manuscript, see ibid., pp. 110–125.

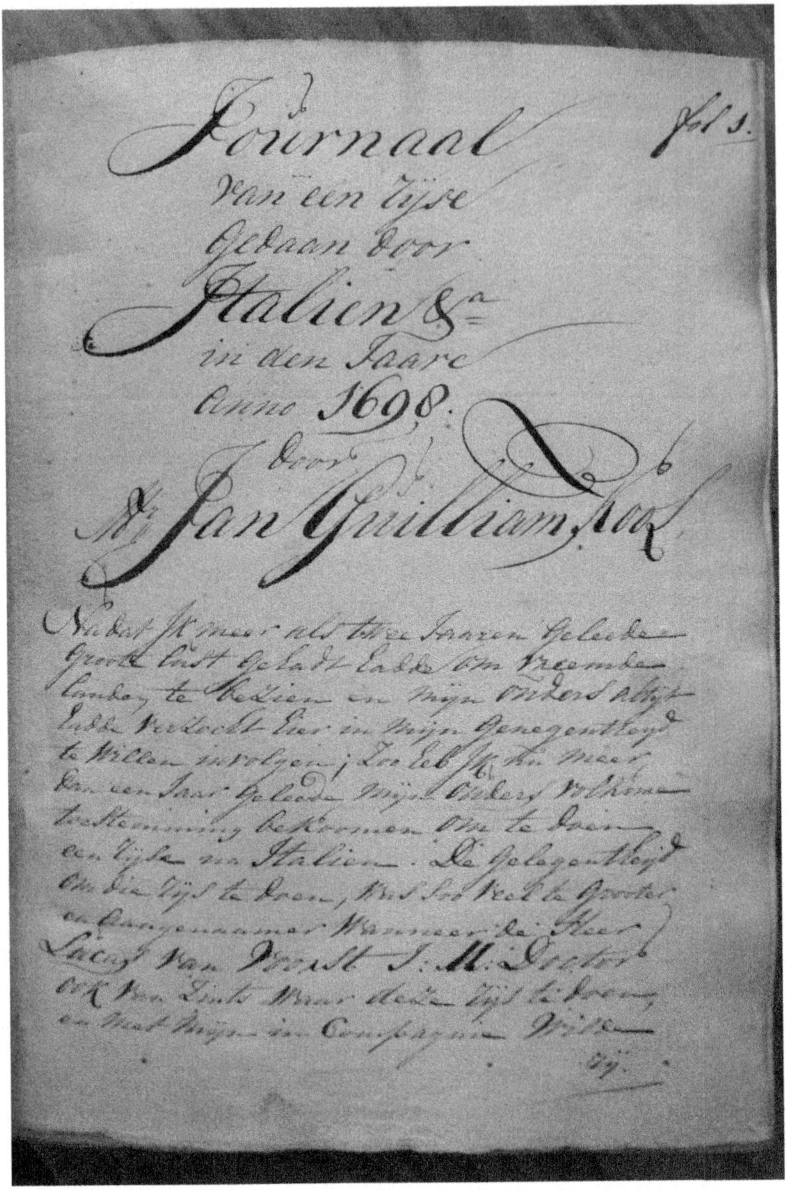

Journaal van een rijse gedaan door Italien &c^a in den Jaare Anno 1698. door Mr Jan Guilliam Kool

fol. 1.

Nadat Ik meer als twee Jaaren geleede groote lust gehadt hadde om vreemde landen te bezien en mijn Ouders altijt hadde verzocht hen in mijn genegentheyd te willen involgen; zoo heb Ik nu meer dan een Jaar geleede mijn Ouders volkome toestemming bekomen om te doen een rijse na Italien. De gelegentheyd om die rijs te doen, was soo veel te grooter en aangenaamer wanneer de Heer Lucas van Voorst J. U. Doctor ook van zints waar deze rijs te doen, en met mijn in compagnie wilde

Figure 4.3. The first folio and title page of Kool's travelogue. Joannes Kool, *Journaal van een rijse gedaan door Italien &c in den Jaare 1698*, BiASA, ms. 34 A 1, fol. 1r. Courtesy of the Ministero dei beni e delle attività culturali e del turismo

The unpublished journal of Joannes Kool offers a unique perspective on travelling scholars at the end of the seventeenth century. The manuscript is an egodocument, chronicling almost every day in detail, with specific attention to the scholarly aspects of the voyage.[97] The travelogue describes his year-long journey through seventeenth-century Europe: its cities, roads, inns, and courts, but more importantly for Kool, its antiquities, music, libraries, books, and learned men. In comparison with other documented travel journals, this one stands out for a variety of reasons. Firstly, Kool travelled after obtaining his doctoral degree from Utrecht University and did not visit any university during his journey. Although universities were ignored by many noble travellers and mature scholars, Kool was neither and he travelled on a *peregrinatio literaria*, much like Vossius and Heinsius before him. Secondly, the manuscript contains an account of almost every single day (with two jumps in time). Thirdly, at 970 folios bound in three volumes, it is the most elaborate Dutch travel journal concerning Europe handed down from the second half of the seventeenth century.[98] And finally, Kool's account offers a unique insight into scholarly practices and encounters, due to the diligence and consistency with which he recorded his experiences. The travelogue—unknown to historians before 2017—thus represents a unique, detailed perspective on the life and thoughts of a young, ambitious scholar on his Grand Tour. Compared to other travelogues of young scholars, Kool's travel journal offers significantly more detail in his accounts of scholarly meetings and the social and cultural aspects of a scholarly voyage.[99]

Local learned circles around Jacob Gronovius, professor at Leiden University, and Johann Georg Graevius, professor at Utrecht University, enabled Kool to study and build his own network. Kool had started to establish himself as a learned man in the Dutch Republic based on his academic work, such as his disputation and dissertation, and more importantly, his connections with widely known scholars such as Jacob Gronovius, Johann Georg Graevius, Perizonius (1651–1715), and Theodorus Janssonius van Almeloveen (1657–1712). Consequently, Kool was able to present himself as a disciple of Graevius and Gronovius in order to gain access to libraries and local learned men along his journey.

Kool's journey was typical of a Grand Tour to Italy (see Figure 4.4). After an eighteen-day journey, during which he stopped a few days in Dordrecht,

97 For an introduction to egodocuments, see Rudolf Dekker, 'Egodocuments in the Netherlands from the Sixteenth to the Nineteenth Century', in *Envisioning Self and Status: Self-Representation in the Low Countries, 1400–1700*, ed. Erin Griffey, 255–285 (Hull: ALCS, 1999).

98 This is based on a comparison with the travel journals registered in the list in Ruud Lindeman, Yvonne Scherf, and Rudolf Dekker, *Reisverslagen van Noord-Nederlanders van de zestiende tot begin negentiende eeuw: een chronologische lijst* (Haarlem: Stichting Egodocument, 1994).

99 Cf. Frank-van Westrienen, *Groote Tour*; Verhoeven, *Europe Within Reach*; Moss, 'Gemaakt op reis'.

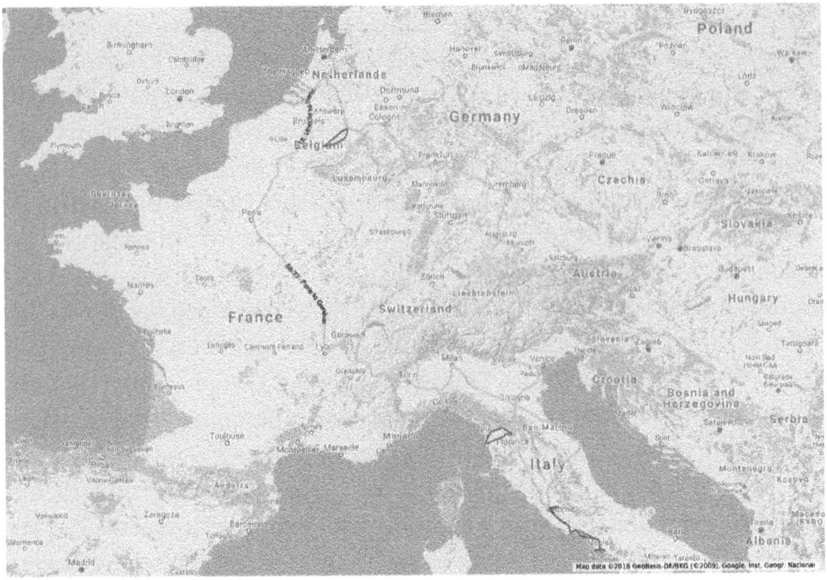


Figure 4.4. An overview of the route taken by Joannes Kool in 1698–1699, interpolated onto a modern map based on the locations mentioned in Kool, 'Journaal', BiASA, mss. 34 A 1–3

Antwerp, and Brussels, Kool spent forty-nine days in the centre of learning and culture in France: Paris. Continuing through Lyon and Geneva, Kool entered Italy, and took multiple day stops in Turin, Genova, Milan, and Bologna, ultimately arriving in Florence. There, he spent forty-nine days revelling in Tuscan culture, but also collating and copying manuscripts in the Medici library, the Bibliotheca Laurentiana Medicea.[100] On 10 October 1698, he reached his ultimate destination: Rome.[101] During his ninety-day stay in Rome, he visited palaces, churches, and libraries. Venice was the only other city where Kool stayed for a prolonged period (in this case, twenty-four days) is Venice.[102] From there, he slowly headed back to Utrecht over fifty-two days, travelling though Padua, Verona, Innsbruck, Augsburg, Nuremberg, Frankfurt am Main, Aachen, Cologne, Liège, and Maastricht. In the matter-of-fact tone characteristic of Kool's travel journal, he notes on the final pages:

> At five in the morning, we sailed with an express yacht [...] to Gorinchem, arrived there at nine, rode with an express carriage from there at ten, paid five guilders and a half for the carriage, and arrived at four in Utrecht.

100 Kool, 'Journaal', BiASA, ms. 34 A 2, fols. $342^{\mathrm{r}}$–$419^{\mathrm{r}}$.
101 Ibid., mss. 34 A 2–3, fols. $418^{\mathrm{r}}$–$775^{\mathrm{r}}$.
102 Ibid., mss 34 A 3, fols. $810^{\mathrm{r}}$–$876^{\mathrm{r}}$.

> Des morgens ten vijf uuren seijlden wij met een expres jacht [...] naa Gorcum, arriveerden aldaar te negen uuren, reeden met een expresse waage daarvan daan te 10. uuren, betaalden voor de waage vijf guldens en een half, arriveerden ten vier uuren in Utrecht.[103]

Kool's journey can be characterised as a *peregrinatio literaria*, much like the travels of Vossius and Heinsius. In total, he spent 82 days at libraries and 106 days on travel (he was away for 373 days in total). Thus, 30 per cent of non-travel days were spent in libraries. At first glance, this number might seem relatively low, but we must keep in mind that Kool spent these days working in a library copying and collating manuscripts. On the remaining days, he visited locals, palaces, churches, antiquities, festivities, or stayed at the inn.

Kool approached the material culture of ancient Rome as a humanist and antiquarian. Armed with Alessandro Donati's *Roma vetus ac recens* (Rome, old and new, 1695), a standard guide to Rome for the learned antiquarian, Kool visited sites of Roman antiquity.[104] In the second half of the seventeenth century, Rome was increasingly interpreted as a 'historical' and 'antiquarian' space by travellers. They showed particular interest in Rome's monuments and antiquities, as well as in its countless statues and inscriptions.[105] Kool experienced the antiquarian space described in Donati's *Roma* and often compared his own observations of Rome with the corresponding passages in Donati providing relevant bibliographical information (chapters and page numbers). For example, after visiting the Pantheon, Kool has nothing to add to Donati's description:

> I consider this temple the best remnant of Antiquity. I will not write down any more since Donati Book III, Chapter 16, has everything.

> vorder acht ik deese tempel het beste overblijfsel van de oudheijd, ik zal daar van niet meer aantekenen om dat Donatus L[iber] III C[apitulum] 16 alles heeft.[106]

103 Ibid., fols. 968$^{r}$–969$^{r}$.
104 Kool probably travelled with the 1695 edition printed in Amsterdam: Alessandro Donati, *Roma vetus ac recens utriusque aedificiis ad eruditam cognitionem expositis* (Amsterdam: Janssons van Waesberge & Johannes Watters, 1695).
105 Rosemary Sweet, 'The Changing View of Rome in the Long Eighteenth Century', *Journal for Eighteenth-Century Studies* 33, no. 2 (2010): 145–164 (esp. 146–147).
106 Kool, 'Journaal', BiASA, ms. 34 A 3, fol. 660$^{r}$.

In a later passage, Kool makes a similar remark when he visits and describes the Arch of Constantine: 'Donati Book III, Chapter 6, describes this arch so decently, and gives such a good explication of the bas-reliefs on it, that I cannot find anything to add to it.'[107] In most instances, Kool merely notes 'vide Donatum' and refers to the descriptions in Donati's *Roma*. However, in exceptional instances, he offers emendations to or criticism of Donati's descriptions. For example, when visiting the Aventine hill in Rome, he makes an argument for the correct location of the Temple of Diana:

> When we were at the Aventine [hill], we saw the Temple of Diana, currently named Santa Sabina, which contradicts Donati Book III, Chapter 10, and proves that the Temple of Diana could not have been here, where the Santa Sabina is now, but that the Temple of Diana was on the other side of the Aventine [hill].
>
> Als van deeze morgen op de Aventinus waaren, zoo zaagen wij het Templum Dianae, hede genaamt Santa Sabina, het welk even wel Donatus L[iber] III Cap[itulum] 10 tegenspreekt en bewijst, dat het templum Dianae niet kan geweest zijn, daar ter plaatse alwaar tegenwoordigh Santa Sabina is, maar dat den Tempel van Diana aan de andere kant van den Aventinus is geweest.[108]

Kool interacted with Roman antiquity in the same way he approached manuscripts: he collated it against what he had read in Donati. That is, he saw the urban and historical landscape as a historical text that could be 'read'. The Roman past was directly relevant for the present and deserved scrupulous attention.[109] The antiquarian space of Rome needed to be observed and compared against Donati's descriptions. Gronovius and Graevius had sent Kool on a mission to record his observations of aspects of Roman antiquity. After all, both of them had published immense *thesauri* that aimed to contain as much knowledge about antiquity as possible.[110] Kool's

107 Donatus L[iber] III Cap[itulum] 6 beschrijft deeze arcus zoo net, en geeft zoo een goede explicatie van de Bassrelieven, die daarop zijn, dat ik niets kan vinden om daar bij te doen.' Ibid., fols. 696$^r$–697$^r$. The inscription can be found in Donati, *Roma vetus ac recens*, p. 185.

108 Kool, 'Journaal', BiASA, ms. 34 A 3, fol. 700$^r$.

109 For early modern attitudes towards antiquity, see Victor Plahte Tschudi, *Baroque Antiquity: Archaeological Imagination in Early Modern Europe* (New York: Cambridge University Press, 2016).

110 Johannes Georgius Graevius, ed., *Thesaurus antiquitatum romanarum*, 12 vols. (Utrecht: François Halma; Leiden: Petrus vander Aa, 1694–1699); Jacob Gronovius, ed., *Thesaurus Graecarum*

observations were later published in a posthumous work of Marquard Gude (1635–1689), the *Antiquae inscriptiones* (Ancient inscriptions, 1731).[111] Both the time spent in libraries and Kool's antiquarian approach to Rome underline the scholarly aspect of his travel. When compared to the journeys of Isaac Vossius and Nicolaas Heinsius, Kool spent even more time on antiquarian and philological research, and much less on the acquisition of books.

Kool's descriptions of his journey and of the interactions with fellow scholars add considerable depth to what we know of scholarly interactions on a *peregrinatio literaria*. Through Europe, early modern scholars could easily identify each other as colleagues, although they had often never met before. Kool's journal shows us that when a young scholar mastered Latin (or Italian for that matter) and brought evidence of his scholarly acumen in the form of books or a dissertation, as well as letters of recommendation to prove his embeddedness in the learned world, the local learned community would welcome him. Additionally, the local scholar and host would often act as an intermediary to help navigate the difficulties of contacting eminent rulers (for example, the Grand Dukes of Tuscany), but more importantly, help the visiting scholar to gain access to the precious manuscript collections for philological work.

Kool wrote about the struggles in accessing certain libraries and manuscripts in Florence and Rome, problems that he could solve by referring to his identity and status as a scholar. This scholarly identity was constructed through three aspects that will be explained in the following subsections: the material manifestations of scholarly civility in the form of books and letters; the 'othering' of clergymen and 'normal' morally flawed young men; and Kool's self-presentation as a disciple in a recognisable and reputable learned tradition and lineage. The remainder of this section will elaborate on Kool's involvement in two separate learned circles: Florence and Rome. The first case focuses on his meeting with Antonio Magliabechi (1633–1714), librarian to Cosimo III de' Medici, Grand Duke of Tuscany, and his attempts to gain access to the Bibliotheca Medicea Laurentiana. The second case focuses on the relatively effortless attempt to get into the Bibliotheca Vaticana in Rome. Both cases illustrate the three aspects of Kool's scholarly identity mentioned above.

*antiquitatum*, 13 vols. (Leiden: Petrus and Baldvinus vander Aa, 1697).

111 Marquard Gude, *Antiquae Inscriptiones quum graecae, tum latinae* [...] (Leeuwarden: François Hesselius, heir of François Halma, 1731). On Gude, see Conrad Bursian, 'Gudius, Marquard', *ADB*, vol. 10 (1879), pp. 88–89; Dirk van Miert, *Humanism in an Age of Science: The Amsterdam Athenaeum in the Golden Age, 1632–1704* (Leiden; Boston: Brill, 2009), pp. 87–88; and Scholten and Pelgrom, 'Scholarly Identity', pp. 113–117.

### 4.2.1 Florence: Magliabechi and the Bibliotheca Medicea Laurentiana

Throughout the seventeenth century, Florence was a hub for scholarly correspondence and book dispersion, mostly directed by the bibliophile and court librarian Antonio Magliabechi (1633–1714).[112] The ruler of the Tuscan state, Cosimo III de' Medici, had travelled twice to the Dutch Republic: between 1667 and 1668, and again in 1669. During these trips, Cosimo acquainted himself with local learned men, such as Nicolaas Heinsius and Jan Swammerdam (1637–1680), and Kool's intellectual fathers: Johann Friedrich Gronovius and Johann Georg Graevius.[113] Magliabechi was central to the scholarly exchange between the Low Countries and Tuscany, exemplified by an extensive exchange of letters and books, but also by his recurring appearances in most Dutch travel journals of the second half of the seventeenth century. Through numerous contacts between Magliabechi and men such as Kool, Jacob Gronovius,[114] Laurentius Gronovius,[115] Coenraed Ruysch,[116] and the brothers Van der Dussen,[117] relations between the Tuscan court and the intellectual elite of the Low Countries were maintained.[118]

In his diary Kool describes how he negotiated his way into the Laurentiana. In one passage he describes his initial meeting with Magliabechi at his house, the so-called *casa Magliabechi*.[119] In another he mentions

112 Jean Boutier et al., eds., *Antonio Magliabechi nell'Europa dei saperi* (Pisa: Edizioni della Normale, 2017); Van Vugt, 'Using Multi-layered Networks', esp. 43–48. On Magliabechi's peculiar ascension to the Medici court and his atypical behaviour at the court, see Caroline Callard, 'Diogène au service des princes: Antonio Magliabechi à la cour de Toscane (1633–1714)', *Histoire, Économie et Société* 19, no. 1 (2000): 85–103.

113 Hoogewerff, *De twee reizen van Cosimo de' Medici*, pp. l–liii. The travel journal of Cosimo III's travels has been translated into Dutch as Wagenaar, ed., *Een Toscaanse prins bezoekt Nederland*.

114 Hoogewerff, 'Laurentius Theodorus Gronovius I', pp. 42–43.

115 Hoogewerff, 'Laurentius Theodorus Gronovius III', pp. 159–160. For the travel journal, see Laurentius Theodorus Gronovius, 'Reize van Gronovius na en door Italie enz.', 1679–1682, 1693–1695, The Hague, KB, 76 H 27.

116 Coenraad Ruysch, 'Journaal van een reis naar Genève, Italië en Frankrijk [...]', 1674–1677, The Hague, NA, FA Teding van Berkhout, no. 1408, fols. $72^r$–$76^r$; transcription generously provided by Alan Moss.

117 Jacob and Pieter van der Dussen, 'Dagboek van een reis van Den Haag door Duitsland en Oostenrijk naar Italië en Zwitserland', 1699–1700, The Hague, KB, KW 128 D 34.

118 For an overview of the relation between Tuscany and the Low Countries under the rule of Cosimo III, see McCormick and Van Veen, *Tuscany and the Low Countries*. For the role of travelling and the scholarly bond between the Dutch and the Grand Duchy of Tuscany, see Van Vugt, 'Structure and Dynamic', pp. 70–106.

119 For a full transcription of Kool's account of this encounter, see Scholten and Pelgrom, 'Scholarly Identity', Appendix A, pp. 127–128.

the difficulty he had in accessing manuscripts and the 'maliciousness of the monks', who denied him full access to manuscripts in the Bibliotheca Laurentiana.[120]

Kool immediately presented himself to Magliabechi on his first day in Florence, on 19 August 1698. Facing Magliabechi's house, Kool notes the contrast between the inhabitant's greatness and the low esteem of the neighbourhood, 'which many have told me about'. He is also aware that he is travelling in the footsteps of his predecessors.[121] Kool describes his manner of presentation in detail:

> After we had knocked at the door [of Magliabechi], we took care to hold the books in the open, so Magliabechi could see them, to gain access more easily. Magliabechi saw [us] through a hole in the wall—there to see if the people in front of his door would please him—yet we were welcome and [Magliabechi] called from above *Adesso Signori*, after little waiting in front of the door, Magliabechi came to open the door, and after that, he adorned us with compliments, we handed him the books, which were all very pleasant.

> Als wij aan de deur geklopt hadden, zoo gebruijkten wij de voorsichtighijd van de boeken bloot te houden, dat Magliabechi die zien konde, om des te lichter acces te krijgen. Magliabechi zag aan de rechterhand van zijn deur door een vierkant gadt, dat in de muur is, om te zien, of de menschen, die voor de deur zijn hem ook aanstaan, dogh wij waaren welkom en riep van boove Adesso Signori, naa een wijnigh gewacht te hebben voor de deur, quam Magliabechi de deur opdoen, en na dat hij ons met complimenten geaccombleerd had, zoo gaaven wij hem de boeken over, die alle zeer aangenaam waren.[122]

Kool carried a variety of books as gifts from himself and Graevius, Gronovius, and Nicolas Toinard (1629–1706), librarian of the Royal Library in Paris: Graevius's edition of Callimachus's *Hymni*, Gronovius's *Manethonis apotelesmatica*, and Jean Baptiste Dubos's *Histoire des quatre Gordiens*, respectively.[123] These

120 For a full transcription of Kool's account of his visit to the Bibliotheca Laurentiana, see ibid., Appendix B, pp. 129–131.

121 The full sentence reads: 'Als wij voor het huijs van Magliabechi quaamen, zoo bevond ik voor eerst waar te zijn, het welk veele mij gezeght hadden, naamentlijk dat het een klijn vervallen huijs was, staande in een straat van geen aansien, Daar deese groot man in woonde.' Kool, 'Journaal', BiASA, ms. 34 A 1, fol. 344$^{r}$.

122 Ibid., fol. 344$^{r}$–345$^{r}$.

123 Johann Georg Graevius, *Callimachi hymni, epigrammata, et fragmenta* (Utrecht: François Halma & Willem vande Water, 1697); Jacob Gronovius, *Manethonis apotelesmaticorum libri sex:*

books were not only presents, but also material manifestations of Kool's embeddedness in the learned circle of Graevius and Gronovius. Kool thus presented himself as a disciple of those two learned men by handing Magliabechi their latest printed works, which originated in prior research in Florence. Additionally, the *Manethonis apotelesmatica* from Gronovius acknowledged the importance of the Florentine library on its frontispiece: 'now brought to light for the first time from the Medici Library.'[124] Dubos's *Histoire*, the book he offered on behalf of Toinard, whom he had met in Paris at the Royal Library, underlined Kool's membership of the civil community of scholars and the hospitality he had received in Paris. These books testified that Graevius, Gronovius, and Toinard, all celebrated members of the learned world, trusted Kool to perform these duties on their behalf. Along with these three material testimonies of virtue, Kool also offered Magliabechi his own printed oration and doctoral dissertation.[125] These works were material reminders of Kool's own scholarly credentials. With the offer of these five books, he presented himself as a reputable scholar, vouched for by illustrious learned men.

According to the social principle of reciprocity in gift-giving, Magliabechi was now obliged to Kool.[126] Magliabechi fulfilled this obligation by arranging an audience with Cosimo III, as well as arranging full access to the Bibliotheca Laurentiana Medicea. At Magliabechi's request, Cosimo III granted Kool full access to the library and the right to collate and copy manuscripts. However, the monks at the library denied Kool access and Magliabechi decided to put them right. In the journal, Kool describes how Magliabechi said

> that he wanted to go to the palace to advance our case, because this was a great maliciousness of the monks; [...] the Grand Duke has immediately sent an order to grant copying, yet those friars [of the library] have been at the court, [arguing that] it was not advisable to allow me into the library, because I was a Reformed Protestant.

*Nunc primum ex Bibliotheca Medicea editi* (Leiden: Fredericus Haaring, 1698); Jean Baptiste Dubos, *Histoire des quatre Gordiens, prouvée et illustrée par les medailles* (Paris: Florentin & Pierre Delaulne, 1695).

124 'Nunc primum ex Bibliotheca Medicea editi.' Gronovius, *Manethonis apotelesmaticorum libri sex*, frontispiece.

125 Joannes Kool, *Oratio de artibus quibus Augustus callide Rempublicam R. invasit* [...] (Utrecht: François Halma, 1696); Kool, *Disputatio juridica inauguralis ad legem Ateriam Tarpejam* [...], doctoral dissertation, Utrecht University (Utrecht: François Halma, 1698).

126 On the ideal of reciprocity in the learned world, see Anne Goldgar, *Impolite Learning: Conduct and Community in the Republic of Letters, 1680–1750* (New Haven, CT; London: Yale University Press, 1995), pp. 21–26. See also see Chapter 3 on friendship and reciprocity (pp. 193–195).

> zo zeijde zijn Ed$^{e}$ dat hij naa het Palijs wilde gaan om onze zaaken te vervorderen te meer, omdat hier in een groote quaataardighijt van de Munniken was; want verhaalde hij mij, den Groot Hartogh heeft aantstonts naa U Ed: audientie order gezonder om te laaten copieren, dogh die H$^{e}$ Preters hebben aan het hof geweest, dat het niet raatzaam was mij op de Biblioteeq te laaten door dien ik een gereformeerde was.[127]

Afterwards Magliabechi visited Cosimo III, and Kool received a letter 'so there was no doubt anymore, that the entire library would be open to me'.[128] He must have thought that all was well. Yet when Kool retrieved a Manetho codex, he was not allowed to sit in the library, but had to sit on the stairs leading up to the library, while the doors on both sides remained closed. Here, Kool notes, 'I was locked up like a prisoner, this place did not please me, since I did not have enough light'.[129] Two days later, in an attempt to continue his collations and copies of manuscripts, Kool visited the Laurentiana again: 'I was locked up [again] and it was so dark, that I could not see, and [I] progressed very little in the copying of the *Carmina* of the poet Maximus.'[130] This was the limit, so he complained to Magliabechi about the dire circumstances under which he had to work in the library. When Magliabechi visited both Cosimo III and the head librarian, a compromise was reached. Although Kool could still not work *in* the library, the librarian would leave the door open to let in more light. The solution was:

> to have more light the door, which only let through light through small gratings, would remain open, and I would not be locked up. What can I do? I should be satisfied.
>
> op dat ik meerder licht zoude hebben zoo zoude de deur, daar ik alleen door klijn gesneede tralien licht kreeg, open blijven, en ik zoude niet opgeslooten zijn, wat zoude ik doen? ik moest te vreede weezen.[131]

127 Kool, 'Journaal', BiASA, ms. 34 A 2, fols. 380$^{r}$–381$^{r}$.

128 '[Z]oo dat der nu geen twijfel meer was, of de gansche Biblioteeq zoude voor mij oopen weezen'. Ibid., fol. 382$^{r}$.

129 '[D]aar ik eve als een gevange[ne] opgeslooten wierd, deeze plaats behaaghde mij niet zeer, de wjil ik geen licht genoegh hadde.' Ibid., fol. 383$^{r}$.

130 '[I]k wierd opgeslooten en het was zoo donker, dat ik niet konde zien, en vorderde zeer wijnigh in het copieren van de *Carmina* Maximi poetae.' Ibid., fol. 388$^{r}$.

131 Ibid., fols. 393$^{r}$–394$^{r}$.

After this brief conflict between Magliabechi, Cosimo III, and the friars of the abbey presiding over the Bibliotheca Medicea Laurentiana, Kool copied and collated manuscripts in the library—or rather, just outside the library on the stairs—for the remainder of his stay in Florence. This memory of religious intolerance is significant in that Magliabechi (and subsequently Kool in his journal) characterised the monks at the library as the uncooperative 'other'. Even more importantly, Magliabechi characterised the behaviour of the monks as 'evil', since it did not accord with the values of openly sharing knowledge with scholars, regardless of their confession.

This confessional hostility and suspicion were not the only reasons Kool was not fully trusted.[132] Previous trouble with Netherlandish scholars had made the wardens of Italian libraries wary. In particular, Jacobus Tollius (1633/1640–1696) was accused of having stolen books from the collection of Isaac Vossius, as well as manuscripts from libraries in Florence and Milan.[133] Kool complained about the distrust of Protestant scholars due to Tollius's bad reputation. He could not easily gain access to the Laurentiana library,

> because I was a Reformed Protestant. The Italians have this suspicion mixed with maliciousness, especially because of Tollius, who stole manuscripts from the library in Milan, and here in Florence. Therefore, other honest [men] still suffer up till this day.

> door dien in een gereformeerde was. Deeze suspicie vermenght met quaataardighijt heeft de Italianen ten meesten deele gegeeven Tollius, die te Milanen, en hier te Florence Mss: uijt de Biblioteeq gestoolen heeft, waarom andere eerlijke noogh ten deezen daage leijden moeten.[134]

132 On the scholarly exchange between Italian and Dutch scholars across this confessional divide, see Jetze Touber, 'Religious Interests and Scholarly Exchange in the Early Enlightenment Republic of Letters: Italian and Dutch Scholars, 1675–1715', *Rivista di Storia della Chiesa in Italia* 68, no. 2 (2014): 411–436; Touber, ' "I am Happy that Italy Fosters such Exquisite Minds" ': Gijsbert Cuper (1644–1716) and Intellectual Life on the Italian Peninsula', *Incontri* 30, no. 2 (2015): 91–106.

133 This incident is mentioned in the correspondence of Gronovius, Graevius, and Heinsius, see Friedrich Koldewey, 'Tollius, Jakob', *ADB*, vol. 38 (1894), pp. 423–427; also mentioned in W. L. Jansen, 'Appropriating *Peri hypsous:* Interpretations and Creative Adaptations of Longinus' Treatise *On the Sublime* in Early Modern Dutch Scholarship' (PhD diss., Leiden University, 2019), pp. 233–234. Another account of the Tollius affair in Florence has been given by Antonio Magliabechi in his letters, see Van Vugt, 'Structure and Dynamics', pp. 91–94.

134 Kool, 'Journaal', BiASA, ms. 34 A 2, fol. 381^r^.

That Cosimo III allowed Kool to work in the Medici library should thus be seen as a sign of personal trust towards Netherlandish scholars generally, and towards Kool in particular.

Gaining access to libraries was not always that troublesome, however. In Rome, we see similar mechanisms and strategies used to gain access, especially the 'othering' of clergymen, the claim to moral superiority, and Kool's self-presentation as a disciple of Gronovius and Graevius.

### 4.2.2 Rome: Cardinal Enrico Noris and the Bibliotheca Vaticana

The role of scholarly diplomat and gatekeeper as embodied by Antonio Magliabechi and Cosimo III in Florence, had been fulfilled in Rome by Christina of Sweden, an educated patroness for scholars such as Isaac Vossius and Nicolaas Heinsius. Both spent several years at Christina's court in Sweden: Vossius between 1642 and 1645, and Heinsius between 1648 and 1655.[135] Kool, however, arrived in Rome a decade after Christina's death, and relied on the help of two cardinals: Enrico Noris (1631–1704) and Pietro Ottoboni (1667–1740). Cardinal Noris was well acquainted with Magliabechi, whose recommendations had helped Noris to become the librarian of Christina of Sweden in Rome, and subsequently librarian of the Vatican library in 1692 until his death. Noris had also met Kool's mentor, Jacob Gronovius, when Gronovius was a professor in Pisa, although their relationship at the end of the seventeenth century appears to have been tense.[136] Cardinal Ottoboni was widely considered a patron of the arts in the late Baroque period, and was a well-connected Roman clergyman on account of having been the cardinal-nephew of Pope Alexander VIII.[137] Thus, in Rome, Kool could again rely on strong contacts between his two mentors, Gronovius and Graevius, and the local learned world.

Kool's strategy for entering the Roman learned world and its libraries was strikingly similar to the one he used in Florence. The first step was an audience with Cardinal Noris, which took place at the latter's palace on

135 Blok, *Vossius*, pp. 253–503.

136 Jacob Gronovius was appointed professor in Pisa in 1674, but left within a year due to conflicts with fellow scholars. See, e.g., J. W. Bierma, 'Gronovius, Jacob', *NNBW*, vol. 1, pp. 986–989, 986.

137 A cardinal-nephew was a cardinal elevated by the Pope who enjoyed more status and power; this position was abolished after Ottoboni had made him the last cardinal-nephew. For more information on Cardinal Ottoboni and his relation with Pope Alexander VIII, see Edward J. Olszewski, *Cardinal Pietro Ottoboni (1667–1740) and the Vatican Tomb of Pope Alexander VIII* (Philadelphia: American Philosophical Society, 2004).

27 November 1698.[138] Kool offered him the same book that he had offered Magliabechi: another copy of Graevius's *Callimachi hymni* on behalf of the author.[139] Self-consciously, Kool avoided talking about Gronovius or alluding to him, because of his tense relationship with Noris:

> we talked about a variety of books, and also some which are still under the press, we also talked about Magliabechi, and his Eminence asked me if we were good friends with mister Gronovius, we did not answer clearly, knowing all too well that there is bad blood between the cardinal and mister Gronovius, as Magliabechi has told me multiple times [...] The Cardinal said of mister Gronovius that he was too critical.
>
> I told the cardinal how I had obtained permission in Paris and Florence to enter the libraries, as well as use them, with regard to copy and collation, and asking of this permission could be obtained at the Vatican Library. The Cardinal answered that the permission to copy was granted to no one, but one could get the permission to collate authors, and he promised to take care of it[.]
>
> wij discoureerden over verschijde nieuwe boeken en over sommige, die nogh onder de pers waaren, wij spraaken ook van Magliabechi, en zijn Eminentie vroegh mij of wij ook goede vrienden met de Heer Gronovius waaren, wij antwoorden niet resolut op deeze vraagh, wel weetende, dat er geen goet bloedt tusschen den Cardinal, en de Heer Gronovius is, want naa Magliabechi mij verschijde maalen verhaalt heeft [...] De Cardinaal zeijde van de Heer Gronovius, dat hij een tooppo [*sic*] critico was.
>
> Ik verhaalde aan de Cardinaal hoe ik in Parijs en Florence de permissie gehad hadde van op de Bibliotheequen te moogen gaan en dezelve te gebruijken, zoo wat belangde het copieren als confronteeren en voeghde daarbij oft men ook die permissie alhier niet zoude konnen hebben [bij] de Bibliotheca Vaticana, waarop de Cardinaal antwoorde, dat de permissie van copieren aan niemant gegeeven wierdt, dogh die een autheur wilde confronteeren, konde de permissie bekoomen, en beloofde mij die te zullen versorgen[.][140]

138 For the full transcription of Kool's audience with Cardinal Noris, see Scholten and Pelgrom, 'Scholarly Identity', Appendix C, pp. 131–133.

139 Graevius, *Callimachi hymni*.

140 Kool, 'Journaal', BiASA, ms. 34 A 3, fols. 727$^r$–728$^r$.

The conflict between Gronovius and Cardinal Noris had been noticed by previous travellers.[141] Yet both Graevius's book and the fact that Kool had gained access to libraries in Paris and Florence must have convinced Noris to help Kool access the Vatican Library.

Just when he was about to leave Noris's palace, an abbot and secretary to Cardinal Noris questioned Kool's trustworthiness and piety, given his Protestant confession. A local antiquarian, Francesco Ficaroni (1664–1747), who had helped Kool to his audience with Cardinal Noris, came to his defence:

> The abbot asked Signor Ficaroni, if we were Christian, maybe he wanted to ask if we were Catholic, yet Ficaroni answered that we were no brutes. The abbot noticed his mistake and said how it were possible that we spoke such good Italian. Yes, answered Ficaroni, one must not be surprised, these two gentlemen [Kool and Van Voorst] understand and speak all languages [...] they do nothing else but study, do not visit whores and do not engage in debauchery, like our [Italian] young folk. Upon hearing this the abbot made a cross and said: '*Jesus Mary*, so many languages, unbelievable!'

> Deeze Abbate vroegh aan Sig$^{r}$ Ficaroni, of wij ook Christenen waaren, misschien wilde hij vragen, of wij ook Catholijke waaren, Dogh Ficaroni antwoorde, dat wij geen beesten waaren, den Abbate wel merkende, dat hi mis had seijde hoe het mogelijk waar, dat wij zoo goet Italiaans spraaken, jaa antwoorde Ficaroni daarop, hier over moet gij U Ed: niet verwonderen, die twee zijn heeren, welke alle taalen verstaan [...] deeze doen anders niet als studeeren, gaan bij geen hoeren en doen geen debauches, gelijk onze jonge lieden doen, den Abbate dit hoorende cruijsighde zigh, en seijde *Jesu Maria* zoo veel taalen, hoe is het moogelijk[.][142]

In other words, Kool embodied the typical virtues of the scholar that we have also seen in the typology of scholarly virtue (see Chapter 1.2). Ficaroni helped and defended Kool out of friendship (*amicitia*). He presented Kool as a scholar who spoke 'all languages' (*eruditio*), was dedicated to his studies (*industria* and *doctrina*), and did not 'engage in debauchery' (and thus

141 This conflict between Cardinal Noris and Jacob Gronovius was also mentioned by Coenraad Ruysch, see Ruysch, 'Journaal', NA, FA Teding van Berkhout, no. 1408, fol. 72$^{r}$.

142 Kool, 'Journaal', BiASA, ms. 34 A 3, fols. 730$^{r}$–731$^{r}$. Kool mentions that it was common to make a cross and say *Jesu Maria*: 'dit cruijsen, wanneer men zigh over verwondert en Jesu Maria te seggen is bij de Italiaanen zeer gebruijkelijk'. Ibid., fol. 731$^{r}$.

showed *pietas*). This exchange shows how Ficaroni presented Kool as an archetypical trustworthy scholar with the virtues that were common in the *vitae*. The scholarly traveller, backed up by local scholars, could enjoy and wield a scholarly identity that came with an aura of virtue, rendering him morally superior to other young travellers.

After attaining the necessary permissions from Cardinal Noris, Kool enjoyed the Vatican Library, where Lorenzo Alessandro Zaccagni (1657–1712), first librarian of the Vatican Library, helped and welcomed him.[143] Zaccagni offered Kool several manuscripts and shared his knowledge about the quality and whereabouts of certain manuscripts. Kool's self-assumed identity as a disciple of Graevius and Gronovius was not merely a representational trope, but came with the assignment to collate manuscripts for the two professors. Yet Kool could not always fulfil this duty:

> Mister Gronovius had requested me to learn about the manuscript of [Lucas] Holstein's *Notae in Stephanum*, to collate it with the printed version, since it has been printed very badly […] Yet, I cannot serve his Honourable [Gronovius], since I did not have enough time to stay in Rome to collate and undertake such a sizeable author.
>
> De Heer Gronovius had mij versocht alhier te verneemen naa het Mss. van Holstenii *Notae in Stephanum* om dezelve met de gedrukte te collationeeren, door dien dezelve zeer vitieus is gedrukt zijn […] dogh ik kan zijn Ed$^{e}$ daar omtrent niet dienen; door dien ik geen tijt genoegh hadden mij in Roma op te houden, om te collationeeren, en eene zoo grooten autheur te onderneemen.[144]

Although Kool indicated that he did not have enough time to finish the collation of a manuscript of Stephanus of Byzantium against an edition by Lucas Holstein, he did spend many days in Florence collating for Graevius and Gronovius.[145] After having worked for an entire day in a Florentine library, Kool laments: 'I would not have copied this author, if Mister Gronovius would not have requested me to do so, therefore I have worked here for someone

143 Domenico Surace, 'Zaccagni, Lorenzo Alessandro', *DBI*, vol. 100 (2020), pp. 294–295.

144 Kool, 'Journaal', BiASA, ms. 34 A 3, fol. 741$^{r}$.

145 Lucas Holstenius, *Notae & castigationes in Stephanum Byzantium de urbibus* […] (Leiden: Pieter van der Aa, 1692). For a complete, annotated overview of the library of Lucas Holstenius, see Alfredo Serrai, *La biblioteca di Lucas Holstenius*, Scienze bibliografiche 2 (Udine: Forum Editrice Universitaria Udinese, 2000).

else again.'[146] In Rome, he must have felt that he had cleared his scholarly debt to Gronovius and Graevius and instead followed his own interests: he spent days collating and copying several manuscripts by the fourth-century philosopher Sallustius.[147]

These four passages above in Kool's travel journal enable me to discern three ways in which a scholarly identity was shaped on a Grand Tour. First, the books Kool presented to both Antonio Magliabechi and Cardinal Noris are important material manifestations of his scholarly civility. Exchange and presents were an important way of establishing oneself as both a gentleman and a scholar. Secondly, Kool and his friends often used an 'othering' mechanism, whereby the scholar or studious type was contrasted with less morally inclined men. This othering is exemplified by Magliabechi's condemnation of monks and clergy and Ficaroni's appraisal of Kool as a virtuous, studious man, in stark contrast with most pleasure-seeking young men. The third and most important way for Kool to establish his scholarly identity was to stress his connections with his intellectual mentors, professors Graevius and Gronovius. By inscribing himself into the learned lineage of Dutch travelling scholars, he was immediately recognised as a scholar of merit and honour on his journeys. This is further exemplified by the responses of learned men on Kool's trip, among them Hervé Simon de Valhébert (c. 1661–c. 1736), librarian and secretary to abbot Jean-Paul Bignon (1662–1743), the Royal Librarian of Louis XIV.

> Monsieur Valhébert said that I was a disciple of Graevius, of which he, like many others have done, exclaimed, *Jesus; un bon Maistre*, as well as praise one's luck, who has studied under that man [Graevius] and to show, how highly regard Graevius's letters are here. I will add here too what Father Sarboeurs of the Abbey Nicaise told me: that many times when abbot Nicaise had received letters from mister Gronovius, he would show them in the assembly with many signs of joy, and as something rare.

> Mons$^{r}$ Valhebert zeijde, dat ik een discipel van de Heer Graevius was, waar op hij, gelijck veele andere hier gedaan hebben, uijt riep, *Jesus; un bon Maistre*, alsoo prijzende het geluk van een persoon, die onder die

146 'Ik zoude dezen autheur niet gecopieerd hebben, ten waar de Heer Gronovius mij dit niet versocht had, daarom heb ik hier weder voor een ander gewerkt.' Kool, 'Journaal', BiASA, ms. 34 A 2, fol. 396$^{r}$.

147 Ibid., ms. 34 A 3, fols. 745$^{r}$–746$^{r}$. For a full list of the manuscripts Kool copied and collated, see Scholten and Pelgrom, 'Scholarly Identity', Appendix D, pp. 134–136.

> man gestudeert hadden en om te toonen, hoe seer dat de brieven van d'Heer Graevius hier geacht zijn, zoo zal ik hier bijvoegen het geen de Pere Sarboeurs van den Abbe Nicaise verhaalde, den Abbe Nicaise dan soo menighmaal als hij brieven van d'Heer Graevius had ontfangen, zoo toonde hij die in de assemblee met zeer veel teekenen van vreught, en als wat raars.[148]

Apart from the commonplace praise of Graevius and Gronovius, letters from renowned Dutch scholars were considered prized possessions. Kool could conveniently deliver these letters—often on a first meeting—and consequently enjoying 'great civility'. It was not unusual for a recipient to be delighted and react like Magliabechi when he received a letter from Graevius. Kool recalled: 'When I handed the letter from Graevius to Magliabechi, he kissed it, before opening it'.[149] For travelling scholars to have renowned mentors was vital because it enabled them to form connections with local learned men. Since Kool was, as his nuptial poem recounted, 'well fed with learning, suckled from the breast of Graevius for years', he possessed not only demonstrable knowledge and skill, but also a distinct identity as a disciple of Graevius.[150]

Soon after his return from Rome, Kool decided to pursue a career as a magistrate in Amersfoort rather than as a scholar. He moved to Amersfoort and became an alderman, an office he held in 1700, 1701, and again from 1703 to 1706.[151] His reputation as a scholar had spread through the letters he had exchanged with Antonio Magliabechi, Gronovius, and Theodorus

148 Kool, 'Journaal', BiASA, ms. 34 A 1, fol. 117$^{r}$.

149 'Wanneer ik den brief van de Heer Graevius aan Magliabechi overleverde, zoo kuste hij deselve, eer hij die oopen deed.' Ibid., ms. 34 A 2, fol. 441$^{r}$.

150 '[D]oorvoedt met wetenschappen, Gezogen uit de borst van Greve jaaren lang'. *Ter bruiloft van den heere Joan Kool, Raad en Schepen der Stadt Amersfoort, en juffrouw Elizabeth Looten* [...] (Amsterdam: François Halma, 1703), p. 3.

151 On Kool's service to the town of Amersfoort, see Abraham van Bemmel, *Beschryving van de Stadt Amersfoort, dezelver Regeeringsvorm, Burgerye en Gilden, Jaarmarkten, Handel en Negotie, Vryheid en Grondgebied*, 2 vols. (Utrecht: Henricus Spruyt, 1760), 2:674–678; Henrik Theodoor Hormann, 'Kool, Mr J(oh)an', in *Gezagsdragers Amersfoort 1405–1795: gecompileerd alfabetisch register*, ed. Hormann (Woudenberg: Bureau voor Familie-Historie, 1998). No funeral record or testament of Kool appears to have survived. On 27 March 1706, his youngest daughter Aletta was baptised in his presence, see Archief Eemland (hereafter: AE), toegang 13.29h, Amersfoort NH, dopen 1687–1708, p. 153. In 1707 Kool and his wife registered in the Amersfoort Weeskamer, see AE, toegang 0039.2, Weeskamer te Amersfoort 1519–1818, inschrijving 1707. His last known letter was written to Theodorus Janssonius van Almeloveen, dated July 1706, see UBU, Hs 996, vol. I, fols. 293$^{r}$–294$^{v}$. In November 1709, his widow Elisabeth Loten married her second husband; see HUA, toegang 711.100, Utrecht NH trouwen 1692–1712, 787, dated 17 November 1709.

Janssonius van Almeloveen.[152] Kool's work on Roman inscriptions ended up in the posthumous publication of Marquard Gude's *Antiquae Inscriptiones* (Ancient inscriptions, 1731), although he could never fully finish his work, to the dismay of Gronovius. The latter wrote to Magliabechi in 1705:

> I am annoyed with Kool's action, however, and I hate it. I will not doubt to seize the opportunity, as soon as I have it, to deal with the matter. Thus, even if he is not present, he cannot viciously forsake the promise he made.

> Factum *Koolii* enimvero displicet, ac abominor; nec dubitabo captare occasionem, quae prima mihi incidet, de ea re agendi. Ita ne absens potest facinorose satis reiicere memoriam promissi?[153]

It is unclear what Gronovius is specifically referring to in this letter, but Kool had evidently failed to live up to a promise he had made. The catalogue of the auction of Kool's library, the *Biliotheca Kooliana*, referred to Kool's unfinished work as a 'large corpus of various inscriptions collected from all over [Italy], carefully edited by Gude'.[154] The work was eventually published in Gude's *Antiquae Inscriptiones*, and the preface written by François van Hessel (1680–1746) simply stated that 'Kool died in the middle of the work'.[155] Although it remains unclear what precisely happened to Kool after his *peregrinatio*, his travel journal offers us considerable insight into how Kool presented himself as a scholar embedded in the learned world and how he was received by Italian scholars. Kool's travelogue might not yield much insight in his philological research, but it does offer the reader a candid account of the social and cultural aspects of a scholarly travel.

152 Four letters from Kool to Magliabechi sent between December 1698 and January 1700, were printed in Giovanni Targioni Tozzetti, ed., *Clarorum Belgarum ad Ant. Magliabechium nonnullosque alios epistolae* [...], 2 vols. (Florence: Ex Typographia ad Insigne Apollinis in Platea Magni Ducis, 1745), 2:329–340.

153 J. Gronovius to Magliabechi, 27 November 1705, in *Clarorum Belgarum epistolae*, ed. Tozzetti, 2:342, n. 1.

154 'Corpus magnum Variorum Inscriptionum, undique sollicita cura a Gudio collectum'. *Bibliotheca Kooliana Instructa elegantissima ac nitida supellectile librorum, Omnis generis ac facultatis praestantissimorum* [...] (Utrecht: Willem van de Water, 1712), 'Libri Manuscripti in Octavo', no. 20, p. 23 lists: 'Corpus magnum Variarum inscriptionum undique sollicita cura a Gudio collectum, quarum maxima pars in Gruteri nova editione non exstat; instructum complurimis indicibus accuratis, a defuncto ad editionem paratum, nitide descriptum & ipsa marmora eleganter picta exhibens, 2 Vol. fol. indices separatim sunt conscripti in 4. & 8.'

155 'Mortuus est medium Koolius inter opus'. Gude, *Antiquae Inscriptiones*, p. 12.

## 4.3 Conclusion

The sustained journeys by Dutch scholars to England, France, and the Italian peninsula illustrate the value early modern scholars attributed to a formative *peregrinatio literaria*. The ideal Justus Lipsius professed in his *Epistola de peregrinatione* was not just hot air: few indeed took time to 'research and learn; that is, truly travel'.[156] The research of Vossius, Heinsius, and Kool can be seen as the necessary fieldwork for the ambitious philologist.[157] These scholars are examples of the sustained *peregrinationes literariae* in the seventeenth century. These voyages were crucial in the formation and continuation of both the prowess of Dutch classical scholarship and the cooperative relationships between Italian and Dutch scholars. What is more, each traveller could rely and build on the fame and connections of the scholars who had travelled before him: Isaac Vossius relied on his identity as an intellectual descendant of Scaliger and Vossius scion; Heinsius identified as the son of Daniel, and friend of Vossius and Gronovius; the sons of Gronovius could refer to their scholarly father and friendships with Vossius; and finally, Kool presented himself as a disciple of the professors Gronovius and Graevius. Throughout the seventeenth century, these travelling scholars were acutely aware of the value of this tradition. On these scholarly journeys, the scholars would present themselves as subscribing to a distinct scholarly identity, based on the scholarly standing and memory of their professors, family, friends, and travelling predecessors. The lineage is based both on father–son relationships, but even more so on professor–student relationships. Scholars such as Joannes Kool could rely on this professor–student relationship, rather than solely on family ties.[158] This is a social system that enabled the travellers to gain access to scholarly networks and precious manuscript collections, based on this scholarly pedigree, as well as the preservation and presentation of the memory of the generations of travelling scholars.

Travelling was essential for the continued sense of an existing and evolving community. With every *peregrinatio literaria* of a Dutch scholar to Italy, the Dutch-Italian scholarly network and community was revitalised. The travels and personal visits helped build trust between the scholars and

156 '[I]ndagare, discere; id est, vere peregrinari'. Lipsius, 'De Ratione cum fructu peregrinandi', p. 107.

157 Risbjerg Eskildsen, 'Exploring the Republic of Letters'.

158 On the importance of the father-son and uncle-nephew relationship in medieval clerical life, see Julia Barrow, *The Clergy in the Medieval World: Secular Clerics, Their Families and Careers in North-Western Europe, c. 800–c. 1200* (New York: Cambridge University Press, 2016).

helped strengthen their bond of scholarly friendship. The scholarly traveller on his *peregrinatio literaria* experienced, formed, and represented the transnational community of classical scholarship. They experienced the reality of the imagined, transnational community of scholarship in the many personal meetings with fellow scholars and their warm welcomes. They represented the local learned community they came from, which in the cases above was the Dutch scholarly community of classical philology. Consequently, the travellers formed their own network of scholarly colleagues and helped strengthen the transnational scholarly network throughout Europe. These sustained scholarly journeys throughout the seventeenth century helped establish bonds of trust and create a strong sense of a transnational Republic of Letters.

Kool's travelogue shows how the scholarly traveller sought to connect to a collective history and a scholarly ancestry. It made him for instance visit the grave of Virgil like many scholars before him, such as Petrarch and Boccaccio. In his journal, he writes:

> When one rides from the city [of Naples], one first passes the sea on the left side and then arrives at the famous Grotta di Cocceo. When one arrives there, one sees on the left side the grave of Virgil.
>
> eerst uijt de stad [Napels] koomende, rijd men eerst langs de zee, die ter linker hand is tot dan met komt bij de vermaarde Grotta di Cocceo eer men daar in komt ziet men aan de linker hand het graf van Virgilius.[159]

The tomb of Virgil was a well-known site of memory for travelling scholars.[160] Travellers visited such places as hallowed scholarly memory to experience the collective history of the Republic of Letters.[161] These sites of memory acted as a material reminder of the scholarly past that they themselves belonged to. The next two chapters focus on such tangible remnants of a scholarly memory culture: tombstones and grave monuments dedicated to the learned men who came to pass.

159 Kool, 'Journaal', BiASA, ms. 34A 2, fol. 538$^r$.

160 Karl A. E. Enenkel, '"Identitites" in Humanist Autobiographies and Related Self-Perception', in Scholten et al., *Memory and Identity*, pp. 31–88 (pp. 68–71).

161 Hulsenboom and Moss, 'Tracing the Sites of Learned Men'.

# 5. The Basilica di Santa Croce: The Florentine Site of Learned Memory

**Abstract:** This chapter examines the Basilica di Santa Croce in Florence as a site of learned memory. It traces its transformation from a Franciscan church into a mausoleum for scholars, focusing on monuments for figures such as Leonardo Bruni and Galileo Galilei. The chapter shows how funerary monuments and inscriptions celebrated intellectual virtues, emphasising how local memory practices were intertwined with broader European scholarly traditions. Santa Croce became a place where Florentine civic pride and European intellectual ideals converged, demonstrating the role of such spaces for scholarly identity formation.

**Keywords:** Florence, Basilica di Santa Croce, memory cultures, scholarly memory, church history, funerary monuments

The Basilica di Santa Croce in Florence appealed to many travellers on their journey through Florence. When Joannes Kool was in Florence, he visited the Basilica di Santa Croce. He marvelled at the 'beautiful tomb of Michael Angelo Buonarroti' and the commemorative aura of the church.[1] Two other tomb monuments caught his eye:

> On the same side [of the church] is the tomb of Leonardo Bruni, during his life secretary of the Republic. On this tomb they have this epitaph, made by Politian [...] There is also a tomb of Carlo Marsuppini, who was also secretary of the Republic. He was a great poet and Signore Magliabechi still has many of his manuscripts.

> Aan dezelve zijde is de Tombe van Leonardus Brunus Aretinus in zijn leeven secretaris van de Republ: op deeze Tombe heeft men dit

1 '[E]en schoone Tombe van Michael Angelo Buonarotia'. Kool, 'Journaal', BiASA, Ms. 34 A 2, fol. 452$^{r}$.

Scholten, Koen. *Rethinking the Republic of Letters: Memory and Identity in Early Modern Learned Communities.* Taylor & Francis Group, 2025
DOI 10.5117/9789048559855_CH05

> Epitaphium, dat Politianus gemaakt heeft. [...] Nogh is daar een Tombe van Carolus Marsupinus, die ook Secretaris van de Republiq geweest is. Deeze was een Groot Poëet en Sign^r Magliabechi heeft nogh veele van zijne manuscripten.[2]

Kool copied the epitaph on the monumental tombs of Leonardo Bruni (1370–1444) and mentioned the tomb of Carlo Marsuppini (1398–1445). He identified them as the graves of learned poets more than two centuries after the monuments had been placed in the church. Bruni and Marsupinni's monuments were part of the memory culture in Santa Croce in which secretaries of the republic were celebrated as learned men. Moreover, the fact that Kool noted that Antonio Magliabechi owned manuscripts of Marsuppini signals that he saw Marsuppini as part of a learned tradition that was directly linked to the present.

Throughout the early modern period, travelling scholars visited the houses of writers, such as the house of Petrarch, and copied epitaphs of illustrious scholars on their journeys.[3] Scholars visited such learned sites of memory and recognised the collective history of learning they often represented. The continued interest in these tombs speaks volumes about their importance as symbols for scholars and learned men. The tomb monuments of scholars and scientists have hitherto received little attention from historians of science and scholarship, despite their inherent ability to convey contemporary ideals of the honourable man of science and scholarship.

Apart from the perspective offered by travellers such as Kool when encountering monuments, the context of the mounments' placement can inform us about contemporary motivations to honour learned men. To honour predecessors and cherish their memory is central to a community's sense of unity. As Gordon Raeburn has argued, 'the disposal of the dead is an important way for the living members of the community to cement their identities.'[4] Monuments are therefore as much about the deceased as they are about the community who deemed it appropriate to honour them. The question of who was considered an honourable predecessor or example, worthy of a funerary monument, reflects the

2 Ibid., fol. 453^r.

3 Harald Hendrix, ed., *Writers' Houses and the Making of Memory* (New York: Routledge, 2008); Hendrix, 'The Early Modern Invention of Literary Tourism: Petrarch's Houses in France and Italy', in ibid., pp. 15–29.

4 Gordon D. Raeburn, 'The Reformation of Burial in Protestant Churches', in *A Companion to Death, Burial, and Remembrance in Late Medieval and Early Modern Europe, c. 1300–1700*, ed. Philip Booth and Elizabeth Tingle, 156–174 (Leiden; Boston: Brill, 2021), p. 157.

norms and values of a community. The same holds for the ideals such predecessors are meant to represent. Reflections on mortality abounded in the early modern period, with the omnipresent pictorial tradition of *memento mori* ('be mindful of dying') in paintings and on tombstones, and Christian advisory books on the *ars moriendi* ('the art of dying').[5] Funerary monuments thus served to remind the onlooker of mortality, but also of the immortal fame and glory of the deceased person to whom they were dedicated. How a certain community curated, selected, and organised the bestowment of immortal fame can inform us about the ideals a certain community held at the time.

How communities treated corpses, and especially those of their most important people, allows us to see how they conceptualised the relation between the living world, the afterlife, and memory. Research on death and the dead in the early modern period has generally focused on two aspects of death and dying. First, social history has examined the impact of death on the lives of individuals and communities with death as a demographic actor. Secondly, cultural historians generally interpret a community's relationship with the dead, and the corresponding rituals, as a way of representing and reconstructing the social and cultural orders of a community.[6] This chapter will focus on the latter aspect of death and memory. More specifically, it will address how scholars, scientists, and learned men were meant to be remembered in the Florentine Basilica di Santa Croce throughout the early modern period. In the sixteenth and seventeenth centuries, the Basilica was transformed into a site of memory where the Florentine state, family of scholars, and the successors of notable scholars celebrated the scholarly fruits of Florentine society. Moreover, a closer analysis of the inscriptions and iconographical traditions of the funeral monuments in the basilica will show that to be honoured as a particularly learned man was the highest honour a Florentine citizen could receive. The presentation of scholarly memory in the Basilica di Santa Croce reflects

5 On the depiction of death with the motto *memento mori*, see Christina Welch, 'Images of Death in Art and Literature in Late Medieval and Early Modern Europe (1300–1700)', in Booth and Tingle, *A Companion to Death, Burial, and Remembrance*, pp. 272–299. For the medieval tradition of preparing for death and the *ars moriendi*, see Tarald Rasmussen and Jon Øygarden Flæten, eds., *Preparing for Death, Remembering the Dead* (Gottingen; Bristol: Vandenhoeck et Ruprecht, 2015), esp. Volker Leppin, 'Preparing for Death: From the Late Medieval *ars moriendi* to the Lutheran Funeral Sermon', ibid., pp. 9–24; Herman J. Selderhuis, 'Ars Moriendi in Early Modern Calvinism', ibid., pp. 109–122; and Claudia Resch, 'Reforming Late Medieval *ars moriendi*: Changes and Compromises in Early Reformation Manual for Use at the Deathbed', ibid., pp. 153–172.

6 Craig M. Koslofsky, *The Reformation of the Dead: Death and Ritual in Early Modern Germany, 1450–1700* (Basingstoke; New York: Macmillan, 2000), pp. 4–11.

the ideals of learning prevalent at the time, yet also sustained the ideal of the learned man, and thereby helped Florentines to imagine these virtuous men as part of a learned community.

Studies of tombs in art history tend to focus on the production and aesthetics of funerary monuments, largely ignoring their epitaphs. However, epitaphs are one of the most important aspects of a monument, not only because they mention the names of the monumentalised person, but also because they often reveal the reason why a deceased person was eternalised in marble, and by whom. Moreover, epitaphs often mention virtues and deeds for which the deceased person was and should be remembered. The early modern occupation with funerary monuments is often interpreted as a combination of the Christian belief in life after death and the belief in immortality through eternal fame. Much as in scholarly life writing (see Chapters 1 and 2), the purpose of a funerary monument was to honour the deceased and inspire virtuous behaviour in its onlookers. A well-known passage from Leon Battista Alberti's *De re aedificatoria* offers a fifteenth-century perspective on the purpose of monuments for Alberti's ancestors from antiquity and contemporary culture:

> Our ancestors used to publicly give statues as well as sepulchres to those who excellently served the state with their blood and their life, to give thanks and to stimulate others to a similar glory of virtue.
>
> Maiores nostri his, qui de re publica sanguine et vita egregie meriti essent, ut gratias referrent, caeterosque ad parem virtutis gloriam excitarent, cum statuas tum et sepulchra dare publice consueverunt.[7]

Alberti made the point that this ancient tradition was worthy of emulation. The sepulchre was thus considered a symbol of honour that represented glory and public standing. Additionally, the 'eminent life' ('vita egregia') of the deceased could serve as an inspiration and exemplar for those that saw the sepulchre. The sepulchre was a site of memory recognised as a public honour and symbol of outstanding merit.

Studies in cultural history on death and commemoration vary in their focus, from what was considered a good way to die—the *ars moriendi*— to burial and funerary practices, to the remembrance of the deceased through

7 Leon Battista Alberti, *De re aedificare*, book VIII, ch. II, 138. Cited here is Leon Battista Alberti, *L'Architettura* [*De re aedificatoria*], vol. 2, ed. and trans. Giovanni Orlandi (Milan: Edizioni il Polifilo, 1966), p. 673.

monuments, tombs, and commemoration events.[8] Funerary monuments reflect a variety of cultural developments in the early modern period. The monuments are embedded in the early modern appreciation for Roman sculpture and epitaphs, the rise of antiquarianism, and a reconfiguration of the relationship between death, memory, and relics in the wake of the Reformation, as well as the preoccupation with acquiring eternal fame and honour by serving the public good and state. This study of funerary monuments is principally indebted to scholarship on material memory in the context of nation building in the nineteenth century, such as the work of Pierre Nora and Jay Winter (see Introduction).[9] The work of Nora and Winter exemplifies the introduction of the idea that places and spaces such as churches, graveyards, statues, and public squares can convey a representation of a past, and, more importantly, sustain cultural identities and a social group's sense of unity and individuality.

This chapter will apply the insights of Nora and Winter—the concept that sites of memory are rich conveyors of meaning and history—to the memorials and tombs dedicated to learned men in the Basilica di Santa Croce. The memorials, tombs, and epitaphs dedicated to learned men in the Basilica di Santa Croce in Florence allow us to reconstruct how collective learned identities were socially constructed and transmitted to the next generation. The funerary monuments of learned men in the basilica will be analysed on three levels: representation, memory culture, and reception. Firstly, which virtues and qualities are ascribed to learned men in Florence? Secondly, what were the historical context and memory culture in which learned men were honoured and elevated in Florence? And thirdly, what was the reception of, and engagement with earlier funerary monuments? In other words, how did the interpretation of the monuments change over time? The remembrance of scholars in churches can thus reveal *who* was being remembered by *whom*, and *what* was considered an appropriate way to remember and honour the scholarly dead.

The first section of this chapter will introduce the concept of funerary monuments of scholars as sites of memory and discuss the early modern practice of honouring learned men and professors throughout Europe. The second section will elaborate on the memory culture in Florence in the early

8 For an overview of the historiography of death and commemoration, see Philip Booth and Elizabeth Tingle, 'Introduction: Dying, Death, and Commemoration, 1350–1700', in *A Companion to Death, Burial, and Remembrance*, pp. 1–30.

9 Pierre Nora, ed., *Les lieux de mémoire*, 7 vols. (Paris: Gallimard, 1984–1992); Jay Winter, *Sites of Memory, Sites of Mourning: The Great War in European Cultural History* (Cambridge: Cambridge University Press, 2014; originally published in 1995).

modern period and the role of the Basilica di Santa Croce in the Florentine memory culture. The third and main section will discuss a selection of the many funerary monuments, tomb slabs, and epitaphs dedicated to learned men in the basilica. This third section is organised chronologically and is divided into three subsections, each dedicated to a distinct period: the early celebration of Florentine learned poets in the fourteenth century, the ideal of the learned man as a central political figure in the fifteenth century, and finally the ideal of the learned man as a gifted genius in the seventeenth and on into the eighteenth century. The changes in the memory culture in the Basilica di Santa Croce in Florence reflect the malleability of both scholarly identity and memory throughout the centuries. The final section concludes and offers a discussion of the results.

## 5.1 Funerary Monuments as Learned Sites of Memory

Honour and fame were central concepts in the funerary and memory cultures of the early modern elite. The deceased person's *fama* and *memoria* were celebrated in the period directly after one's death, especially during the funeral oration. Based on the classical tradition, funerary orations were usually held before the burial and written by an esteemed orator. Funeral oratory stressed praise of the virtuous life of the deceased. John McManamon argued that the practice of funerary oratory was full of idealism and served to 'enrich the spirit of any commonwealth no matter what its constitutional form'.[10] The funerary orations emphasised the Ciceronian ideal that those who gave their lives in service of the public good were rewarded with grace and honour in the afterlife. This ideal reverberated in nearly all funerary orations in the early modern period. Cicero's narrative of meritocracy abounded in this genre, especially his claim that 'all those who have preserved, aided, or enlarged their fatherland, have a special place prepared for them in the heavens, where they may enjoy an eternal life of happiness.'[11] The genre of funeral oration reflects the prevalent narrative that one should not look merely at *exempla* from antiquity, but also at one's virtuous contemporaries, who could serve as examples of how to live one's life and benefit the public

10 John M. McManamon, *Funeral Oratory and the Cultural Ideals of Italian Humanism* (Chapel Hill; London: University of North Carolina Press, 1989), p. 4.

11 '[O]mnibus, qui patriam conservaverint, adiuverint, auxerint, certum esse in caelo definitum locum, ubi beati aevo sempiterno fruantur.' Cicero, *De Re Publica*, bk. 6, ch. 13; translation from Cicero, *On the Republic; On the Laws*, trans. Clinton Walker Keyes (Cambridge, MA; London: Harvard University Press, 1928), pp. 264–265.

good.[12] McManamon distinguished three types of praise that were common in funeral oratory: first, the praise of the *bona externa* such as birthplace, descent, and education; second, the innate goods of the body such as strength, beauty, and innate brilliance; third, and most prevalent, the goods of the soul, which are the virtues that someone acquired, possessed and emanated.[13] The analysis of the funerary monuments in the second section will also draw on this tripartite division of these types of praise.

In this chapter, honour is taken to mean the public level of distinction in a certain community or society. The word 'honour' comes from the Latin verb *honorare,* which means 'to worship and respect', and amounts to a right to respect of virtuous superiors with a high status.[14] One's position in society, or the height of one's honour, was reflected in the location of one's seat in the church or at the dinner table, one's clothes, and one's funeral and tomb.[15] With regard to funerals and death, different degrees of honour were bestowed on the deceased.[16] Increasingly often in the fifteenth century, exemplary statesmen and clergy received pompous funerary rites in an attempt to mimic the funerals of emperors of antiquity.[17] Honour was not confined to the individual: communities such as guilds and societies were acutely aware of their honour and status, and so were mayors of towns and leaders of states.[18] This preoccupation with accruing and displaying honour and virtue is particularly reflected in early modern sites of memory. Early modern memorials, portraits, and funerary monuments are rich in references to the deeds, positions, and virtues of individuals, but also to their ties to communities, municipalities, and states. Humanist culture cherished the concept of immortality through *fama* (fame or renown), and funerary monuments, epitaphs, and sepulchres were considered a perfect way to ensure that one's *fama* continued to thrive posthumously.[19] To attain

12 McManamon, *Funeral Oratory*, pp. 19–35.

13 Ibid., p. 20.

14 Frank Henderson Stewart, *Honor* (Chicago: Chicago University Press, 1994).

15 Keith Thomas, *Ends of Life: Roads to Fulfilment in Early Modern England* (Oxford: Oxford University Press, 2009), pp. 147–186; Andrew Butterfield, 'Social Structure and the Typology of Funerary Monuments in Early Renaissance Florence', *RES* 26 (1994): 47–67.

16 Thomas, *Ends of Life*, pp. 149–150, 154–155.

17 For example, the funeral of Francis I of France, see Roy Strong, *Art and Power: Renaissance Festivals, 1450–1650* (Woodbridge: Boydell, 1984), p. 67; for the funeral of Cosimo de' Medici, see Sharon T. Strocchia, *Death and Ritual in Renaissance Florence* (Baltimore; London: Johns Hopkins University Press, 1992), pp. 180–188.

18 Thomas, *Ends of Life*, pp. 164–165.

19 See McManamon, *Funeral Oratory*, ch. 2, 'A Humanist Conviction: "Virtue Increases When Praised" ', pp. 5–35.

'eternal life in marble' was a trope among humanists and scholars, attested to by numerous epitaphs.

Early modern learned communities had sites of memory to secure their collective identity. Recently, historians of science and scholarship have turned to such forms of community formation. For example, book catalogues were used to remember a person's library.[20] Rare book and manuscript collections were often named after the benefactor to honour his generosity and erudition, such as the Vossius and Scaliger collections at Leiden University, or the Bibliotheca Machliabechiana at the current Biblioteca Nazionale Centrale di Firenze.[21] Exemplary scholars were also memorialised in portrait galleries that often adorned libraries to inspire those who worked there with the images of their illustrious predecessors.[22] Of course, during the funeral, the orations would have been a central act of commemoration, aimed at consolidating the memory of the outstanding life of the deceased.[23]

The church was a unique site of memory in the early modern period. It fulfilled many roles: it was a house of worship, the place travellers would visit to see relics or works of art, and first and foremost a centre for a religious community. Additionally, churches inspired their visitors to live a virtuous life. The religious artwork reminded visitors to be mindful of death (*memento mori*) and the funerary monuments invited emulation. After all, to enter a church one often had to tread on the tombstones of many virtuous predecessors. All these factors made a church an ideal location for the remembrance of scholars as exemplary members of the religious community, but also as reminder for visitors to the church that the families in their specific congregation brought forth illustrious men of science and learning. From a religious standpoint, a burial in the holy ground of the church was considered an extraordinary honour. The closer one's grave was to the main altar, the more prayers and 'spiritual power' from mass the deceased would receive.

20 See, e.g., Philippe Schmid, 'Catalogues in Catalogues: Imitation and Competition in Early Modern Book Collecting', in *Book Trade Catalogues in Early Modern Europe*, ed. Arthur der Weduwen et al., 399–425 (Leiden; Boston: Brill, 2021).

21 David McKitterick, *The Invention of Rare Books: Private Interest and Public Memory, 1600–1840* (Cambridge: Cambridge University Press, 2018); Eric Garberson, 'Libraries, Memory and the Space of Knowledge', *Journal of the History of Collections* 18, no. 2 (2006): 105–136.

22 Ludmilla Jordanova, *Defining Features: Scientific and Medical Portraits, 1660–2000* (London: Reaktion Books, 2000); Martin Warnke, 'Das Bild des Gelehrten im 17. Jahrhundert', in *Res Publica Litteraria: Die Institutionen der Gelehrsamkeit in der frühen Neuzeit*, vol. 1, ed. Sebastian Neumeister and Conrad Wiedemann, 1–31 (Wiesbaden: Harrassowitz, 1987).

23 See McManamon, *Funeral Oratory*, ch. 6, 'Academic Ideals: "Perfected in the Arts Appropriate to Humanity"', pp. 123–152.

In the Catholic and Protestant parts of Europe learned men, professors, and scholars were eternalised in marble in churches in the early modern period. These learned memory cultures were both part of a larger tradition as well as a reflection of the local and civic context. In general, scholars and professors were honoured in the context of a lineage of university professors or in the context of a civic lineage of illustrious men. For example, Stefanie A. Knöll has studied the memory culture at the universities of Oxford, Leiden, and Tübingen, and asked how funerary monuments helped shape the collective identity of the university community.[24] According to Knöll, the funerary monuments in churches conveyed a collective identity, since the 'academics fashioned themselves in opposition to the uneducated' who 'sought to impress [...] through their use of the Latin language and sophisticated symbols.'[25] The tombs and monuments are ideal conveyors of the memory and identity of the university, since they build on the tradition of honouring family members in the local church. Universities, however, often remembered their professors in a specific church that was associated with the university. Knöll argued for the interpretation of funerary monuments as 'active mediators of the self-image their subjects wanted to convey in order to influence and shape society'.[26] In Oxford, Leiden, and Tübingen, the professorial elite represented themselves as a vital and powerful part of civic society through their funerary monuments. Knöll's study focused on the representational aspect of funerary monuments, but similarly showed how the monuments helped elevate the status of the professors and the university.

On the Italian peninsula, many early modern university cities honoured their scholars, scientists, and professors in churches. For example, Jill E. Carrington studied the sepulchres of professors in Padua from the fourteenth up until the sixteenth century.[27] Her study delves deep into the motivations, patronage structures, and intended meaning of the monuments. Carrington argued that the tombs of professors from the Paduan Studio expressed 'a sense of well-defined corporate identity as an intellectual elite, elevated and set apart from anyone outside the professorial caste.'[28] The elaborate monuments dedicated to professors helped to elevate the

24 Stefanie A. Knöll, 'Creating Academic Communities: Funeral Monuments to Professors at Oxford, Leiden and Tübingen, 1580–1700' (PhD diss., University of Sussex, 2003).

25 Ibid., p. 249.

26 Ibid., p. 11.

27 Jill Emilee Carrington, 'Sculpted Tombs of the Professors of the University of Padua, c. 1358–c. 1557' (PhD diss., Syracuse University, 1996).

28 Ibid., p. 201.

social standing of the professor in Paduan society.[29] A similar tradition of honouring university professors as a distinct social class can be observed in thirteenth- and fourteenth-century Bologna.[30] Renzo Grandi's study shows that in the seventh decade of the thirteenth century, Bologna *dottori* began a monumental self-celebratory program by erecting sepulchral monuments, thereby expressing a collective identity.[31] These monuments of professors in Padua and Bologna were a substantial part of the cultural memory of the universities. The professors' tombs helped elevate the virtues of learning and teaching, as well as visualise the importance of the universities for the cities.

Finally, the University of Pisa had a tradition of erecting tombs for their professors. In the fifteenth and sixteenth centuries, many Pisan professors and scholars were honoured with tomb monuments.[32] The monuments were meant to reflect the prosperity of the university in Pisa, the city, and the Medici rule in Tuscany. In the seventeenth century, the popularity of the tomb monument faded, and professors were more often honoured with an epitaph and a bust.[33] It is safe to state that throughout the early modern period the cultural symbol of the professor was ubiquitous, and the learned man was recognised as an honourable figure. The universities of Oxford, Tübingen, Padua, Bologna, and Pisa all presented a collective history and lineage in their funerary monuments. Throughout Europe, the tomb monuments helped elevate the social and cultural standing of the professor and learned man in society at large.

29 Ruth Wolff, 'Le tombe dei dottori al santo: considerazioni sulla loro tipologia', in *Cultura, arte e committenza nella Basilica di S. Antonio di Padova nel trecento: Atti del Convegno internazionale di studi Padova, 24–26 maggio 2001*, ed. Luca Baggio and Michela Benetazzo, 277–297 (Padua: Centro Studi Antoniani, 2003).

30 Renzo Grandi, *I monumenti dei dottori e la scultura a Bologna (1267–1348)* (Bologna: Commune di Bologna / Istituto per la Storia di Bologna, 1982); Grandi, 'Dottori, scultori, pittori: ancora sui monumenti bolognesi', in *Skulptur und Grabmal des Spätmittelalters in Rom und Italien: Akten des Kongresses 'Scultura e Monumento sepolcrale del tardo medioevo a Roma e in Italia' (Rom, 4.–6. Juli 1985)*, ed. Jörg Garms and Angiola Maria Romanini, 353–365 (Vienna: Verlag der Österreichischen Akademie der Wissenschaft, 1990). For addenda to Grandi's study, see Martin Bertram, 'Mittelalterliche Gelehrtengräber in Bologna: Anmerkingen zu dem neuen Werk von Renzo Grandi', *Quellen und Forschungen aus italienischen Bibliotheken und Archiven* 65 (1985): 427–435.

31 Grandi, *I monumenti dei dottori*, pp. 51–53.

32 Elena Carrara, '*Exequiae, funera, sepolcra*: Rituale e iconografia nei monumenti dei professori dello Studio tra Quattro e Cinquecento', in *Scultura a Pisa nell'età moderna: Le sepolture dei docenti dello Studio*, ed. Cinzia Maria Sicca, 21–73 (Pisa: Edizioni Plus / Pisa University Press, 2007), pp. 46–57.

33 Cristiano Giometti, 'Professori e cavalieri: Cerimonie funebri e monumenti del Seicento nel segno della continuità', in Sicca, *Scultura a Pisa nell'età moderna*, pp. 75–90.

At the same time, learned men and professors were honoured with funerary monuments in the context of civic and patriotic histories of illustrious men. In London, for example, Westminster Abbey became the pantheon of eminent poets and learned men.[34] Over the course of the seventeenth and eighteenth centuries, the abbey was slowly filled with eminent national symbols, but also eminent poets and learned men, such as Edmund Spenser (1552/1553–1599) and Isaac Casaubon.[35] The most important and elaborate monument is dedicated to Isaac Newton (1642–1727), whose monument reflects the contemporary praise for eminent scientists.[36] These monuments were also of interest to eighteenth-century scholars, most importantly the Royal Society of Antiquaries who published a series of these monuments in the second half of the eighteenth century.[37] These publications further promoted these scholars and scientists as cultural icons, figures who reflect the greatness of a state and city.

Thus, the monumentalisation of learned men was increasingly organised in large churches that were framed as national pantheons. They were less remembered in churches, basilicae, and abbeys in the context of their parish or local civic community. They were often not memorialised in the context of other learned men, but rather in a lineage of great men stemming from a city, state, or nation.[38] The late eighteenth century saw the development of the ideal of the pantheon, a type of temple where the greatest men of the nation were celebrated to invoke emulation.'[39]

These developments of a memory culture can be seen throughout Europe and the changes described above happened roughly at the same time. However, if one looks more closely at the individual funerary monuments, their intended meaning, and the context in which they were placed, generalisation on the European level becomes problematic. While many funerary monuments of learned men frequently appeal to a wider European learned world, they are often deeply entrenched in local traditions and memory culture. Funerary monuments should therefore be understood as a cultural

34 Alfred Rupert Hall, *The Abbey Scientists* (London: R. & R. Nicholson, 1966).

35 Matthew Craske, 'Westminster Abbey 1720–70: A Public Pantheon Built Upon Private Interest', in *Pantheons: Transformations of a Monumental Idea*, ed. Richard Wrigley and Matthew Craske, 57–79 (Aldershot; Burlington, VT: Ashgate, 2004).

36 Bernard Cohen, 'Commemorations & Memorials: Isaac Newton, 1727–1977', *Vistae in Astronomy* 22 (1979): 381–394.

37 Joseph Ayloffe and John Nichols, *An account of some ancient monuments in Westminster Abbey* (London: Printed by J. Nichols, 1780).

38 Wrigley and Craske, *Pantheons*.

39 Eveline G. Brouwer, *Public Pantheons in Revolutionary Europe: Comparing Cultures of Remembrance, c. 1790–1840* (Basingstoke; New York: Palgrave Macmillan, 2012), p. 9.

and social phenomenon in their specific geographical location and place in time. Social historian Andrew Butterfield reminds us that they should be understood as a reflection of social hierarchies:

> In celebrating individuals, monuments implicitly commemorate individuals as members of groups, albeit as special or distinguished members. As such, a monument not only serves to depict the features of a deceased individual; it also preserves a record of the social relations that surrounded its creation.[40]

Funerary monuments reflect a contemporary social hierarchy and what was considered laudable in a historical setting. In late medieval Padua and Bologna, for example, the elite status of the university professors was reflected in the sheer amount and lustre of their funerary monuments.[41] The tomb monuments from the thirteenth and fourteenth centuries depicted the professors as teachers, with illustrations of classroom scenes and lecturing professors, stressing the virtue of teaching. This iconography of the professor teaching from his *cattedra*, underscoring his position as the highest official in a university, was also present on a tomb monument of the Pisan professor Ligo Ammannati († 1359).[42] In the sixteenth and seventeenth centuries, the professor was generally presented on epitaphs as a learned man of unique virtue and learning. The professor was less seen as a teacher in a university community, but rather as an individual scholar who exhibited outstanding virtue and learning.[43] The iconographical changes reflect the remoulding of the collective identity of the professor and the university. The significance of their iconographical elements become apparent in the social and cultural contexts of the funerary monuments and those who are represented. The many examples of memory cultures surrounding professors and learned men exemplify the importance of taking into consideration both the local cultural context as well as the larger European trends in memory cultures. In Chapter 2, we saw that the collections of *vitae* were often a local appropriation of a wider ideal of learning; the funerary monuments of learned men in churches allow us to see in more detail how learned men were remembered in a local context.

40 Butterfield, 'Social Structure and the Typology of Funerary Monuments', p. 67.

41 Carrington, 'Sculpted Tombs'; Grandi, *I monumenti dei dottori*.

42 Carrara, '*Exequiae, funera, sepolcra*', p. 46.

43 This development can also be observed in the portraits of writers, see, e.g., Lieke van Deinsen, *The Panpoëticon Batavûm: The Portrait of the Author as a Celebrity* (Amsterdam: Rijksmuseum, 2016).

The history of remembrance and memory cultures in churches outlined above is also visible in the Basilica di Santa Croce in Florence. Ultimately, the basilica became a pantheon of illustrious and learned men in the eighteenth century. Since Florence did not represent itself as a university city, the learned men in the Santa Croce were rarely professors. The remembrance and representation of erudition in the Basilica di Santa Croce in Florence is thus an expansion of the study of funerary monuments beyond the well-studied university settings of Oxford, Leiden, Tübingen, Padua, Bologna, and Pisa. The Basilica di Santa Croce offers an interesting case study for understanding the meaning, construction, and representation of the cultural symbol of the learned man outside a university city in the early modern period. In this chapter, the memory of learned men in the Basilica di Santa Croce will be seen as a representation of the ideals of learned men, yet also as a mediator who had an influential role in cementing the ideals of learned men in Florentine culture and beyond. To better understand the distinct context of the Basilica di Santa Croce, the next section will elaborate on the history of the basilica and the memory culture in Florence.

## 5.2 The Florentine Memory Culture and the Basilica di Santa Croce

For Florence and the Florentine state, the Basilica di Santa Croce was, and remains to this day, a true pantheon of illustrious men. From the fifteenth century onwards, learned men were celebrated with elaborate sepulchres, especially in the northern and southern naves of the church. This emanated from a distinct fifteenth-century memory culture in which statesmen were remembered as poets. From the sixteenth century onwards, under the influence of Giorgio Vasari (1511–1574) and the Grand Duke Cosimo I de' Medici (1519–1574), the memory culture shifted, initiating the celebration of men who graced the Florentine state. In the second half of the eighteenth century, the Santa Croce was considered a true pantheon of illustrious men, sparking a revived interest in its Republican memory. At this point, omissions in the collection of illustrious men were remedied, for example with a new sepulchre for Niccolò Machiavelli (1469–1527) in 1787, and there was also a large increase in marble epitaphs and sepulchres for contemporary men.

The Basilica di Santa Croce was originally built as a Franciscan church and convent. The building of the basilica started in 1294 and finished in 1385. A cloister and chapterhouse were added in 1470. The church consists of a relatively large Franciscan nave, with fourteen chapels, most of which are decorated with frescoes and reliefs by well-known artists such as Giotto and Donatello.[44] The basilica was the head church of the Santa Croce quarter in Florence, one of the four districts of the city, and thus served as the main religious building for all the residents of this quarter.[45] In the face of the Counter-Reformation, between 1566 and 1584, Giorgio Vasari completely redesigned the interior of the church according to the wishes of Cosimo I de' Medici.[46] The medieval *tramezzo* or dividing wall, which separated the church from the choir, was broken down (see Figure 5.1).[47] The southern and northern naves were completely redesigned: the walls were cleared and symmetrical altars were placed in each bay. This led to the movement and destruction of fresco's, statues, tomb monuments, tomb slabs, and sculptures. The precise extent of the destruction is not known.[48] Nevertheless, this architectural intervention has implications for any study of the memory culture in the church. All tomb monuments which predate Vasari's renovation were consciously selected to remain in the church. It is possible to look back at monuments from before the second half of the sixteenth century only through a Vasarian lens. All the tomb monuments and funerary monuments that remained in the basilica after 1584 reflect Vasari's historical consciousness and sensibility.

Despite Vasari's redesign, Santa Croce is one of the few churches in Florence that still has tomb slabs: around three hundred, inscribed with family

44 For a general history of the Santa Croce, see Monica Maffioli, ed., *Santa Croce nell'800* (Florence: Alinari, 1986); Walter Paatz and Elisabeth Paatz, *Die Kirchen von Florenz: Ein kunstgeschichtliches Handbuch*, 6 vols. (Frankfurt a.M.: Vittorio Klostermann, 1940–1954), 4:497–701. For a closer interpretation of the funerary monuments, see Luciano Berti, ed., *Il pantheon di Santa Croce a Firenze* (Florence: Cassa di Risparmio di Firenze, 1993). For an inventory of the church, see Umberto Baldini and Bruno Nardini, eds., *Il complesso monumentale di Santa Croce: la basilica, le capelle, i chiostri, il museo* (Florence: Nardini Editore, 1983).

45 The other church in the Santa Croce quarter was the San Pier Maggiore. This church was broken down in 1784 due to its unstable structure; see Paatz and Paatz, *Die Kirchen von Florenz*, vol. 3.

46 See Marcia B. Hall, *Renovation and Counter-Reformation: Vasari and Duke Cosimo in Sta Maria Novella and Sta Croce, 1565–1577* (Oxford; New York: Oxford University Press, 1979), esp. 16–32.

47 For a reconstruction of the *tramezzo*, see Marcia B. Hall, 'The *Tramezzo* in Santa Croce, Florence, Reconstructed', *The Art Bulletin* 56, no. 3 (1974): 325–341; and more recently Andrea De Marchi, 'Relitti di un naufragio: affreschi di Giotto, Taddeo Gaddi e Maso di Banco nelle navate di Santa Croce', in *Santa Croce: Oltre le apparenze*, ed. De Marchi and Giacomo Piraz, 33–71 (Pistoia: Gli Ori, 2011), esp. pp. 34–35.

48 For a complete inventory of the church, see Paatz and Paatz, *Die Kirchen von Florenz*, 4:544–591; for the list of inventory that has been moved or lost, see ibid., pp. 591–613.

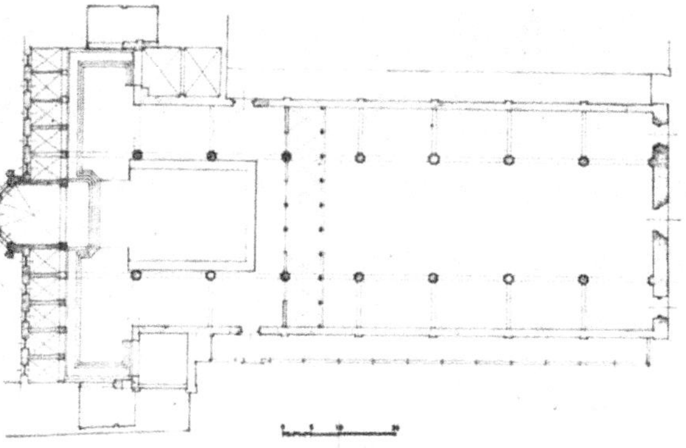

Figure 5.1. The church design of Santa Croce before Vasari's redesign. Note the closed choir before the main altar, as well as the *tramezzo* (marked with eight dots in the middle) dividing the church in two parts for the laity, and the clergy and friars. From Marcia B. Hall, 'The Tramezzo in Santa Croce, Firenze, Reconstructed', *The Art Bulletin* 56, no. 3 (1974), p. 329

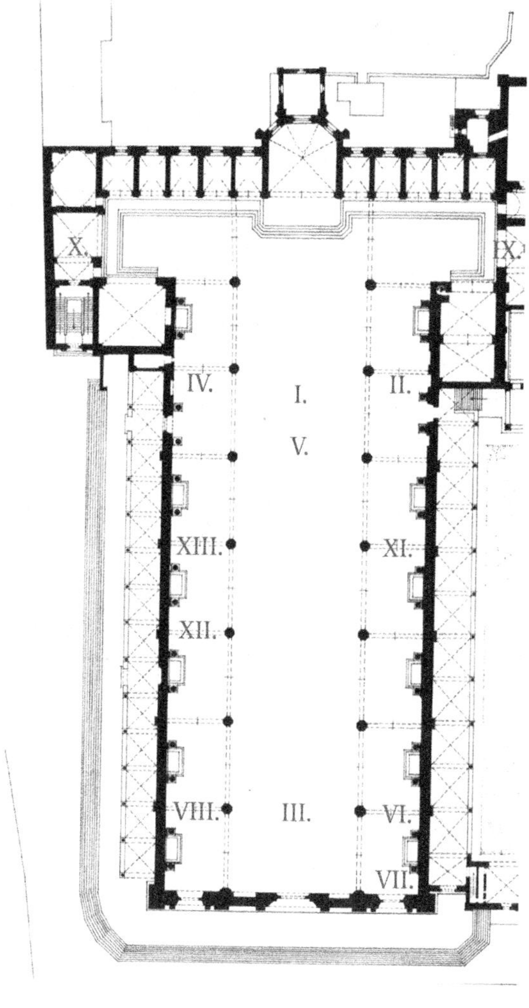


Figure 5.2. Current church design of Santa Croce. Based on, and adapted from, the schematic from Lando Bartoli, 'La Basilica di Santa Croce al di là dello spazio e del tempo', in *Il complesso monumentale di Santa Croce: la basilica, le capelle, i chiostri, il museo*, ed. Umberto Baldini and Bruno Nardini, 33–49 (Florence: Nardini Editore, 1983), pp. 48–49. The Roman numerals refer to the positions of monuments that are discussed below

names and heraldic symbols, most of which predate the redesign.[49] These family graves became increasingly common in the fourteenth and fifteenth centuries. Before 1229, canon law dictated that only saints, popes, royalty, and benefactors of the church could be buried in the sacred ground inside a church.[50] From 1229 onwards, floor tombs for affluent families steadily increased. Up until 1400, memorialisation took place in the context of religiosity and piety, particularly in the context of family, genealogy, and nobility.

The end of the fourteenth century, the start of the so-called Florentine Renaissance, signals a shift in the memory culture in Florence. From this moment on, power and prestige were continuously on display in Florence by way of festivals, grandiose marriages, court spectacle, and funerary processions.[51] The Florentine state—or rather, its ever-changing elite regime—continuously struggled for power, legitimacy, and a distinct identity, in the midst of multiple wars, conflicts, and calamities.[52] First, plague outbreaks severely diminished the number of inhabitants between 1348 and 1400, effectively cutting the population to one third of its former size, causing political and economic tensions.[53] The Black Death led to what Samuel Cohn dubbed a 'cult of remembrance': burials, perpetual remembrance masses, commemorative meals, commissions for art and symbols of familial pride, and testamentary investment in monuments all significantly increased after the vicious plague outbreak of 1362–1363.[54] Secondly, the regime struggled

49 Stefano Rosselli, *Sepoltuario Fiorentino ovvero descrizione delle chiese capelle e sepolture loro armi & inscrizioni della citta di Firenze e suoi contorni*, 1657, ASFi, mss. 624–625, fols. 21$^{r}$–23$^{r}$, cited in Paatz and Paatz, *Die Kirchen von Florenz*, 4:547. A digital version of the original manuscript is attached to Michelina Di Stasi, *Stefano di Francesco Rosselli: antiquario fiorentino del XVII sec. e il suo sepoltuario* (Florence: Edizione Polistampa, 2014). For more information on the family insignia and symbols on the tomb slabs, see Rita Sabelli et al., eds., *Le tombe terragne della Basilica di Santa Croce* (Florence: Città di Vita, 2001).

50 Butterfield, 'Social structure', p. 59.

51 Strong, *Art and Power*, esp. ch. 4, pp. 126–152.

52 For a general history of the Florentine state, see Gene Brucker, *The Civic World of Early Renaissance Florence* (Princeton, NJ: Princeton University Press, 1977). For a history of the later Grand Ducal period, see Eric W. Cochrane, *Florence in the Forgotten Centuries, 1527–1800: A History of Florence and the Florentines in the Age of the Grand Dukes* (Chicago; London: University of Chicago Press, 1973).

53 Gene Brucker argues that the political system lost cohesiveness and vitality in the wake of the Black Death, allowing for a new aristocratic spirit and corporatism in the fourteenth century, see Brucker, *Civic World*, pp. 14–59. The population of Florence fell from 120,000 to 36,909 between c. 1300–1427; for population estimates of Florence, Perugia, Siena, Pisa, Arezzo, and Assisi, see Samuel Kline Cohen, Jnr., *The Cult of Remembrance and the Black Death: Six Renaissance Studies of Central Italy* (Baltimore; London: Johns Hopkins University Press, 1992), pp. 5–6.

54 Ibid.

for power and territory, both during the wars for papal territory (1375–1378) and against Florentine exiles and political enemies in Pisa, Lucca, Siena, and Milan, most notably the Visconti family (1389–1402).[55] Even without external pressure, the governing elite experienced changes of power and internal pressure. Under these circumstances, for much of the fourteenth and fifteenth centuries Florence had a need for regional and pan-European recognition of its stature and ability as a state.[56]

In this period funerary practices and rituals changed accordingly. A competition between established patricians and newly arrived men in combination with a relatively high social mobility in Florentine society helped create a culture in which social perception was key. Since funerals were strategic moments to display status and influence its perception, the period between 1360 and 1400 saw an increase in pompous funerals and rituals. It was at this point, Sharon Strocchia argued, that the concept of an honourable burial was redefined, changing from a pious and solemn occasion to a publicly visible and pompous affair.[57] Strocchia's thesis is strengthened by the steady increase in monumental graves and the increasing popularity of humanism and the classical notion of praise throughout the fifteenth century.[58]

The cultural figure of the learned man had a central place in the Florentine memory culture. The emergence of a memory culture celebrating learned men started in the late fourteenth century. In 1396, the Signoria, the governing council of Florence, ordered the construction of monumental sepulchres in the cathedral for Accursius (1182–1263), a celebrated jurist, and for Florence's most illustrious poets: Dante (1265–1321), Petrarch (1304–1374), Giovanni Boccaccio (1313–1375) and Zanobi da Strada (1312–1364). While these plans never materialised, the decree offers insights into the values which bound these men together. The Signoria

> deliberated on the 22nd of December, 1396 [...] to carry out the movement to the city of Florence of the bones, that could conveniently be found and saved from the once illustrious men of famous memory, Florentine citizens such as: Dominus Accursius, doctor of law and glossator of the entire corpus of sacred civil law. [And] the poets Dante Alighieri, Francisco Petrarca, Zenobio de Strada and Boccaccio.

55 The so-called Visconti wars (1389–1402) were a continuous brawl for military, diplomatic, and political power between powerful central-Italian families; see Brucker, *Civic World*, pp. 125–186.

56 Strocchia, *Death and Ritual*, pp. 75–82.

57 Ibid., pp. 56–67.

58 Ibid., pp. 143–148; McManamon, *Funeral Oratory*, ch. 2, pp. 5–35.

> deliberaverunt die 22 mensis decembris 1396 [...], et fecisse conduci ad civitatem florentie ossa, que poterunt commode reperiri et haberi de olim illustribus et celebris memoriae viris civibus florentinis, videlicet: Domino Accursio legum doctore et glosatore ordinario totius corporis sacrarum legum civilium[que]. Dante Alleghieri, Domino Francisco Petrarcha, Domino Zenobio de Strata et Domino Johanne Boccaccii de Certaldo Poetis.[59]

As is stressed in this decree, one important requisite was the return of the mortal remains of both Dante and Petrarch to restore the honour of these men, and more importantly, the honour of Florence. From the fourteenth century onwards, both Dante and Petrarch were the great pride of Florence: Dante, because he was considered the founder of the Florentine language, and a great poet;[60] Petrarch, because he was credited as the restorer of classical Latin and the *studia humanitatis* in Florence.[61] The Signoria's many requests were to no avail. In 1429, the Signoria still requested the bones of these two men to be returned to Florence from Ravenna:

> It has already been decided by our Republic, that sepulchres be made for the famous poets Dante Alighieri and Francis Petrarch, with fitting magnificence, in our city, the *patria* of these very poets.

> Fuit iam pridem per n[ost]ram remp[ub]l[icam] constitutum, ut Dantis Alaghieri [sic] et Francisci Petrarce, inclitorum poetarum sepulcra cum ea, qua decet, magnificentia in urbe nostra, hoc est in patria ipsorum poetarum, construentur.[62]

59 Decree of the Signoria, 22 December 1396. The first transcription and reference to this decree appears in Francesco Moisè, *Santa Croce illustrazione storico-artistica: Con note e copiosi documenti inediti* (Florence: Tipografia Galileiana, 1845), pp. 484–485. The reference in Moisè refers to the discontinued Archivio delle Riforme, *Provvisione*, filza 87. Later references and transcriptions, most recently in Elisabeth Oy-Marra, *Florentiner Ehrengrabmäler der Frührenaissance* (Berlin: Gebr. Mann Verlag, 1994), p. 123, erroneously refer to the suspended Archivio delle Riforme. The document now resides at ASFi, *Provvisione*, filza 87, unknown fols.

60 Hans Baron, *The Crisis of the Early Italian Renaissance: Civic Humanism and Republican Liberty in an Age of Classicism and Tyranny*, 2 vols. (Princeton, NJ: Princeton University Press, 1955), 1:38–40.

61 Ibid., 1:229–240.

62 Signoria to Ostasio III da Polenta, 1 February 1429. Similar to the previously mentioned decree of the Signoria, the first transcript of this document appeared in Moisè, *Santa Croce*, p. 485. Anne Markham Schulz discusses these documents in a footnote yet fails to mention the continued requests of the Signoria for the return of the mortal remains of Dante: see Markham Schulz, *The Sculpture of Bernardo Rossellino and his Workshop* (Princeton, NJ: Princeton University

Figure 5.3. Domenico di Michelino (1417–1491), *Dante and His Poem*, 1465. Santa Maria del Fiore, Florence. Tempera on panel, 290 × 232 cm. Copyright Marie-Lan Nguyen / Wikimedia Commons. The caption has three elegiac couplets: 'Qui coelum cecinit mediumque imumque tribunal • Lustravitque animo cuncta poeta suo • Doctus adest Dantes sua quem Florentia saepe • Sensit consiliis ac pietate patrem • Nil potuit tanto mors saeva nocere poetae • Quem vivum virtus carmen imago facit'

The Florentine memory culture was filled with representations of learned poets from the late fourteenth century onwards. A fresco depicting Claudian, Dante, Petrarch, Boccaccio, and Zanobi da Strada was displayed at the 'Aula Minor' in Palazzo Vecchio, the town hall of the city of Florence. Only fragments of this so-called *Uomini Famosi* cycle, dating from around 1375, remain. Several depictions of these five illustrious men allegedly adorned walls and ceilings of palaces in Florence at that time.[63] Apart from these frescos, two wooden doors to the Sala dei Gigli were adorned with two wood reliefs of Dante and Petrarch, both pointing to an opened book, made in 1480.[64] In the same vein, there are a few frescos depicting learned poets in Florence's cathedral, the Santa Maria del Fiore. One memorial painting depicts Dante

Press, 1977), p. 33, n. 9. The reference in Moisè, *Santa Croce*, refers to the suspended Archivio delle Riforme, *Lettere della Signoria*, filza 30. My efforts to find the original document were in vain.

63 Oy-Marra, *Florentiner Ehrengrabmäler*, pp. 17–18.

64 See Claudio Paolini, 'Francesco di Giovanni detto il Francione, *Dante, Petrarca e libri; Vasi di fiori e sportelli aperti*', in *Inganni ad arte: Meraviglie del* trompe-l'œil *dall'antichità al contemporaneo*, vol. III.4, ed. Annamaria Giusti, 141–144 (Florence: Mandragora, 2009).

standing next to the city of Florence with his *Divine Comedy* (see Figure 5.3). This fifteenth-century pictorial tradition allows us to see how the laureate poet—that is, Dante, Petrarch, Boccaccio, and Zanobi da Strada—served as an *exemplum virtutis* of honourable Florentine citizenship.[65]

To be laureate poet continued to be the highest form of honour in Florentine society. The depiction of Dante and other laureate poets became one of the most persistent symbols in Florence.[66] The depictions often evoke the sense of a literary debate and the fictional meeting of poets in Dante's *Inferno*, where Dante comes across Virgil, Homer, Horace, Ovid, and Lucan.[67] In the sixteenth century, Giorgio Vasari painted a portrait that depicted Dante in such a lively debate with five other men of learning: the laureate poets Petrarch, Guido Cavalcanti (c. 1255–1300), Giovanni Boccaccio (1313–1375), and (probably) two later humanist commentators on their work, Cristoforo Landino (1424–1498) and Marsilio Ficino (1433–1499) (see Figure 5.4).[68]

The special commemoration of Boccaccio, Petrarch, and Dante as laureate poets continued well into the seventeenth century. Francesco Bocchi (1548–1613 or 1618), a Florentine writer, in his collection of learned men titled *Elogiorum; quibus Viri Doctissimi nati Florentiae decorantur liber primus* (First book of eulogies with which the most learned men born in Florence are honoured, 1609), dedicated numerous pages to these three poets. Bocchi wrote:

> Three illustrious men have so placed their feet on the summit of elegance: Boccaccio, Petrarch, and Dante that they are celebrated with glorious words in the opinion of everyone around the world. They are mentioned by such a reasoning that each one for himself excels in the knowledge of the best arts, thereby enlightening their and our *patria*, [which is] one and the same.

65 Ibid., pp. 25–38, 75–76.

66 Jonathan Nelson, 'Dante Portraits in Sixteenth Century Florence', *Gazette des Beaux-Arts* 119 (1992): 59–77.

67 Dante Alighieri, *The Divine Comedy*, vol. 1: *Inferno*, trans. and comm. by John D. Sinclair (New York: Oxford University Press, 1961), Canto IV, ll. 85–90, pp. 62–63; led by Virgil, 'the good Master', Dante writes: 'I saw four great shades coming to us; their looks were neither sad nor joyful. The good Master began: "Mark him there with sword in hand who comes before the three as their lord; he is Homer, the sovereign poet. He that comes next is Horace the moralist, Ovid is the third, and the last Lucan" '.

68 Deborah Parker, 'Vasari's *Portrait of Six Tuscan Poets*: A Visible Literary History', *Lectura Dantis* 22–23 (1998): 45–62; Edgar Peters Bowron, 'Giorgio Vasari's Portrait of Six Tuscan Poets', *The Minneapolis Institute of Arts Bulletin* 60 (1971–1973): 42–53. The identity of the two men without laurel crowns is disputed; see Nelson, 'Dante Portraits', pp. 60–62.

Figure 5.4. Giorgio Vasari, *Portrait of Six Tuscan Poets*, 1544. Oil on panel, 132 × 131 cm. Copyright Minneapolis Institute of Arts / Wikimedia Commons. From left to right: Cristoforo Landino (?), Marsilio Ficino (?), Petrarch, Giovanni Boccaccio, Dante, Guido Calvacanti

> In summi decoris possessionem ita pedem posuerunt tres viri clarissimi, Boccaccius, Petrarca, Dantes, ut orbis terrae iudicio ubique; gloriosis vocibus celebrentur. ii ea ratione nominantur, ut unus, quisque optimarum artium scientia excellat, & suam, nostramque hanc eandem patriam illustret.[69]

Bocchi's account of these three poets takes up a disproportionately large part of the volume: sixty-three pages for seventeen learned men, and a total of twenty pages for the poets. For centuries, the cultural symbol of

69 Francesco Bocchi, *Elogiorum; quibus Viri Doctissimi nati Florentiae decorantur, Liber Primus* (Florence: apud Iuntas, 1609), p. 70.

the laureate poet was foundational for the remembrance and presentation of learned men in the memory culture in Florence. The laureate poet became, and long remained, the ultimate example of a Florentine man of virtue.

The collective identity of the laureate poet was a distinct Florentine cultural icon of learning. As we will see in the next section, the memory culture of learned men in the Basilica di Santa Croce up until the fifteenth century was strongly influenced by the memory cult around the laureate poets in the preceding centuries. In this period of state building, one can see the different roles the learned man came to occupy in the memory culture of a city. In the university cities of Bologna and Padua, the cultural icon of the benevolent professor came to represent their civic identities as accomplished cities.[70] The memory culture in the Basilica di Santa Croce in Florence does not stand in such a university tradition but, rather, represents a civic identity where humanist learning was considered a unique trait and pride of the Florentine state.

This historical context outlined above is important to study the individual funerary monuments in the Basilica di Santa Croce. The next section will look closely at the iconography, textual representation, and collective identity conveyed by the sepulchres, epitaphs, and tombstones dedicated to learned men in the basilica from the fourteenth up until the eighteenth century.

## 5.3 Typology and Chronology of Learned Memory in the Basilica di Santa Croce

The Basilica di Santa Croce contains many funerary monuments and a fair share of them are dedicated to learned men. Yet the iconography, symbolism, virtues, and modes of remembrance changed considerably in the early modern period. These changes reflect the malleability of the ideal of a learned man, as well as the ways and goals of honouring them with a marble tombstone or monument. The remainder of this chapter will discuss a range of these funerary monuments to illustrate the different communities that wanted to immortalise learned men, such as the Florentine state and the specific Florentine learned communities, and the virtues and scholarly qualities they represented.

70 Carrington, 'Sculpted Tombs'; Sicca, *Scultura a Pisa nell'età moderna*.

**Table 5.1. Monuments discussed in this chapter in Basilica di Santa Croce, Florence, ordered by date of death**

| Name | † | Birth | Type | Placement | Map Fig. 5.2 | Fig. |
|---|---|---|---|---|---|---|
| Verrocchio, Guliano | 1442 | 1400 | Tomb slab | c. 1442 | I | 5.7 |
| Bruni, Leonardo | 1444 | 1370 | Memorial tomb | 1446–1450 | II | 5.5 |
| Galilei, Galileo di Giovanni | 1447/ 1451 | 1369/ 1370 | Tomb slab | c. 1447–1451 | III | 5.8 |
| Marsuppini, Carlo | 1453 | 1399 | Memorial tomb | 1445–1456 | IV | 5.6 |
| Sansone, Francesco | 1499 | 1414 | Tomb slab | c. 1499 | V | |
| Michelangelo | 1564 | 1475 | Memorial tomb | 1568 | VI | 5.9 |
| Buonarroti, Francesco | 1632 | 1574 | Hanging mon. | c. 1632 | VII | 5.10 |
| Galilei, Galileo | 1642 | 1564 | Memorial tomb | 1737 | VIII | 5.11 |
| Giugni, Vincenzo Maria | 1718 | 1758 | Hanging mon. with bust | 1758 | IX | |
| Buonarroti, Filippo | 1733 | 1661 | Hanging mon. with bust | c. 1733 | VII | 5.10 |
| Galileo, Alessandro | 1737 | 1691 | Hanging mon. with bust | c. 1737 | X | |
| Micheli, Pier Antonio | 1737 | 1679 | Hanging mon. with bust | 1737 | XI | |
| Cocchi, Antonio | 1758 | 1695 | Hanging mon. with bust | 1773 | X | |
| Lami, Giovanni | 1770 | 1697 | Memorial tomb | 1775 | XII | 5.13 |
| Tavanti, Angelo | 1782 | 1714 | Memorial tomb | c. 1782 | XIII | |

### 5.3.1 Learned Poets and Orators as the Pinnacle of Virtue, c. 1400–1564

The title of poet laureate continued to be the highest form of praise for a Florentine and the desire to publicly display and honour great Florentine men as poets dominated the memory culture. As a result, some chancellors of the Republic were framed and memorised as learned poets, rather than as statesmen. Coluccio Salutati (1331–1406), the first chancellor of the Republic, set a precedent. He received a Latin funeral oration, and the Florentine state wanted to pay for his sepulchre, although this seems never to have materialised.[71] Poggio Bracciolini (1380–1459) wrote an epitaph to adorn Salutati's tomb that praised his qualities as a learned man. Salutati was

71 Strocchia, *Death and Ritual*, pp. 106–120; Oy-Marra, *Florentiner Ehrengrabmäler*, pp. 76–79.

'like a common father to learned men, [and] it was mostly thanks to him that Greek learning first came to Florence.'[72] During his funeral, he was crowned with a laurel crown: 'After his death he was presented with a laurel wreath by public decree as a sign of the learning and the virtues in which he excelled.'[73] After all, he was 'the most eloquent man of his age', and 'who left behind many monuments of his innate brilliance.'[74] The contribution of Salutati to the politics and common good of the Florentine state was honoured by remembering him as the most eloquent laureate orator of his time. He received extraordinary honour and became a point of reference for how to adequately honour a chancellor.

The continuation of this tradition can clearly be seen in grave monuments of Leonardo Bruni (1370–1444) and Carlo Marsuppini (1399–1453) in the Basilica di Santa Croce. These two succeeding chancellors of the Florentine state were not praised and remembered because of their piety, or even their statesmanship, but rather on account of their scholarly and public merit. Bruni was the first person to be buried in the Santa Croce not because of his religious virtues or noble descent, but as a statesman and man of learning. This started a tradition of sepulchres honouring those who had served the Florentine public good particularly well. Bruni and Marsuppini were not remembered the context of family, nobility, or even religious virtues, but rather in the context of their acceptance by the Florentine people as learned men and poets.

Leonardo Bruni was a pupil of Coluccio Salutati, whom he succeeded as the chancellor of Florence, the highest public office. He served as chancellor

72 'Doctorum virorum quasi com[m]unis parens, huius pr[a]ecipuo opere / gr[a]ec[a]e litter[a]e primum Florentiam commigrarunt.'

73 'Post obitum corona laurea donatus est iussu populi in doctrin[a]e virtutumque quibus excelluit insigne.'

74 'Vir fuit [a]etatis su[a]e optimus ac eloquentissimus, qui sui ingenii multi reliquit monumenta'. The full inscription Poggio wrote reads: 'Sepulchrum Colucii Pieri Salutati / Hic opido Stignani bonis parentibus ortus cum ab ipsa adolescentia / eloquenti[a]e et bonarum artium studiis operam dedisset / cancellarius florentinus factus est. Quod officium XL / ferme annos summa cum integritate ac laude administravit. / Doctorum virorum quasi com[m]unis parens, huius precipuo opere / gr[a]ec[a]e litter[a]e primum Florentiam commigrarunt, quibus rebus om- / nium civium benevolentiam est consecutus. LXXV [a]etatis / anno excessit e vita, summo civitatis merore. Post obitum / corona laurea donatus est iussu populi in doctrin[a]e vir- / tutumque quibus excelluit insigne. Vir fuit [a]etatis su[a]e / optimus ac eloquentissimus, qui sui ingenii multa re- / liquit monumenta laude et gloria digna ad memoriam / posteritatis.' Florence, BCNF, ms. Magliabechiano VIII. 1445, fol. 207$^{v}$; cited in, and translation based on, Stefano U. Baldassarri, 'Poggio Bracciolini and Coluccio Salutati: The Epitaph and the 1405–1406 Letters', in *Poggio Bracciolini and the Re(dis)covery of Antiquity: Textual and Material Traditions*, ed. Roberta Ricci, 71–87 (Florence: Firenze University Press, 2020), pp. 73, 85.

Figure 5.5. Bernardo Rossellino, Sepulchre of Leonardo Bruni, 1446. Basilica di Santa Croce, Florence. The monument is an asset of the Fondo Edifici di Colto administered by the Direzione Centrale degli Affari dei Culti and the Ammistrazione del Fondo Edifici di Colto del Ministero dell'Interno. Photograph by author, courtesy of Santa Croce Opera. Marked as I in Figure 5.2

between 1410–1411, and again from 1427 until his death in 1444.[75] Bruni is often seen as a central figure in the development of the strong relation between humanistic scholarship and politics in Florence.[76] As a scholar, he was well-known for his Latin translations of Greek works, such as Aristotle's *Politics* and *Nicomachean Ethics*. Additionally, Bruni was an influential historian, who wrote an extensive history of Florence and a biography of Cicero. Bruni wielded much political power in the city of Florence, which makes it striking that the memory his monument in the Santa Croce conveys only alludes to his qualities as an erudite man.

Bruni's marble sepulchre is placed near the altar, on the southern wall in the central nave, just before the onset of the southern nave (see Figure 5.2, marker I). The sepulchre is of considerable dimensions and depicts Leonardo Bruni in a supine position—commonly called a *gisant* position—on top of a casket (see Figure 5.5). Bruni's head is facing the entrance of the church. He wears a laureate crown, just as his predecessor, and holds a book close to his chest. The top of the sepulchre is ornamented with a depiction of Madonna and child. The marble casket is decorated with a prose epitaph that invites the onlooker to remember Leonard Bruni as a learned man:

> After Leonardo has departed from life,
> *Historia* mourns, *Eloquence* is mute.
> It is said that the Greek,
> as well as the Latin muses, could not hold their tears.
>
> Postquam Leonardus e vita migravit
> Historia luget Eloquentia muta est
> Ferturque musas tum graecas tum
> Latinas lacrimas tenere non potuisse

The prose epitaph, written by his successor Carlo Marsuppini, equated the loss of Bruni's life to a loss of eloquence and learning.[77] This point is

75 Cesare Vasoli, 'Bruni, Leonardo, detto Leonardo Aretino', *DBI*, vol. 14 (1972), pp. 618–633; Lauro Martines, *The Social World of the Florentine Humanists, 1390–1460* (Toronto; Buffalo, NY; London: University of Toronto Press, 2011), pp. 117–123, 165–175.

76 John R. Najemy, 'Civic Humanism and Florentine Politics', in *Renaissance Civic Humanism: Reappraisals and Reflections*, ed. James Hankins, 75–104 (Cambridge: Cambridge University Press, 2000); Ronald G. Witt, *'In the Footsteps of the Ancients': The Origins of Humanism from Lovato to Bruni* (Leiden; Boston: Brill, 2000), pp. 392–442; and for the foundational monograph on civic humanism, see Baron, *Crisis of the Early Italian Renaissance*.

77 For a more detailed analysis of the intertextual references in the epitaph, see Victor Schmidt, 'A Humanist's Life Summarized: Leonardo Bruni's Epitaph', *Humanistica Lovaniensia* 47 (1998): 1–14.

reinforced by the iconographical design of the monument. Bruni's marble likeness clenches a book, probably his *Historiarum Florentini populi liber* (History of the Florentine people) to his chest. The display of books on tombs symbolised erudition. For example, on the tombs of professors from the university of Padua in the fourteenth and fifteenth centuries, the book was always present to represent the professor's high standing in the learned world.[78] In the first volume of the *Historiae*, Bruni provided the Florentine state with a story of its genesis. In later volumes, he introduced the Medici as important historical actors in the formation of the Florentine Republic.[79] Bruni was thus central in establishing a civic identity for Florence and bestowed an honourable collective memory on the city. The ideal of the city of Florence as a place where politics and scholarship thrive can also be found in his famous panegyric *Laudatio Florentinae urbis* (Praise of the City of Florence).[80] Additionally, in his *Life of Dante* Bruni praised Dante as a virtuous man who was dutiful to the Republic of Florence; as such, Bruni helped to sustain the image of the poet as a virtuous citizen and political actor.[81] The sepulchre refers to Bruni's status as a learned man with the depiction of the book, as well as the Florentine tradition of memorialising exceptional statesmen as laureate poets. The tomb also praises Bruni as a learned man, whose lasting achievements were his works of history and erudition. Notwithstanding the fact that as a chancellor, Bruni was a man of political power, honour and fame were awarded on account of his scholarly and literary work. This is the first example of how the Florentine humanist elite cultivated the amalgamation of the profession of statesmanship and learning. This elevated the position of the learned man as one of the loftiest positions in Florentine society and its memory culture.

A few years later, in 1453, the tomb of Carlo Marsuppini was constructed (see Figure 5.6; for the position see Figure 5.2, marker II).[82] Marsuppini

78 Carrington, 'Sculpted Tombs', pp. 184–186.

79 Gary Ianziti, 'Leonardo Bruni, the Medici, and the Florentine Histories', *Journal of the History of Ideas* 69, no. 1 (2008): 1–22.

80 Leonardo Bruni, *Laudatio Florentinae urbis*, ed. Stefano U. Baldassarri (Florence: SISMEL edizioni del Galluzzo, 2000); or in Italian, see Leonardo Bruni, *Panegirico della città di Firenze: testo italiano a fronte di frate Lazaro da Padova*, ed. Giuseppe de Toffol (Florence: Nuova Italia Editrice, 1974).

81 On Bruni's *Life of Dante*, see Najemy, 'Civic humanism and Florentine politics', pp. 90–93. For the role of Bruni's *Historiae* in political thought, see James Hankins, *Virtue Politics: Soulcraft and Statecraft in Renaissance Italy* (Cambridge, MA: Harvard University Press, 2019), pp. 218–237.

82 For general information on the Marsuppini tomb, see Antonio Natali, 'I sepolcri di Leonardo Bruni e Carlo Marsuppini', in Berti, *Il pantheon di Santa Croce*, pp. 17–55. For the restoration of

came from a noble family in Arezzo and became the first secretary of the Florentine chancellery, from 1444 until his death in 1453, as the successor of Leonardo Bruni.[83] A decade earlier in 1430, he had been appointed at the Studio Fiorentino as the professor of poetry, rhetoric, and Greek, a position he held for the rest of his life. Marsuppini was the private tutor of the children of Cosimo de' Medici (1389–1464) and a well-respected scholar. He worked extensively on the translation of Greek authors, especially the works of Homer, and he was a prolific Latin poet.[84]

There are striking similarities between the sepulchres of councillors Bruni and Marsuppini that reaffirmed Marsuppini's position as successor of Bruni. Not only are they placed symmetrically, facing each other in the southern and northern naves, respectively (see Figure 5.2, markers II and IV), but even their construction, iconography, and epitaphs are strikingly similar. Bruni and Marsuppini are directly facing each other in the same *gisant* position, both with their feet pointing towards the main altar. Each effigy is positioned in a supine funerary position on decorated drapery, on a bed atop the sarcophagus. Their heads rest on pillows and are tilted to face onlookers, and consequently each other. Furthermore, both sepulchres feature a Madonna and child enclosed by a rounded upper arch. The two marble men both have their hands wrapped around a book. Marsuppini's epitaph also alludes to his qualities as a learned man:

> Stand and look at this marble preserving a great poet. Go now, for whose *ingenium* the world was not enough. Carlo, the great glory of his age, knew everything that nature, the heavens, and custom have to tell. Ausonian and Greek muses, now hang loose your hair. Alas, the fame and honour of your choir [of muses] has fallen.
>
> Siste vides magnum quae servant marmora vatem
> I ingenio cuius non satis orbis erat
> Quae natura Polus quae mos ferat omnia novit
> Karolus aetatis gloria magna suae

the tomb, see Christopher Weeks, 'The Restoration of Desiderio Da Settignano's Tomb of Carlo Marsuppini in S. Croce, Florence', *The Burlington Magazine* 141, no. 1161 (1999): 732–738.

83 Paolo Viti, 'Marsuppini, Carlo', *DBI*, vol. 71 (2008), pp. 20–21; Martines, *Social World*, pp. 127–130.

84 For the Latin poetry of Marsuppini, see Ilaria Pierini, ed., *Carlo Marsuppini. Carmini Latini: Edizione critica, traduzione e commento* (Florence: Firenze University Press, 2014). For the translations of Homer, see Alessandra Rocco, *Carlo Marsuppini traduttore d'Omero: la prima traduzione umanistica in versi dell'Iliade* (*primo e nono libro*) (Padua: Il poligrafo, 2000).

Figure 5.6. Desiderio da Settignano, Sepulchre in honour of Carlo Marsuppini, 1455–1456. Basilica di Santa Croce, Florence. The monument is an asset of the Fondo Edifici di Colto administered by the Direzione Centrale degli Affari dei Culti and the Ammistrazione del Fondo Edifici di Colto del Ministero dell'Interno. Photograph by author, courtesy of Santa Croce Opera. Marked as II in Figure 5.2

Ausoniae 7Graiae[85] crines nunc solvite musae
occidit heu vestri fama decusque chori

Marsuppini wrote the epitaph for Bruni and it therefore stands to reason that the successor of Marsuppini, Poggio Bracciolini (chancellor in 1453–1459), wrote his epitaph.[86] The epitaph's language is complex and dense, showcasing many allusions to Marsuppini's standing as a scholar and poet. Marsuppini is referred to as a 'vates', which means both poet and prophet and is a direct reference to illustrious Roman poets, such as Virgil and Ovid. The Roman and Greek muses ('Ausoniae et Graiae musae') who are ordered to let go of Marsuppini refer to his achievements as a Latin and Greek author. The epitaph is written in dactylic hexameters alternating with dactylic pentameters, forming elegiac distichs that were typical of elegies and lyrical poetry. While Marsuppini's epitaph is much more lyrical compared to Bruni's prose text (in line with Marsuppini's reputation as a poet and Bruni as an orator), they both imply a strong affinity with Greek and Latin literature and mention the dedicatee's innate brilliance and learning. The muses figure in both epitaphs, too: the muses could not restrain their tears in the case of Bruni and are reluctant to let Marsuppini go.

The spatial position of the monuments of Bruni and Marsuppini in the church adds considerably to their significance. The two grave monuments were originally behind the *tramezzo* (see Figure 5.1). The will of the politician Francesco Bruni (1315–c. 1385) gives an indication of the role of the *tramezzo*, since he asked to be buried 'facing the portal, close to the choir, which divides the space for men from the space for women' ('iuxta portam que est propinqua coro et dividit locum hominum a loco mulierum').[87] As was the

85 '7Graiae' is taken to mean 'et Graiae' where the 7 is interpreted as an ampersand; see Adriano Cappelli, *Lexicon Abbreviaturarum: Wörterbuch lateinischer und italienischer Abkürzungen wie sie in Urkunden und Handschriften besonders des Mittelalters gebräuchlich sind, dargestellt in über 14000 Holzschnittzeichen*, 2nd ed. (Leipzig: Verslagsbuchhandlung von J. J. Weber, 1928), pp. xxi, xxx, and 408. The term *Ausonia* is sometimes used as a general descriptor for the region surrounding Rome or the Italian peninsula: see, e.g., Virgil referring to 'Ausonia terra' in *Aeneid*, IV, l. 349; VI, l. 807; and VII, l. 55. The phrase 'Ausoniae et Graiae musae' appears again in the second half of the fifteenth century in Guarino Veronese, *Carmina*, ed. Aldo Manetti (Bergamo: Istituto universitario di Bergamo, 1985), 3, l. 4.

86 Paul Shaw, 'Poggio Bracciolini, an Inscription in Terranuova, and the Monument to Carlo Marsuppini: A Theory', in *Poggio Bracciolini and the Re(dis)covery of Antiquity*, ed. Ricci, pp. 149–162.

87 ASFi, Diplomatico, Santa Croce, 1386; cited in Moisè, p. 108, n. 2, and subsequently in Ena Giurescu Heller, 'Access to Salvation: The Place (and Space) of Women Patrons in Fourteenth-Century Florence', in *Women's Space: Patronage, Place, and Gender in the Medieval Church*, ed.

case in many Franciscan churches, the choir in Santa Croce was a closed-off space, mostly reserved for clergy, friars, men of good standing, as well as patrons who leased a family chapel.[88] Bruni and Marsuppini are placed close to the choir, which further stresses their perceived importance, since a burial in the choir was considered to be an exceptional honour, often exclusive for eminent clergy. The reference to the choir of muses in Marsuppini's epitaph can also be seen as an implicit reference to the honourable burial place: 'The fame and honour of your choir' ('vestri fama decus que chori').[89] Additionally, the picture of Dante, the archetypical, Florentine poet laureate, graced the wall behind the *tramezzo*, 'in the middle of the church'.[90] From the end of the fourteenth century onwards, the laureate statesman became a vivid example and symbol of outstanding virtue and honour.

The sepulchres of Bruni and Marsuppini signal a break with tradition. It was unprecedented to honour orators and poets in the holy confines of the choir alongside ecclesiasts. Earlier memorial tombstones behind the *tramezzo* were dedicated to important Franciscan theologians and ecclesiasts. Among them is, for example, the tombstone of Francesco Sansone (1414–1499), an influential Franciscan minister, learned friar and patron of the arts.[91] His tombstone was cast in bronze and states that Sansone was of 'virtuous life and religion' ('probita vita et religio'). His tomb was placed there on behalf of the Franciscan order to honour his virtuous and pious

Virginia Chieffo Raguin and Sarah Stanbury, 161–184 (Albany, NY: State University of New York Press, 2005), p. 176.

88 Donal Cooper, 'Experiencing Dominican and Franciscan Churches in Renaissance Italy', in *Sanctity Pictured: The Art of the Dominican and Franciscan Orders in Renaissance Italy*, ed. Trinita Kennedy, 47–61 (Nashville; London: Frist Center for the Visual Arts and Philip Wilson Publishers, 2014), pp. 57–59.

89 This term 'choir' gains additional meaning in the context of the existence of the rood screen. Cf. the translation in John Pope-Hennessy, *Italian Renaissance Sculpture* (London: Phaidon, 1996), p. 305, which renders the last line of the epitaph as 'Alas, the fame and splendour of your choir is dead'. The reference to the choir of muses that has fallen can be construed as a subtle reference to the placement of the tomb facing the choir of the church.

90 This reference comes from *Libro di Antonio Billi*, 1487–1537: 'Taddeo Gaddi, ilquale dipinse nella Mercantia di Firenze sopra il banco, nelquale luogo disse essere discepolo di Giotto il gran' maestro, dipinse nella chiesa di Santa Croce circa al mezo la chiesa il miraculo del fanciullo resuscitato. Dove è la figura di Dante Alinghieri, dove sono tre figure al naturale insieme; et la sua è quella del mezo.' In Antonio Billi, *Il libro di Antonio Billi esistente in due copie nella Bibliotheca Nazionale di Firenze*, ed. Carl Frey (Berlin: G. Grote'sche Verlagsbuchhandlung, 1892), p. 8.

91 Giovanna Baldissin Molli, 'Francesco Sansone', *DBI* 90 (2017), pp. 325–327. For an extensive study of Sansone's influence, see Giovanna Baldissin Molli, ed., *Frate Francesco Sansone 'de Brixi' Ministro generale OFMConv (1414–1499): Un mecenate francescano del Rinascimento* (Padua: Centro Studi Antoniani, 2000).

Figure 5.7. Unknown sculptor, Tomb slab and grave of Guliano Verrocchio, 1442. Basilica di Santa Croce, Florence. The monument is an asset of the Fondo Edifici di Colto administered by the Direzione Centrale degli Affari dei Culti and the Ammistrazione del Fondo Edifici di Colto del Ministero dell'Interno. Photograph by author, courtesy of Santa Croce Opera. The full inscription reads: 'F[rater] Julianus Verrocchius theologus insignis hic sepul[-] • tus est F[rater] Antonius de Medicis eiusdem ordinis theol[ogus] • et Etruriae provinciae minister ne iaceret incultus • qui aetatis suae fuerat decor hoc monumentum • donavit vix[it] a[nnos] XXXXII men[ses] VI dies X obiit MCCCCXXXXII'

life. In a similar vein, friar Antonio de' Medici ordered a marble tombstone for the now largely forgotten theologian and friar Guliano Verrocchio (1400–1442), praising his role as a theologian at the convent and university (see Figure 5.7).[92] The likeness of Verrocchio on the tomb slab reveals he was depicted holding a book, much like Bruni and Marsuppini, as well as the marble renditions of erudite professors in Padua and Bologna. The book was an important iconographical symbol of learning. Verrocchio was honoured on account of his formal role and his connection to the Franciscan order, without any resort to pompous virtues, honour, or fame. The grand sepulchres of Bruni and Marsuppini therefore represent a departure from this tradition, since they have no connection with the Franciscan order, but were learned men who received honours because of their statesmanship in service of the Florentine state. The Florentine humanist ruling elite had found a unique way to honour their chancellors: as erudite, public orators, and laureate poets.

The meaning of the sepulchres of Bruni and Marsuppini changed in the sixteenth century due to the renovations to the church ordered by Giorgio Vasari. The *tramezzo* was removed and the Basilica became an open space that emphasised the imposing memorial tombs on the northern and southern wall of the church, including those of Bruni and Marsuppini. From Vasari's point of view, Bruni, and Marsuppini should be seen together as a comparison and analogy, something Vasari called a *paragone* in his life stories, the *Vite*.[93] The symmetrical position of these sepulchres in the church invited an early modern onlooker to assess the virtue and honour of these men in relation to each other. Bruni and Marsuppini's symmetrical tombs reinforced this sense of comparison (*paragone*) and succession: Bruni represented the art of prose and Marsuppini the art of poetry. The symmetrical positioning of sepulchres in opposing naves would become an example for the final resting places of exceptionally honourable men. From the end of the sixteenth century onwards, the Santa Croce became a growing site of learned memory, where

92 For the attribution of the tombstone to Andrea del Verrocchio and the relation between Guliano Verrocchio and Antonio de' Medici, see Dario A. Covi, 'An Unnoticed Verrocchio?', *The Burlington Magazine* 110, no. 778 (1968): 4–9, 11.

93 For the theory of *paragone* in sixteenth-century art, see, e.g., Alessandro Nova, 'Paragone-Debatte und gemalte Theorie in der Zeit Cellinis', in *Benvenuto Cellini: Kunst und Kunsttheorie im 16. Jahrhundert*, ed. Nova and Anna Schreurs, 183–202 (Cologne: Böhlau, 2003); for a consideration of Leonardo da Vinci's *Paragone*, see Monica Azzolini, 'In Praise of Art: Text and Context of Leonardo's *Paragone* and Its Critique of the Arts and Sciences', *Renaissance Studies* 19, no. 4 (2005): 487–510; for the importance of *paragone* in the *Icones* genre, see Dorigen Caldwell, 'The *Paragone* Between Word and Image in *Impresa* Literature', *Journal of the Warburg and Courtauld Institutes* 63 (2000): 277–286.

visitors could experience a grand open church design, with sepulchres that offered a vivid overview of Florentine scholarly virtue.

One tomb slab from the second half of the fifteenth century shows the continuation of the tradition of honouring great men as both statesmen and learned men. The tombstone in the middle of church (close to the main entrance) of Galileo di Giovanni Galilei (1369/1370–1447/1451), a forefather of the well-known astronomer, illustrates this dual honour both on account of civil service and learning (see Figure 5.8). Galilei was part of a higher governmental council, the so-called council of Twelve Good Men (*Dodici Buonomini*) as well as a scholar and teacher.[94] The inscriptions reads:

> Galileo Galilei, once a Bonaiuti, was the acme and master of philosophy and medicine during his time, who also in the highest magisterial position, loved the public good [*res publica*] in a wonderful way; his son, Benedictus, in sacred memory of him and his well-lived pious life, put forth this tomb to his father, himself, and their descendants.

> Temporibus hic suis phylosophye · Atq[ue] medicine culmen fuit et magister · Galileus DeGalileis olim Bonaiutis qui · Etiam summo in magistratu miro · Quodam modo rem publicam dilexit · Cuius sancte memorie bene acte · Vite pie Benedictus filius hunc tumu[-] · lum patri sibi suis q[ue] posteris edidit

The inscription dedicated to Galileo di Giovanni Galilei refers to the fact that he served the public good (*res publica*) well, both in his capacity as teacher and magistrate. The tradition of burying teachers, scholars, and statesman in the Santa Croce thus extended beyond the pompous sepulchres for Bruni and Marsuppini. For statesmen and scholars, the highest honour was to be buried with a sumptuous sepulchre. Second were the theologians and clergy who were buried behind the *tramezzo* near the choir, and finally the teachers such as Galileo Galilei who were buried in the church itself. Like Verrocchio, Galilei is depicted on his tomb as an erudite man, holding a book against his chest with both hands (see Figure 5.8).

The dedication of elaborate funeral monuments to secretaries Bruni and Marsuppini as learned men is unique and points to the specific Florentine memory culture, in which high magistrates were honoured as scholars. In comparison, the tombstone of Lorenzo Valla (1407–1457), humanist scholar and once a student of Bruni, which was once located in the church of St John

94 Francesco Bausi, 'Galilei, Galileo', *DBI* 51 (1998), pp. 471–473.

Figure 5.8. Unknown sculptor, tomb slab and grave of Galileo di Giovanni Galilei (1369/1370–1447/1451), 1447–1451. Basilica di Santa Croce, Florence. The monument is an asset of the Fondo Edifici di Colto administered by the Direzione Centrale degli Affari dei Culti and the Ammistrazione del Fondo Edifici di Colto del Ministero dell'Interno. Photograph by author, courtesy of Santa Croce Opera

Lateran in Rome, is much humbler.[95] The tombstone does not have a poetic Latin inscription (Valla was not known as a poet), bears no reference to muses, nor is there an allusion to a succession. Valla's tombstone was taken care of by his mother, rather than by a successor or with public funds. The inscription on the tombstone is dedicated to Valla, an 'apostolic writer, who during his life exceeded all others in eloquence' ('Apostolicoque scriptori, qui sua aetate omneis eloquentia superavit').[96] The Florentine memory culture of the laureate orators and poets Bruni and Marsuppini—in the tradition of Dante, Boccaccio, and Petrarch—had a lasting impact. Both statesmen and scholars received an exceptional honour in Basilica di Santa Croce, and theirs was the honour of the choir. The tomb slabs dedicated to Verrocchio and Galilei likewise reflect the high status of the learned man in Florentine society in the fifteenth century.

### 5.3.2 From Laureate Statesmen to Gifted Geniuses, 1568–1737

From the second half of the sixteenth century onwards, the memory culture in the Basilica di Santa Croce changed in terms of the representation of learned men, as well as the memory brokers who wanted to honour those men with monuments. As said, Vasari's redesign of 1568 transformed the church into one big, open space. He is best known as the author of the influential history of artists, his *Vite*, but as a painter, he created portraits that conveyed a sense of literary and intellectual history. As mentioned before, in his 1544 group portrait of Tuscan poets (see Figure 5.4), Vasari used his art to portray intellectual history and revive the Florentine memory cult around the laureate poets.[97] Vasari redesigned the Basilica di Santa Croce to reflect his ideal of a site of memory: a symmetrical hall of illustrious men who were gifted with innate brilliance.

Shortly after the redesign, Vasari designed a monument in honour of Michelangelo (1475–1564). This monument is both a testament to

95 For a detailed history of the Valla tombstone, see Jan L. de Jong, 'De Sepulcro Laurentii Vallae quid veri habeat: Tracing the Tomb Monument of Lorenzo Valla in St. John Lateran, Rome', *Quellen und Forschungen aus italienischen Archiven und Bibliotheken* 94, no. 1 (2015): 94–128.

96 Lorenzo Valla, *Opera, nunc primo* [...] *in unum volumen collecta, & exemplaribus variis collatis, emendata* [...] (Basel: Henrich Petri, 1540), epitaph on the frontispiece; cited in and English translation from De Jong, 'De Sepulcro Laurentii Vallae', p. 97. The full transcription of the inscription on the frontispiece of Valla's *Opera* reads: 'Laurentio Vallae harum aedium sacrarum canonico, Alphonsi regis et Pontificis Maximi secretario, Apostolicoque scriptori, qui sua aetate omneis eloquentia superavit, Caterina mater filio pientissimo posuit. Vixit annos L. Ob[iit] An[no] Domini M.CCCC.LXV. [1445] Calendis Augusti.'

97 Parker, 'Vasari's *Portrait of Six Tuscan Poets*'.

Figure 5.8. Unknown sculptor, tomb slab and grave of Galileo di Giovanni Galilei (1369/1370–1447/1451), 1447–1451. Basilica di Santa Croce, Florence. The monument is an asset of the Fondo Edifici di Colto administered by the Direzione Centrale degli Affari dei Culti and the Ammistrazione del Fondo Edifici di Colto del Ministero dell'Interno. Photograph by author, courtesy of Santa Croce Opera

Lateran in Rome, is much humbler.[95] The tombstone does not have a poetic Latin inscription (Valla was not known as a poet), bears no reference to muses, nor is there an allusion to a succession. Valla's tombstone was taken care of by his mother, rather than by a successor or with public funds. The inscription on the tombstone is dedicated to Valla, an 'apostolic writer, who during his life exceeded all others in eloquence' ('Apostolicoque scriptori, qui sua aetate omneis eloquentia superavit').[96] The Florentine memory culture of the laureate orators and poets Bruni and Marsuppini—in the tradition of Dante, Boccaccio, and Petrarch—had a lasting impact. Both statesmen and scholars received an exceptional honour in Basilica di Santa Croce, and theirs was the honour of the choir. The tomb slabs dedicated to Verrocchio and Galilei likewise reflect the high status of the learned man in Florentine society in the fifteenth century.

### 5.3.2 From Laureate Statesmen to Gifted Geniuses, 1568–1737

From the second half of the sixteenth century onwards, the memory culture in the Basilica di Santa Croce changed in terms of the representation of learned men, as well as the memory brokers who wanted to honour those men with monuments. As said, Vasari's redesign of 1568 transformed the church into one big, open space. He is best known as the author of the influential history of artists, his *Vite*, but as a painter, he created portraits that conveyed a sense of literary and intellectual history. As mentioned before, in his 1544 group portrait of Tuscan poets (see Figure 5.4), Vasari used his art to portray intellectual history and revive the Florentine memory cult around the laureate poets.[97] Vasari redesigned the Basilica di Santa Croce to reflect his ideal of a site of memory: a symmetrical hall of illustrious men who were gifted with innate brilliance.

Shortly after the redesign, Vasari designed a monument in honour of Michelangelo (1475–1564). This monument is both a testament to

95 For a detailed history of the Valla tombstone, see Jan L. de Jong, 'De Sepulcro Laurentii Vallae quid veri habeat: Tracing the Tomb Monument of Lorenzo Valla in St. John Lateran, Rome', *Quellen und Forschungen aus italienischen Archiven und Bibliotheken* 94, no. 1 (2015): 94–128.

96 Lorenzo Valla, *Opera, nunc primo* [...] *in unum volumen collecta, & exemplaribus variis collatis, emendata* [...] (Basel: Henrich Petri, 1540), epitaph on the frontispiece; cited in and English translation from De Jong, 'De Sepulcro Laurentii Vallae', p. 97. The full transcription of the inscription on the frontispiece of Valla's *Opera* reads: 'Laurentio Vallae harum aedium sacrarum canonico, Alphonsi regis et Pontificis Maximi secretario, Apostolicoque scriptori, qui sua aetate omneis eloquentia superavit, Caterina mater filio pientissimo posuit. Vixit annos L. Ob[iit] An[no] Domini M.CCCC.LXV. [1445] Calendis Augusti.'

97 Parker, 'Vasari's *Portrait of Six Tuscan Poets*'.

Michelangelo's perceived status as a legendary *Florentine* artist, as well as to Vasari's personal devotion and admiration for the man and his art. Vasari dubbed Michelangelo the master and founding father of art.[98] In his *Vite*, he praised Michelangelo in embellished prose as a divine artist and a perfect culmination of all the arts:

> But the man who wins the palm among artists both living and dead, who transcends and surpasses them all, is the divine Michelangelo Buonarroti, who reigns supreme not merely in one of these arts but in all three [painting, sculpture, and architecture] at once. This man surpasses and triumphs over not only all those artists who have almost surpassed Nature but even those most celebrated ancient artists themselves, who beyond all doubt surpassed Nature: and alone he has triumphed over ancient artists, modern artists, and even Nature herself, without ever imagining anything so strange or so difficult that he could not surpass it by far with the power of his most divine genius through his diligence, sense of design, artistry, judgement, and grace.[99]

Vasari's life of Michelangelo emanates civic pride as well as an appeal to Michelangelo's divine achievements in the arts, especially sculpture, painting, and architecture. The praise of Michelangelo was a panegyric on Florence too, 'the most worthy among all other cities'. Vasari claimed Michelangelo gave his genius to the city, 'so that the perfection Florence justly achieved with all her talents might finally reach its culmination in one of her own citizens.'[100] The tomb monument in the Santa Croce acknowledged the divinity of Michelangelo, as well as stressing his relation to 'his native city, Florence'.[101] Michelangelo, after all, embodied the perfection of Florence. This widely published life reflects how Michelangelo was remembered in the Santa Croce.

Michelangelo's tomb monument is certainly not a sepulchre dedicated to a scholar. Yet the monument fundamentally changed the context in which learned men would be remembered in Santa Croce in the seventeenth century. The monument of Michelangelo started a tradition of seeing artists, as well as scholars and learned men, as gifted geniuses. Self-proclaimed

98 Alessandro Cecchi, 'L'estremo omaggio al "padre e maestro di tutte le arti": Il monumento funebre di Michelangelo', in Berti, *Il pantheon di Santa Croce*, pp. 57–82.

99 Giorgio Vasari, *The Lives of the Artists*, trans. with introduction and notes by Julia Conaway Bondanella and Peter Bondanella (Oxford; New York: Oxford University Press, 1991), pp. 281–282.

100 Ibid., p. 415.

101 Ibid.

successors and family members of, for example, the Buonarroti family would seek to associate themselves with their illustrious masters. Three marble statues on the sepulchre represent the key divine arts Michelangelo mastered: the muses of sculpture, painting, and architecture. The epitaph has a matter-of-fact tone and conveys little humanistic flair:

> For Michelangelo Buonarroti, from the old Simonian family, sculptor, painter, and architect, widely known to everyone because of his fame. Leonardo, for his most beloved and well-deserved uncle, after moving his bones from Rome to this greater temple, [and] placing them in the sepulchre of his [kinsmen], encouraged by the most serene Cosimo de' Medici, Grand Duke of Tuscany, protector of the city, in the year 1570. He lived 88 years, 11 months, and 15 days.

> Michaeli Angelo Bonarotio
> E vetusta Simoniorum familia
> Sculptori pictori et architecto
> Fama omnibus notissimo
> Leonardus patruo amantiss[imo] et de se optime merito
> Translatis Roma eius ossibus atque in hoc templo maior[e]
> Suor[um] sepulcro conditis cohortante sereniss[imo] Cos[i]mo Med[icis]
> Magno Hetruriae Duce P[atrono ] C[ivitatis]
> Ann[o] Sal[utis] M.D.LXX.
> Vixit Ann[os] LXXXVIII M[enses] XI D[ies] XV

Michelangelo's tomb is an example of virtues and characteristics deemed important for great men in the sixteenth century. First, descent and genealogy: that Michelangelo is named here both a Buonarroti and stemming from an 'old Simonian family' attests to this sensitivity. Secondly, the humanistic fixation on *fama*, as we have seen on Bruni's and Marsuppini's sepulchres, has been made explicit on this sepulchre. Thirdly, while the benefactors of tombs were usually not mentioned, Michelangelo's tomb makes them explicit: Leonardo Buonarroti and Cosimo I de' Medici (1519–1574), Grand Duke of Tuscany. That is how they associated themselves with Michelangelo's posthumous *fama*.

In the following centuries, several illustrious men from the Buonarroti family would eagerly associate themselves with the grandeur the name carried. The house of Michelangelo, which he left to his nephew Leonardo who took care of the sepulchre, was turned into a museum by Michelangelo Buonarroti the Younger (1568–1646). He dedicated this museum, the Casa Buonarroti, to his granduncle and perpetuated the name and fame of the

Figure 5.9. Giorgio Vasari, Sepulchre of Michelangelo, 1568. Basilica di Santa Croce, Florence. The monument is an asset of the Fondo Edifici di Colto administered by the Direzione Centrale degli Affari dei Culti and the Ammistrazione del Fondo Edifici di Colto del Ministero dell'Interno. Photograph by author, courtesy of Santa Croce Opera

Buonarroti family. When he died, his body was placed close to the tomb monument of the 'divine' Michelangelo in the Basilica di Santa Croce.[102] The Vasari design elevated just a few tomb monuments, one behind every

102 Cecchi, 'L'estremo omaggio al "padre e maestro di tutte le arti"'.

grand pillar. Behind every grand pillar, family and friends who wanted to associate themselves with the illustrious deceased would be buried here too.

The epitaphs of two additional Buonarroti family members, Francesco (1574–1632) and Filippo Buonarroti (1661–1733), were strategically placed next to Michelangelo's grand tomb. Francesco Buonarroti was an architect and a knight of the Order of St. John.[103] His epitaph praised him as the secretary of Antoine de Paule (1551–1636), Grand Master of the order, extolling his faith, his candour of spirit, and his outstanding knowledge of the accomplishments and deeds of his order.[104] Francesco was no artist, but rather showed excellent learning. All these modes of remembrance are familiar and to the point. Finally, however, the epitaph connected Francesco to the famous Buonarroti family name: Michelangelo Buonarroti the Younger placed this 'so that for his beloved brother there would be a place near the bones or near the name of his ancestors'. To be buried in a family grave or close to one's family was a normal affair for noble families in the Basilica di Santa Croce throughout the early modern period. However, to be placed next to an honourable funerary monument and associate oneself with that monument on the epitaph appears to have started with Michelangelo's tomb.

This Buonarroti family tradition was still very much alive more than a century after Michelangelo's death. Filippo Buonarroti was a court official, as well as an avid antiquarian and collector of Etruscan art.[105] His epitaph alludes to his learning and qualities as a scholar, and he, too, wanted to be honoured with a burial and remembrance as closely as possible to Michelangelo. The epitaph dedicated to Filippo reads:

> To Filippo Buonarroti, Senator of Florence, herald of the majesty of the Etruscan Kings. Fair, sharp, prudent, distinguished interpreter of antiquity, conspicuous on account of his understanding of law, his knowledge of nature and history, the monuments of his innate brilliance and the examples of his singular modesty and probity. The mourning wife Fausta

103 Ludovica Sebregondi Fiorentini, 'Francesco Buonarroti, cavaliere gerosolimitano e architetto dilettante', *Rivista d'arte* 38, no. 2 (1986): 49–86.

104 The full epitaph reads: 'D[eo] O[ptimo] M[aximo] / Fr[atri] Francisco Bonarrotio Leonardi F[ilio] / Eq[uiti] hieros s[ancti] Ioan[nis] in fonte Commend[atori]. / Fr[atris] Antonii de Paula M[agistri] M[ilitum] pro lingua ital[ica] a secr[etis] / Prudentia, fide, animi candore / Institutorum ac rer[um] gest[arum] sui ordinis / Eximia cognitione praecellenti / Michaelangelus B[uonarroti the Younger] ut penes ossa maiorum / Vel nomini locus esset fratri suaviss[imo] P[osuit] / Obiit meritae IIII non[arum] Oct[obris] / Ann[o] Sal[utis] MDCXXXII aetatis suae LVIII'.

105 For an impression of his Etruscan art collection and work as an antiquarian, see Daniela Gallo, ed., *Filippo Buonarroti e la cultura antiquaria sotto gli ultimi Medici* (Florence: Cantini Edizioni d'Arte, 1986).

Figure 5.10. The hanging monuments and epitaphs of Francesco Buonarroti (above) and Filippo Buonarroti (below), c. 1632 and c. 1733, respectively. Basilica di Santa Croce, Florence. The monument is an asset of the Fondo Edifici di Colto administered by the Direzione Centrale degli Affari dei Culti and the Ammistrazione del Fondo Edifici di Colto del Ministero dell'Interno. Photograph by author, courtesy of Santa Croce Opera

Malvoltia and son Leonardo executed [this epitaph]. He lived 62 years, 20 days. He died December 8, 1733.

Philippo Bonarrotio Senatori Flor[entiae]
Maiestatis Etruscorum regum adsertori
Iusto sagaci prudenti
Summo antiquitatis interpreti iuris scientia
Naturae atq[ue] historiae cognitione ingenii monumentis
Singularis modestiae probitatisque exemplis conspicuo
Fausta Malvoltia uxor et Leonardus Filius
Maerentes p[osuerunt] vix[it] a[nnos] LXII d[ies] XX
Obiit VI eid[us] dec[embris] a[nno] MDCCXXXIII.

The epitaph frames Filippo as a virtuous man ('fair, sharp, prudent') and as a productive scholar, because of his knowledge and 'the monuments of his *ingenium*'. The family connection to Michelangelo is not made as explicitly clear on this epitaph as on the epitaph of Francesco Buonarroti, but was underlined in this case due to the proximity to the Francesco Buonarroti epitaph and Michelangelo's sepulchre.

The family name was the standard context for burial and funerary practices throughout the seventeenth and eighteenth centuries. It was for that reason that Stefano Rosselli (1598–1663) organised all the grave monuments in Florence according to their family arms and ties in his 1657 *Sepoltuario Fiorentino*.[106] Another trend the Buonarroti monuments point out is the emergence of a memory cult around a certain sepulchre, creating the sense of lineage and association. Both Filippo and Francesco wanted to be buried close to the 'genius' Michelangelo, showing that the Buonarroti family represented a lineage of exceeding virtue. The epitaphs, however, do refer to their own scholarly achievements: Francesco's wide-ranging knowledge of the history of his order and Filippo's knowledge of antiquity and law, who left us monuments of his innate brilliance, especially his antiquarian work on the history of the Etruscans, the alleged forefathers of the Tuscans. The fruits of science and scholarship are presented on these epitaphs as worthy of honour and praise. The learned man is praised on account of his innate brilliance that has bestowed gifts upon the learned world. To be buried close to the great Michelangelo was to place oneself in the family tradition of excellence of the Buonarroti name.

106 Rosselli, *Sepoltuario Fiorentino*, 1657, ASFi, mss. 624–625. A digital version of the original manuscript is attached to Di Stasi, *Stefano di Francesco Rosselli*.

The century-long struggle and discussion about how the Medici astronomer and mathematician Galileo Galilei (1564–1642) should be honoured in Santa Croce shows that new generations of scholars and rulers desired to associate themselves with prominent 'geniuses'. Much like the Bruni–Marsuppini *paragone*, followers of Galileo Galilei wanted to present him as the living continuation of Michelangelo's innate brilliance and genius. This was made even more likely since Galileo was born within days of Michelangelo's death. Michelangelo excelled in sculpture, painting, and architecture; Galileo was a master of geometry, astronomy, and philosophy. One of Galileo's followers, Vincenzo Viviani (1622–1703), considered himself to be the true disciple of Galileo and insisted on giving Galileo the posthumous honour and memory he deserved.[107] The placement of a monumental tomb, however, was problematic since the Catholic church had sentenced Galileo to house arrest on a 'vehement suspicion of heresy' charge.[108] Owing to this conviction, Galileo's name was associated with heretical views and actions, and a burial in the sacred grounds of the Santa Croce was therefore deemed unfitting. Although the Roman and local clergy did not permit an official honorary burial and tomb, Viviani managed to arrange a small tomb behind the Capella del Noviziato of the Santa Croce, right between the convent and the sacristy, just outside the basilica. Here, Galileo found a resting place out of public sight. The burial there was visited by Viviani and Evangelista Torricelli (1608–1647), who succeeded Galileo as court mathematician, while the Grand Duke and clergy shied away.[109] Cardinal Francesco Barberini (1597–1679) warned the Florence inquisitor in a letter about the risks of publicly honouring the heretical Galileo:

> His Beatitude decided that you, with your usual adroitness, should be sure to let it come to the Grand Duke's ears that it is not good to erect mausoleums to the corpse of someone who was sentenced by the Tribunal of the Holy Inquisition, and died while enduring the sentence: because decent people might be scandalised, to the detriment of His Highness's piety. But in case you could not prevent this plan, you should pay attention

107 Anna-Luna Post, 'Claiming Fame for Galileo: Reputation and Scholarly Credibility in Early Modern Italy' (PhD diss., Utrecht University, 2020).

108 For Galileo's trial documents, see Sergio Pagano, ed., *I documenti vaticani del processo di Galileo Galilei, 1611–1741*, new. ed. (Vatican City: Archivio segreto Vaticano, 2009); for a compilation in English, see Thomas F. Mayer, ed., *The Trial of Galileo, 1612–1633* (Toronto: University of Toronto Press, 2012).

109 For the appropriation of Galileo by Viviani and Torricelli, see Michael Segre, *In the Wake of Galileo* (New Brunswick, NJ: Rutgers University Press, 1991).

> that the epitaph, or inscription, which will be placed on the tomb, should not include words that might offend this Tribunal's reputation. And you should treat him who will deliver the funeral eulogy similarly, too, paying attention to read and examine it carefully, before it is delivered or printed.[110]

The lengths to which Viviani went to restore the honour of Galileo knew few bounds. He edited and published Galileo's collected work, *Opere*, wrote his life story, and worked hard to build a tomb for Galileo in the Santa Croce. Throughout all these endeavours, however, Viviani received pushback from the Inquisitor in Florence and clergy. He could not, for example, include Galileo's *Dialogo* in the *Opere*, since it was put on the *Index of Forbidden Books* in 1633.

Viviani gave up on his plan to build a monumental tomb for Galileo in Santa Croce and move him from a backroom in the Medici chapel to the bay facing Michelangelo, his *paragone*; instead, he left this to the next generation in his last will and testament.[111] He did, however, ensure that the backroom of the Medici chapel, Galileo's burial place, was decorated with an elaborate inscription and a bust.[112] Determined to bestow posthumous honour on Galileo, Viviani decided to adorn the façade of his house with a fitting memorial (see Figure 5.12).[113] With the assistance of his friend and

110 Cardinal Barberini to Giovanni Muzzarelli, 25 January 1642, in Galileo Galilei, *Opere di Galileo Galilei*, ed. Antonio Favaro, vol. 18 (Florence: Tipografia Barbèra, 1906), no. 4197, pp. 379–380. The original reads: 'S. Beatitudine, col parere di questi miei Eminentissimi, ha risoluto che ella, con la sua solita destrezza, procuri di far passare all'orecchie del Gran Duca che non è bene fabricare mausolei al cadavero di colui che è stato penitentiato nel Tribunale della Santa Inquisitione, et è morto mentre durava la penintenza, perchè si potrebbono scandalizzare i buoni, con pregiuditio della pietà di S. Altezza. Ma quando pure non si potesse distornare cotesto pensiero, dovrà ella avvertire che nell'epitafio o inscrittione, che si porrà nel sepolcro, non si leggano parole tali, che possano offendere la riputazione di questo Tribunale. La medesima avvertenza dovrà pur ella havere con chi reciterà l'oratione funerale, procurando di vederla e considerarla ben, prima che si reciti o si stampi.' This translation from Stefano Gattei, 'Galileo's Legacy: A Critical Edition and Translation of the Manuscript of Vincenzo Viviani's *Grati Animi Monumenta*', *The British Journal for the History of Science* 50, no. 2 (2017): 181–228 (p. 182).

111 SCArch, *Opera di Santa Croce*, vol. 446, 'Viviani's testament', fol. 143$^{v}$–144$^{r}$. The testament describes how Viviani's remains should be buried as close as possible to Galileo.

112 Paolo Galluzzi, 'I sepolcri di Galileo: le spoglie "vive" di un eroe della scienza', in Berti, *Il pantheon di Santa Croce*, pp. 145–182; an abridged version of this chapter has been translated into English: see Paolo Galluzzi, 'The Sepulchers of Galileo: The "Living" Remains of a Hero of Science', trans. Michael John Gorman, in *The Cambridge Companion to Galileo*, ed. Peter Machamer, 417–447 (Cambridge: Cambridge University Press, 1998).

113 For Viviani's memorial of Galileo on his *palazzo*, see Roberto Lunardi and Oretta Sabbatini, eds., *Il rimembrar delle passate cose. Una casa per memoria: Galileo e Vincenzo Viviani*, Testi

Figure 5.11. Sepulchre of Galileo Galilei, 1737. Basilica di Santa Croce, Florence. The monument is an asset of the Fondo Edifici di Colto administered by the Direzione Centrale degli Affari dei Culti and the Ammistrazione del Fondo Edifici di Colto del Ministero dell'Interno. Photograph by author, courtesy of Santa Croce Opera

e Studi 18, vol. 3 (Florence: Polistampa, 2009). This edition contains full transcriptions and translations to Italian. For transcriptions with an English translation, see Rufus Suter, 'The Galileian Inscriptions on the Façade of Viviani's House in Florence', *Osiris* 12 (1956): 225–243. For transcriptions of the inscriptions relating to Galileo's finger, a broken glass object, and the

Figure 5.12. The memorial to honour Galileo Galilei on the façade of the so-called Palazzo di Cartelloni (Palace of Scrolls) in the Via Sant'Antonino 11, Florence. The façade has two large scrolls on either side of the door that recount Galileo's life story. Photograph by author

architect Giovan Battista de' Nelli (1661–1725), the work started in 1693. Both Viviani and Nelli died before the façade was finished, and the work was finished by Nelli's son, Giovanni Battista Clemente de' Nelli (1725–1793). The two huge stone 'scrolls' alongside the door tell of Galileo's life and describe him as a *pious* reinventor of philosophy: Galileo, 'man of celestial innate brilliance', the 'delight of his Fatherland, of Tuscany, of Italy, indeed of all Europe, for reigniting the torch of philosophy', who 'first showed others the more certain road to truth, and traversed it fruitfully, while always being accompanied on this difficult path by piety'.[114] While Galileo took on international fame as a martyr of science, Viviani presented Galileo as the forefather of a radically new philosophy. Galileo's supporters and followers tried to control his memory by questioning Galileo's sentence; they criticised the treatment he had received; implicitly resisted the Inquisition's verdict; changed the narrative of Galileo's life, especially with Viviani's

inscription on the façade of Galileo's villa in Arcetri, see Rufus Suter, 'Four Galileian Inscriptions', *Isis* 43, no. 3 (1952): 231–236.

114 '[C]oelestis plane ingenii Vir'; 'Patriae, Etruriae, Italiae, imo Europae totius delicio, Philosophiae renascentis faci.'; 'viam ad veritatem certius indagandam alios primus docuit, feliciterque peregit, comitante semper per tam arduum iter pietate[.]' Translations based on Suter, 'The Galileian Inscriptions', pp. 237, 226, and 227, respectively.

life of Galileo; and finally, erased the story of the trial from the memory of Galileo.[115]

Just before his death, Viviani published a large part of the inscriptions as an addendum to an edition of Aristaeus's geometry, titled *De locis solidis* (On conic sections, 1701).[116] The façade in its current state was only finished in 1793 by Nelli Jnr. It was a tribute by Nelli's son to his father, Viviani, and to Galileo. The complete texts of the inscriptions were published by Nelli's son in a small booklet titled *Grati animi monumenta Vincentii Viviani* (Monuments of the beloved soul of Vincenzo Viviani, 1791).[117] Nelli Snr. was the student of Galileo's successor as court mathematician, Evangelista Torricelli, and Giovanni Battista Nelli was committed to perpetuating Galileo's name and fame, as well as the name and fame of his father. Nelli Jnr. took care of building the monumental tomb in Santa Croce, in the first place for Galileo, but also for his father and Viviani. In a legal document, lawyers who worked for Nelli presented legal arguments to the ministers of the Santa Croce for the placement of the tomb in the basilica.[118] In 1737, almost a century after Galileo's death, thirty-four years after Viviani's death, and twelve years after Nelli's father's death, the sepulchre in the Santa Croce was finally finished (see Figure 5.11). The tomb is a complex site of memory, paying homage to Galileo, his disciple Viviani, and the student of Galileo's successor, Giovan Battista Nelli. The inscription on the lavish tomb, placed directly facing Michelangelo's sepulchre (see markers III and IV), reveals the connections between Galileo's successors who wanted to associate themselves with him:

> Galileo Galilei, noble Florentine, the great revivor of geometry, astronomy, and philosophy, comparable to none of his time, may he rest peacefully here. He lived 78 years and passed away in the year 1641. Ten patricians, prefects of this holy temple, took care of the eternal gem of their *patria*, a monument by Vincenzo Viviani, for the ash of his master [Galileo], as well as for himself. Carried out according to the testament [of Viviani], by his heir Giovanni Battista Clemente Nelli, son of senator Giovan Battista, and finished with willing spirit, in the year 1737.

115 Post, 'Claiming Fame for Galileo', pp. 205–222.

116 Vincenzo Viviani, *De locis solidis secunda divinatio geometrica in quinque libros* [...] (Florence: Hyppolyto Navesi, 1701).

117 Giovanni Battista Clemente De' Nelli, ed., *Grati animi monumenta Vincentii Viviani* (Florence: Franciscus Moücke, 1791).

118 SCArch, *Opera di Santa Croce*, 'Galileo', vol. 446, fols. 168$^{r}$–170$^{v}$.

Galilaeus Galileius Patric[ius] Flor[entinus]
Geometriae astronomiae philosophiae maximus restitutor
Nulli aetatis suae comparandus
Hic bene quiescat
Vix[it] a[nnos] LXXVIII. obiit a[nno] M.D.C.XXXXI.
Curantibus aeternum patriae decus
X. viris patriciis sacrae huius aedis praefectis
Monimentum a Vincentio Viviano magistri cineri sibique simul
Testamento f[ieri] i[ussit]
Heres Io[annes] Bapt[ista] Clemens Nellius Io[annis] Bapt[istae] Senatoris F[ilius]
Lubenti animo absolvit
An[no] M.D.CCXXXVII

The praise for Galileo's work as a mathematician, astronomer, and philosopher is rather brief. More than half of the lines on the epitaph are dedicated to the placement history of the monument. Nelli Jnr. at this point was twelve years old, which is why 'ten patricians' helped him realise the monument. To recap, Nelli Jnr. was carrying out the wishes of his father, Nelli Snr., who was the testator of Viviani, who wanted to redeem Galileo's wish to be buried in Santa Croce.[119] Nelli Jnr. had a vested interest in creating a connection to Galileo, through his father and Torricelli. At the end of his life, he wrote and published the second biography of Galileo as well, titled *Vita e commercio litterario di Galileo Galilei* (The life and learned exchange of Galileo Galilei, 1793), a considerable expansion of Viviani's first 'Raccontro storico'.[120] The monument dedicated to Galileo is thus a representation of an intellectual lineage, from Galileo all the way to Giovanni Battista Clemente Nelli.

These origin stories of the monuments of Michelangelo and Galileo clarify the importance of the memory and honour of the deceased to later generations. Francesco and Filippo wanted to associate themselves with Michelangelo; Viviani, Giovan Battista Nelli, and his son Giovanni Battista Clemente Nelli all wanted to be associated with Galileo Galilei. In the ideal of a *paragone*, the monuments to Michelangelo and Galileo are facing each other, much like Bruni and Marsuppini's. Both Michelangelo and Galileo's marble monuments are adorned with three muses. Michelangelo is

119 SCArch, *Opera di Santa Croce*, 'Viviani's testament', vol. 446, fols. $142^r$–$157^r$.
120 Giovanni Battista Clemente De' Nelli, *Vita e commercio litterario di Galileo Galilei nobile e patrizio fiorentino mattematico e filosofo sopraordinario de' Gran Duchi di Toscana Cosimo e Ferdinando II*, 2 vols. (Lausanne: stamperia Moücke, 1793).

accompanied by the muses of architecture, painting, and sculpture; Galileo by the muses of geometry, astronomy, and philosophy. These parallels are no coincidence. Viviani helped to cement the myth that the innate brilliance of Michelangelo lived on in Galileo. In his life story of Galileo, the 'Racconto istorico', Viviani falsified Galileo's birth date from 15 to 19 February, 1564, to bring it closer to the date of Michelangelo's death, 18 February, 1564, and help propagate the idea that Galileo was Michelangelo reborn.[121] Viviani's testament clearly expresses the desire 'that his body will be put down in a casket close to that of Signore Galileo' ('che il suo cadavere si ponga in deposito vicino a quello del medesimo Sig: Galileo'), and if a monument was placed in the church, 'on the side, or below his bones' ('et a canto, o sotto alle di lui ossa').[122]

The Galileo monument became a central point in the Santa Croce for the remembrance of his intellectual successors, such as Viviani and Nelli, but also those who claimed to be descendants of the noble family of Galileo Galilei.[123] The monument to Alessandro Galileo (1691–1737), architect, was initially placed next to that of Galileo Galilei in 1739, in an attempt to connect him with the great Galilean heritage.[124] Alessandro wanted to become part of the Florentine nobility, but it seems that his claim to descent from the family of Galileo Galilei was not accepted in Florentine society.[125] The monument to

121 William E. Wallace, 'Matters of Life and Death: Galileo in the Afterlife of Michelangelo', *Source: Notes in the History of Art* 17, no. 3 (1998): 20–24; Segre, 'Viviani's Life of Galileo', *Isis* 80, no. 2 (1989): 206–231 (pp. 212–214).

122 SCArch, *Opera di Santa Croce*, vol. 446, 'Viviani's testament', fol. 144r. The full sentence reads: '& leggendo la sepoltura il suo proprio cadavere nella detta Chiesa di Santa Croce sotto alla detta statua, e memoria del predetto Grand Galileo, et a canto, o sotto alle di lui ossa quando saranno ivi trasportate, et in tanto, che non sara adempito il sub suo Concetto, vuole, et ordina, che il suio cadavere si ponda in deposito vicino a quello del medesimo Sig: Galileo, e che tutto quanto sopra la disposto intorno al detto sepolcro, e memoria, et al proprio cadavere.'

123 Silvia Cusmano, 'Galilei, Alessandro', *DBI*, vol. 51 (1998).

124 SCArch, 'Suo monumento eretto luogo ove etistevano i despositi senatori Giugni, e Galilei'. The monument for Alessandro Galileo was later moved to the Bardi chapel in the Basilica di Santa Croce, probably in the nineteenth century; see Paatz and Paatz, *Die Kirchen von Florenz*, 4:578, 607. The epitaph reads: 'D[eo] O[ptimo] M[aximo] / Alexandro Galilei / Patricio Florentino / Matematico et architecto clarissimo / Qui septem annis apud Britannos / Summa cum laude versatus / A Cosmo III. et Io[anne] Gastone I. Magnis Etruriae Ducibus / In patria regiis munimentis praepositus / A Clemente XII Pont[ifice] Max[imo] Romam evocatus / Faciem templi Florentinae nationis / Sacellum Corsinae gentis in Lateranensi basilica / Augustamque basilicae frontem cum porticu excitavit / Quibus vix absolutis morte praereptus / Magnum sui desiderium civibus exterisque reliquit / Obiit Romae XI. Kal[endis] Ian[uariis] Anno MDCCXXX VII aetatis suae XLVI / Ibique iacet in ecclesia D. Nicolai in Arcione / Ad radices montis Quirinalis / Galileus et Antonius / Patri suo optime merito moestissimi posuere'.

125 Cusmano, 'Galileo, Alessandro'.

Galileo Galilei shows the Galilei family emblem, a ladder, prominently on top of the sepulchre; the Alessandro Galilei monument shows the same emblem, but this time in the hands of a weeping muse. The location and the display of the family emblem underscore Alessandro's last attempt to connect himself via a familial lineage to the great genius of yore. The monument was later placed in a chapel at the back of the church, further detaching Alessandro's from the Galilean legacy. Regardless, the monuments to both Michelangelo and Galileo Galilei became central points of a memory culture in which later generations wanted to associate themselves with the family lineage or the intellectual succession. Vasari's redesign highlighted a few monuments that changed the meaning and representation of learned men. First, learned men such as Filippo and Francesco Buonarroti associated themselves with the grandeur of Michelangelo as a gifted mind. The monument dedicated to Galileo also aims to show a continuation of Michelangelo's genius and was immediately built as an intellectual family grave. The two tombs suggest that Michelangelo's innate brilliance lived on in his family members as well as in Galileo and his followers.

The desire of followers and descendants to place themselves in a direct line from Michelangelo and Galileo resonates with the presentation of a scholarly pedigree on the *peregrinatio literaria* (see Chapter 4). Scholarly travellers such as Kool, as well as the Buonarroti and followers of Galileo, created a learned memory culture in which they presented themselves as part of a lineage of scholarship and learning. The general framework of the Republic of Letters based on classical authors and collective virtues, which was present in early funerary monuments in Santa Croce as well as the sixteenth-century collections of *vitae* (see Chapter 1), made way for a distinct, local—Florentine, in the case of Santa Croce—history of learning. This history of learning involved a lineage of innate brilliance and individual genius that was passed down along the family line (in the case of the Buonarroti) as well as a master–follower line (in the case of Viviani and Nelli). The funerary monuments of Michelangelo and Galileo show that learned men did not inscribe themselves into a general pan-European history of learning and fame, but rather into a local memory culture that emphasised a personal transmission of learning and fame.

### 5.3.3 A New Wave of Memory and Grandeur, 1737–c. 1800

In the eighteenth century, the number of epitaphs in the Basilica di Santa Croce steadily increased. This was part of a wider European increase in memorial monuments and epitaphs in the eighteenth century. In this era,

many European cities constructed or refashioned old churches into public pantheons to celebrate illustrious men.[126] For example, Westminster Abbey in London was slowly turned into a site of memory for illustrious poets and thinkers.[127] The Basilica di Santa Croce also became a true pantheon for famous Italians, and for Florentines in particular. The Galileo monument and some of the epitaphs and graves discussed above were erected in the 1730s. A new wave of memorialisation kicked off in the second half of the eighteenth century and the number and character of the epitaphs and hanging monuments changed.[128] The second half of the eighteenth century saw a more profound nostalgia, with a boom in publications of eulogies of illustrious Tuscan men in Florence. The learned man was now increasingly seen in a Florentine history of learning.

The memory boom gave rise to many compilations of Tuscan or Florentine illustrious men, such as Giammaria Mazzuchelli's edition of Filippo Villani's fourteenth-century collection of lives of illustrious Florentine men.[129] The Italian printer and engraver Giuseppe Allegrini published *Serie di ritratti d'uomini illustri Toscani*, a collection of eulogies and icons of illustrious men in the Plutarchan historical tradition.[130] The anonymous author of the preface wanted to present a history of the rebirth of the arts and sciences in Florence:

> Collected together, the *Elogi degli Illustri Toscani* will form a history of a rebirth [*rinascimento*] of the sciences and the arts, almost entirely due to the genius of the nation, and to the magnificence and taste of the Sovereign House of the Medici. Thereafter, in times closer to us under the auspices of the same [Medici], it will be possible to observe how the truth began to appear clearer on our horizon, and to spread so much light in the world by means of our discoveries, as to make reason triumph, and dispel the ancient ghosts of ignorance.

126 Eveline G. Bouwers, *Public Pantheons in Revolutionary Europe: Comparing Cultures of Remembrance, c. 1790–1840* (Basingstoke; New York: Palgrave Macmillan, 2012).

127 Craske, 'Westminster Abbey 1720–70'.

128 Arjan de Koomen, 'L'età dei Lumi e il sublime: I monumenti a Niccolò Machiavelli e a Vittorio Alfieri', in Berti, *Il pantheon di Santa Croce*, pp. 183–220.

129 Filippo Villani, *Le vite d'uomini illustri florentini*, ed. Giammaria Mazzuchelli (Venice: Giambatista Pasquali, 1747).

130 Giuseppe Allegrini, ed., *Serie di ritratti d'uomini illustri Toscani con gli elogi istorici dei medesimi*, 4 vols. (Florence: Allegrini, 1766–1773), and a second edition: *Elogi degli uomini illustri Toscani*, 4 vols. (Lucca: [n.p.], 1771–1774). For a short biographical overview, see Anita Mondolfo, 'Allegrini, Giuseppe', *DBI*, vol. 2 (1960).

> Tutti insieme raccolti gli Elogj degli Illustri Toscani formeranno un'istoria del rinascimento delle scienze, e delle arti, quasi unicamente dovuto al genio della Nazione, e alla magnificenza, ed al gusto della Casa Sovrana dei Medici. Quindi sotto i medesimi auspizi in tempi più a noi vicini si potrà osservare, come cominciasse sul nostre Orizonte a comparir più chiara la verità, e diffondersi tanta luce nel mondo per mezzo delle nostre scoperte, da farne trionfar la ragione, e dissipare gli antichi fantasmi dell'ignoranza.[131]

In the eighteenth century, the Basilica di Santa Croce became a site of memory to show that Florence, the Medici, and their learned men had become a beacon in the world of learning and science. Jesuit Giuseppe Richa (1693–1761), in his ten-volume overview of all the churches in Florence, eloquently introduced the Santa Croce as a temple of memory, where many writers (*scrittori*) are part of the congregation. Richa claimed, 'there are more sepulchres here than in all of the other big churches in Florence together'. What is more, they have 'increased the light of sacred history, and the glory of the temple: here one finds saintly remains, sacred remains, profane remains, and erudite remains.'[132] Increasing the light of sacred history, the continuation of the biblical account of history, and adding glory to the Santa Croce are unreservedly portrayed as positive results. When he refers to *ceneri*, literally meaning ashes and figuratively glowing or smouldering remains, Richa alludes to both the mortal remains as well as to the fading, but still glowing light the illustrious men emitted. Richa's account shows that the Santa Croce was widely known as a pantheon of illustrious men in the second half of the eighteenth century.

The tombs of Galileo and Michelangelo, as well as those of Bruni and Marsuppini, gave the Santa Croce an appeal of importance, fame, and honour. Florentine men who considered themselves to be 'uomini illustri' wanted to be buried in line with their perceived fame. In the eighteenth century, the ornamental epitaphs, and sepulchres of Giovanni Lami (1770), Pier Antonio Micheli (1737), Vincenzo Maria Giugni (1758), Antonio Cocchi (1773), and Angelo Tavanti (1782) were placed in the church. All these epitaphs remember them as erudite men who served the common good

131 Allegrini, *Elogi degli uomini illustri Toscani* (Lucca, 1771), 1:x.

132 '[E]sse sola più sepolture, che molte altre insieme della maggiori chiese di Firenze, [...] per l'impegno di accrescere lume alla Sacra Storia, e gloria al Tempio medesimo: ove ravviseremo Ceneri Sante, Ceneri Sacre, Ceneri Profane, e Ceneri Erudite.' Giuseppe Richa, *Notizie istoriche delle chiese fiorentine*, vol. 1 (Florence: Pietro Gaetano Viviani, 1754), p. 73.

Figure 5.13. Sepulchre of Giovanni Lami, c. 1770. Basilica di Santa Croce, Florence. The monument is an asset of the Fondo Edifici di Colto administered by the Direzione Centrale degli Affari dei Culti and the Ammistrazione del Fondo Edifici di Colto del Ministero dell'Interno. Photograph by author, courtesy of Santa Croce Opera

well. Micheli's epitaph boasts that he was 'known everywhere because of his wisdom' ('ubique notus ac propter sapientiam') and because he wrote works of natural history and was the botanist of the Grand Duke of Tuscany.[133] It is noteworthy that the connection with Florence and the Medici family is highlighted, while the epitaph does not mention that Micheli was a professor of botany at the University of Pisa.[134]

The monument and tomb built for Giovanni Lami (1698–1770) illustrates the central place learned men came to occupy in the Basilica di Santa Croce in the second half of the eighteenth century (see Figure 5.13). The executor of Lami's testament, Giovanni Filippo Elmi (fl. 1740–1788), informed the headmaster of the basilica that he wanted to build 'a sepulchre in the church of Santa Croce where there are other graves of Galilei, of the Buonarroti, and of other illustrious and well-known men'.[135] According to the placement document in the archive of Santa Croce, Lami's sepulchre should 'perpetuate the memory of such a great man, adornment and splendour of the Republic of Letters.'[136] Lami wanted to stand among other illustrious men to strengthen and perpetuate his own memory as a great man of letters. His work for the Republic of Letters was considered honourable and worthy of a monumental tomb (see Figure 5.13). The tomb monument shows Lami as a charismatic orator. Both columns that sustain the temple in which he stands are festooned with laurel crowns, and a small owl representing wisdom sits on a branch at the bottom of the monument. The epitaph alludes to Lami's standing as a professor, theologian, and scholar, a 'polyhistor, the most learned in Greek and Latin of his age', who wrote the most erudite commentaries and defended the Catholic faith against the impiety of Arianism (that is, anti-trinitarianism).[137] The epitaph clearly stresses the scholarly works he

133 The full epitaph of Pier Antonio Micheli reads: 'Petrus Antonius Michelius / Vixit annos LVII dies XXII in tenui re / Beatus omnis historiae naturalis / peritissimus magnorum Etruriae / ducum herbarius inventis et scriptis / Ubique notus ac propter sapientiam / Suavitatem pudorem optimis / Quibusque aetatis suae egregie carus / Obiit IV nonas Ianuarias MDCCXXXVII / Amici aere conlato titulum posuere'.

134 Alessandro Ottaviani, 'Micheli, Pier Antonio', *DBI*, vol. 74 (2010).

135 '[U]n Deposito nella chiesa di Santa Croce ovè sono altri depositi di Galileo, de Buonarruoti, e di altri uomini illustri e celebri'. SCArch, *Opera di Santa Croce*, vol. 440, 'Letter concerning Lami's monument', 2 fols., fol. $1^{r-v}$.

136 '[P]erpetua La memoria di un' uomo così grande, ornamento e splendore della Republica Letteraria.' Ibid., fol. $2^{r}$.

137 The full epitaph reads: 'I X Θ C / Ioanni Lamio ecclesiasticae historiae in Florentino Atheneo professori / A divo Francisco Lotharingio Rom. Imp. inter consiliarios theologos adscito / Graece et Latine doctissimo polyhistori aetatis suae nemini secundo / De catholica religione quam eruditis eruditissimis commentariis adversus / Arianorum impietatem defendit optime merito morum candore veritatis / Tuendae studio et in egenos charitate quos supremis tabulis

brought forth and his renown as a scholar. At the end of the eighteenth century, the epitaphs on the tombs of learned men stressed the relevance of both the public good and the commonwealth of learning.

Bequeathing one's library to the Florentine people also became a strategy to gain fame in the eighteenth century. Antonio Magliabechi's personal collection was turned into a renowned public library, and from its inception in 1737, it became normal practice for learned men to donate books as a way to perpetuate their memory.[138] Lami, who was the librarian of the large Biblioteca Riccardiana, left his 114 manuscripts and 5,000 volumes to the Biblioteca Magliabechiana.[139] The official documents describe his bequest as a unification of the Biblioteca Magliabechiana and 'Biblioteca di Giovanni Lami' which attests to the grand size of Lami's library. The library, however, was never renamed and therefore remembered as a gift to the Magliabechiana in public memory.[140] Bequests to the Magliabechiana helped to elevate men of learning and sustain their memory for future generations.[141]

The epitaph of Antonio Cocchi (1695–1758) in the Santa Croce basilica reflects a similar mnemonic strategy. Cocchi was a prominent cultural and intellectual figure in eighteenth-century Florence. He was known for his knowledge of medicine, philology, botany, and antiquities.[142] A monument in his honour was only erected twelve years after his death. In the epitaph, Cocchi is praised as 'the most outstanding philosopher and physician', who excelled in the virtues of 'prudence, honesty, and kindness'.[143] The monument was erected in the church after his death in 1773 by his friend and Pisan professor of theoretical medicine Domenico Brogiani.[144] Brogiani was a former pupil of Antonio Cocchi, and in the 1750s a professor of his

heredes / Ex asse instituit bonis omnibus probatissimo / Philippus Elmius hereditatis curator ex auctoritate principis M[onumentum] P[osuit] / Vixit annos LXXII menses IL dies XXIX obiit postridie nonas Februarias / Anni a partu virginis MDCCLXX / O[ssa] h[ic] s[ita] s[unt]'.

138 See, e.g., BNCF, L'Archivio Magliabechiano, filze III, V, VIII, IX, X, and XI.

139 Maria Pia Paoli, 'Lami, Giovanni', *DBI*, vol. 63 (2004), pp. 226–233.

140 BNCF, L'Archivio Magliabechiano, filza X, fols. $50^r$–$91^r$.

141 Maria Mannelli Goggioli, *La Biblioteca Magliabechiana: libri, uomini, idee per la prima biblioteca pubblica a Firenze* (Florence: Leo S. Olschki, 2000).

142 Ugo Baldini, 'Cocchi, Antonio', *DBI*, vol. 26 (1982), pp. 451–461. For a biography of Cocchi, see Luigi Guerrini, *Antonio Cocchi naturalista e filosofo*, Biblioteca di medicina e storia 2 (Florence: Polistampa, 2002). On his role as an antiquarian, see Miriam Fileti Mazza and Bruna Tomasello, eds., *Antonio Cocchi primo antiquario della Galleria Fiorentina, 1738–1758* (Modena: F. C. Panini, 1996).

143 The full epitaph of Antonio Cocchi reads: 'Antonio Cocchio / Philosopho et medico praestantissimo / Francisci caesaris antiquario / Doctrinarum linguarum que peritia / Prudentia probitate beneficentia / Suis exteris que probatissimo / Dominicus Brogianius / Amico et praeceptori gratus P[osuit] / A[nno] S[alutis] MDCCLXXIII. ab eius obitu XV.'

144 SCArch, *Opera di Santa Croce*, vol. 430, 'Antonio Cocchi', 3 fols.

son, Raimondo Cocchi.[145] In the case of Cocchi, it was not his family but his successor in Pisa who arranged the monument. Yet Cocchi's appointment as a professor in Pisa is not mentioned on the epitaph at all. To further secure Cocchi's heritage as a scholar, his son donated his manuscripts to the Biblioteca Magliabechiana in 1774.[146]

Around the end of the eighteenth century, the Santa Croce had become one of the most important sites of memory on the Italian peninsula that represented the genius of the Tuscan state. This was further emphasised by the project to give Machiavelli a more prominent place in the pantheon of illustrious Tuscans with a new memorial tomb.[147] In the nineteenth century, prominent Florentines wished to turn the Santa Croce into a *complete* pantheon. In 1829, a final attempt was made to bring Dante back to Florence for his eighth centenary, which can be seen as part of the European-wide 'centenary fever', in which national memory cultures flourished.[148] More than four centuries after the initial requests by the Signoria of Florence, the remains were still not moved from Ravenna. This lacuna was filled with a grandiose marble tomb, which now stands proudly halfway up the southern nave.[149] Additionally, in 1865, Enrico Pazzi (1818–1899) put up a large statue of Dante, crowned with a laurel and looking stern, in front of the church on the Piazza Santa Croce to further cement Dante's place in the line of illustrious Tuscan men. In the nineteenth century, Dante was seen as an exile who had longed for unity, a symbol from the past to help guide the unification of Italy. The nineteenth-century memory cult around Dante was positioned in and around Santa Croce because the basilica represented Tuscan heritage and civic pride.[150] The Basilica di Santa Croce had become a symbol of Tuscan, but also of national greatness. At this point, the monuments to the many scholars and learned men were increasingly framed as monuments to Italian illustrious men.[151] The memory of learned men was

145 Ugo Baldini, 'Cocchi, Raimondo', *DBI*, vol. 26 (1982), pp. 477–483, esp. p. 478.

146 BNCF, L'Archivio Magliabechiano, filza IX, fols. 382r–384r.

147 De Koomen, 'L'età dei lumini e il sublime'.

148 Joep Leerssen and Ann Rigney, eds., *Commemorating Writers in Nineteenth-Century Europe: Nation-Building and Centenary Fever* (Basingstoke; New York: Palgrave Macmillan, 2014).

149 Anne O'Connor, 'Dante Alighieri—From Absence to Stony Presence: Building Memories in Nineteenth-Century Florence', *Italian Studies* 67, no. 3 (2012): 307–335; Ettore Spalletti, 'Gli esordi del pantheon romantico dal monumento a Vittorio Alfieri al cenotafio di Dante', in Berti, *Il pantheon di Santa Croce*, pp. 221–243 (pp. 239–240).

150 O'Connor, 'Dante Alighieri', pp. 329–330.

151 Carlo Sisi, 'Il cimitero romantico di Santa Croce', in *Santa Croce nell'800*, ed. Maffioli, pp. 125–133; Anne O'Connor, *Florence: City and Memory in the Nineteenth Century* (Florence: Città di Vita, 2008).

malleable throughout the ages, and their remembrance in Santa Croce has become an open-ended story.

## 5.4 Conclusion

The Basilica di Santa Croce was a learned site of memory throughout the early modern period. The memory brokers who had a vested interest in securing the memory, honour, and fame of the deceased were often family or successors. A successor could be someone who followed the deceased in an official capacity or someone who presented themselves as an intellectual heir or successor. The idea of a succession of power and learning was stressed by the design and placement of the tombs: two identical tombs facing each other in opposite naves.

The Basilica di Santa Croce in Florence is an amalgamation of learned sites of memory in the form of funerary monuments that each have their historical context and representation of learning and virtue. The funerary monuments of professors and learned men in Padua, Bologna, Pisa, Oxford, Leiden, and Tübingen show that they helped elevate the status of the professor in these cities.[152] Florence was not a university town, yet the Santa Croce shows that the social standing of the learned man was very high in the city throughout the early modern period. The highly visible memory culture in the Santa Croce helped sustain the ideal of the learned men throughout the early modern period and the interrelations between the monuments created a sense of a Florentine community of illustrious men of learning and erudition, although the characteristics and virtues associated with learned men changed over time.

The learned memory culture in Basilica di Santa Croce in Florence is an example of how the cultural symbol of the learned man was reinforced in a civic audience by both Florentine rulers and the intellectual elite. The funerary monuments in the basilica helped visualise the existence of a transhistorical learned community in Florence, making it clear for travellers such as Joannes Kool that the tombs of Bruni and Marsuppini reflected a rich Florentine tradition of learning and scholarship. To be honoured in Santa Croce with a monumental tomb was a clear indicator that one earned a place in the succession of learning in Florence. For example, Giovanni Lami's testament (1770) clearly states he wanted a sepulchre in Santa Croce

152 Carrington, 'Sculpted Tombs'; Grandi, *I monumenti dei dottori*; Sicca, *Scultura a Pisa nell'età moderna*; Knöll, 'Creating Academic Communities'.

in the midst of 'other illustrious and well-known men' ('altri uomini illustri e celebri'),[153] because Lami was considered an 'adornment and splendour of the Republic of Letters' ('ornamento e splendore della Republica Letteraria').[154] The learned memory in Santa Croce thus appealed to local learned men and together they presented a historical narrative of Florentine learning to travellers and visitors, as well as the local churchgoers. The Santa Croce as a site of memory helped two groups to *imagine* a learned community via the culture of memory. First, the students, scholars, and men of learning in Florence, who could imagine a transhistorical learned community and were given a sense of tradition; second, the students and scholars who would visit the basilica and recognise not only the transhistorical Florentine learned community, but also its prominent place in a pan-European learned community, to which they would sometimes refer as the Republic of Letters (*respublica literaria*) or the learned world (*orbis literatus*). Two learned successions came together when Kool visited the Santa Croce: the lineage of Dutch scholarly travellers and the Florentine succession of learned men represented by the funerary monuments.

The next chapter will introduce the Pieterskerk in Leiden as a site of learned memory. While Santa Croce was a basilica of a quarter of Florence, the Pieterskerk became the main church associated with Leiden University in the seventeenth century. Both churches provide case studies for how the ideals of the Republic of Letters were appropriated and given meaning in a local context. While the characteristics and contexts of the remembrance were very different, the fact that both cities had a church to honour learned men attests to the ubiquity and persistence of the ideal of the virtuous man of learning. The learned communities the monuments in the churches in both Florence and Leiden represented and *imagined*, however, were local. Yet, as the response of Joannes Kool showed at the start of this chapter, these forms and places of remembrance were recognisable throughout Europe as learned sites of memory.

153 SCArch, *Opera di Santa Croce*, vol. 440, 'Letter concerning Lami's monument', 2 fols., fol. $1^{r-v}$.

154 Ibid., fol. $2^{r}$.

# 6. The Pieterskerk: Representing the Learned Community of Leiden University

**Abstract:** This chapter analyses the Pieterskerk in Leiden as a memorial space for scholars, particularly those from Leiden University. Starting in the late sixteenth century, the church became a place to honour professors and intellectual figures. The chapter explores how these funerary monuments conveyed virtues such as learning, wisdom, and erudition, aligning local scholarly pride with the transnational ideals of the Republic of Letters. It also discusses how Leiden's academic community sought to memorialise its contributions to European scholarship, creating a link between local and international intellectual traditions.

**Keywords:** Leiden University, Pieterskerk, memory cultures, church history, funerary monuments

You who trample on me, you too will be trampled on; think of this and pray for me
Qui me calcas calcaberis et tu id cogita et ora pro me[1]

Funerary monuments invite onlookers to reflect on their life and death. As outlined at the start of the previous chapter, the funerary monuments, tomb stones, and epitaphs of learned men can be considered learned sites of memory that convey a sense of the past to the following generations. The Pieterskerk in Leiden became a site of memory for Leiden University and the Leiden scholarly community in general. This chapter will argue that the church and its monuments helped create and *imagine* a learned community

1 Tomb slab of Knight Alselme de Cais, 17 June 1710; in the exit to the annex of St. John's Co-Cathedral, Valletta, Malta.

Scholten, Koen. *Rethinking the Republic of Letters: Memory and Identity in Early Modern Learned Communities.* Taylor & Francis Group, 2025
DOI 10.5117/9789048559855_CH06

by presenting an image of a flourishing university with a collective history. The remembrance of learned men in the Pieterskerk reveals who was being remembered, but also what was considered an appropriate way to remember and honour the scholarly dead. Consequently, the funerary monuments reflect the prevailing ideals of learning, erudition, and the learned man in a wider culture. The Pieterskerk provides an example of how pan-European ideals of the learned persona were appropriated and given meaning in the context of a Netherlandish university city.

From the last quarter of the sixteenth century onwards, the Pieterskerk was the place where many scholars and professors were buried and honoured with a tombstone, hanging monument, or a monumental tomb. In contrast with the Basilica di Santa Croce, the monuments in the Pieterskerk represented the university community of Leiden University, rather than a mausoleum of eminent citizens in general. In the eighteenth century, the Pieterskerk had attained the status of a *mausoleum academicum*, because by that point many eminent professors had been buried or honoured there. The Pieterskerk is thus a repository of learned memory, where we can distinguish learned memory cultures from the late sixteenth up until the eighteenth century. The community and ideals the learned memory was meant to represent crystallised in the early seventeenth century, when Leiden University was starting to develop and propagate an institutional identity as the 'Athens of the North'.

The Pieterskerk was the centre of a Protestant religious community from 1572 onwards, but also functioned as a public space for visitors and students. As such, the monuments in the church have different audiences: ordinary members of the parish, members of the academic community, as well as travellers and visitors.[2] The learned memory made the past of the academic and religious communities visible for future generations. The funerary monuments in the Pieterskerk are thus conveyors of learned memory that helped make palpable the entangled histories of the university and the city of Leiden.

The Pieterskerk has mainly been studied from the perspective of art history, most notably in the monograph *De Pieterskerk in Leiden*.[3] This study provides the history of the church, as well as several case studies that discuss its graves, tombs, and monuments. However, the book does

2 In this study, the travel accounts of the Pieterskerk have not been studied besides Arnoldus Buchelius and Gotfried Hegenitius's descriptions of the church.

3 Elizabeth den Hartog, ed., *De Pieterskerk in Leiden: Bouwgeschiedenis, inrichting en gedenktekens* (Zwolle: WBooks, 2011).

not provide a systematic overview of the tombs and monuments in the church, nor does it include transcriptions of the epitaphs and inscriptions on those memorials. Some of the passages hint at the relationship between Leiden University and the Pieterskerk, yet the monograph does not systematically address the Pieterskerk as a site of memory. Stefanie Knöll argued that the Pieterskerk is an example of a wider culture of remembrance at universities. As I already mentioned, Knöll compared the graves and tombstones dedicated to scholars and professors in the Protestant university towns of Leiden, Oxford, and Tübingen. She concluded that the funerary monuments in these three cities represented and helped shape the collective identity of the university community.[4] Knöll argued that the hanging monuments for professors in the Pieterskerk should be interpreted as a way for the academic community to distinguish itself from other citizens and social groups. In Oxford and Tübingen, where the academic community also managed to carve out an elite status for itself, university professors were honoured in a similar way. This status was represented in their funerary monuments, which explicitly presented them as professors and men of learning. Knöll's main conclusion is that the university professors used 'their funerary monuments as a means not just of personal commemoration, but equally to create a communal identity amongst their professional colleagues.'[5]

This chapter will expand on Knöll's study, both in terms of its source material and its theoretical approach. Knöll compiled a helpful list of the monuments and tombstones that can still be found in the Pieterskerk, yet did not take into consideration the many lost monuments and tombstones that have been recorded in seventeenth-century sources. This study considers two additional sources: the inventory drawn up by the antiquarian Arnoldus Buchelius (1565–1641) and Gotfried Hegenitius's (1596/1598–c. 1646) published travelogue.[6] Buchelius's *Inscriptiones monumentaque in templis et monasteriis Belgicis inventa* (Inscriptions and monuments found in Netherlandish churches and monasteries) and Hegenitius's 'Iter Frisio-Hollandicum' will be used to analyse the tombstones and funerary monuments dedicated to professors and learned men in the Pieterskerk, which were lost or moved.

4 Stefanie A. Knöll, 'Creating Academic Communities: Funeral Monuments to Professors at Oxford, Leiden and Tübingen, 1580–1700' (PhD diss., University of Sussex, 2003).

5 Ibid., p. 247.

6 Arnoldus Buchelius, 'Inscriptiones monumentaque in templis et monasteriis Belgicis inventa', c. 1609–1632, UBU: hs. 1648 (Hs 4 L 16); Gotfried Hegenitius, 'Iter Frisio-Hollandicum', in Buchelius and Abraham Ortelius, *Itinerarium Frisio-Hollandicum, et Itinerarium Gallo Brabanticum* [...], 11–177 (Leiden: Elzevier, 1630).

Additionally, all the inscriptions and locations of the monuments and tombstones that can still be found in the Pieterskerk have been conferred with Buchelius's catalogue and Hegenitius's travelogue, to ensure that the current design is original and unchanged. As such, these sources add new information to the list of monuments and tombstones compiled by Stefanie Knöll in her dissertation in 2003 and the 1864 inventory by Kneppelhout van Sterkenburg.[7]

This study will also take a different theoretical approach. Whereas Knöll saw the monuments as a reflection of the social standing of the professors, or, as she put it, 'the idea that professors constituted the academic and moral elite of the country',[8] this study will be more concerned with how the memory culture in the Pieterskerk appropriated and related to pan-European ideals of learning and scholarship. This chapter will show how the ideals and community the monuments represented changed over time, following personal considerations and the university context. Additionally, this chapter will identify trends in the memory culture in Leiden and the close relationship between the modes of remembrance and the history of the university more generally.

The chapter at hand will consider the memorials, tombs, and epitaphs dedicated to learned men and especially professors in the Pieterskerk in Leiden. Particular attention will be paid to the representations and historical contexts of the individual monuments and tombstones. Two leading questions inform the analysis. First, which virtues and qualities are ascribed to learned men in Leiden? Second, what is the cultural historical context and memory culture in which learned men were honoured and how did this affect the representation of learned men? Key to this analysis is the language of the epitaphs. Epitaphs are one of the most important and revealing aspects of a monument. In these short texts, the merit and worth of the deceased learned man were presented to posterity, offering us an insight into the perceived virtues and values with which learned men were associated. Moreover, epitaphs often mention virtues and deeds for which the deceased person was and should be remembered.

The first section of this chapter will introduce the academic memory culture in Leiden the Pieterskerk was embedded in and the church's entanglement with the history of the university. The remainder of the chapter has been organised chronologically. The second section of this chapter will

7 Knöll, 'Creating Academic Communities', pp. 383–423; K. J. H. C. Kneppelhout van Sterkenburg, *De gedenkteekenen in de Pieters-kerk te Leyden* (Leiden: P. H. van den Heuvel, 1864).

8 Knöll, 'Creating Academic Communities', p. 192.

Figure 6.1. Christiaan Hagen, *View of the Pieterskerk in Leiden*, part of the wall map *Grote Hagen*. Amsterdam, 1675. Etch and engraving, 250mm × 296mm. Rijksmuseum: RP-P-AO-10-25

address the memory culture and funerary monuments in the first two decades after the foundation of Leiden University in 1575. This section will show how professors were buried and memorialised as excellent teachers in the Pieterskerk from 1575 onwards. The third section of this chapter will address the heyday of the professorial memory culture in the Pieterskerk, in which professors were honoured as ornaments of the university and the city, and the university started to assume an institutional identity as *the* university of the northern Netherlands. The professors were remembered less on account of their teaching but rather for their scientific research and standing in the learned world. In the fourth section, the tradition of honouring professors from the 1640s onwards will be addressed. Around that time, the hanging monument lost its popularity and many of the tombstones of professors adopted a standard form and epitaph. The final section draws conclusions, makes a comparative analysis with the memory culture in the Basilica di Santa Croce in Florence, and comments on the importance of learned memory in churches for the perpetuation of pan-European ideals of scholarship and the Republic of Letters.

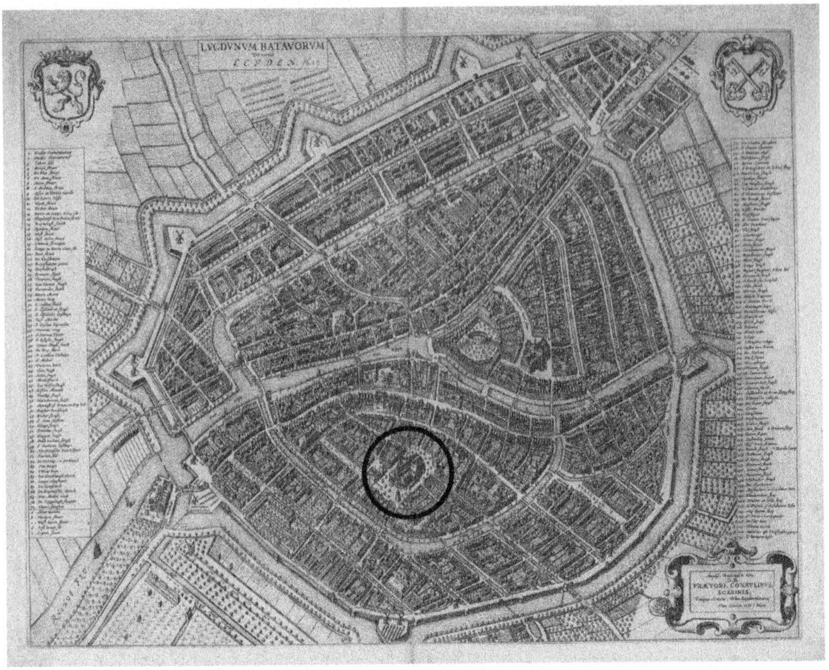


Figure 6.2. *Map of Leiden*. The print appeared in Johannes Willemszoon Blaeu, *Toonneel der steden van de Vereenighde Nederlanden*, 2 vols. (Amsterdam: Johannes Willemszoon Blaeu, 1652), vol. 1. Etch and engraving, 422mm × 542mm. Rijksmuseum: RP-P-AO-10-11-2. The location of the Pieterskerk is indicated by the circle

## 6.1 Leiden University and Its Memory Culture

The Pieterskerk is essentially a freestanding cross basilica in Gothic style. The first traces of the Pieterskerk can be dated to approximately the twelfth century. On top of the foundations of this old church, a bigger Pieterskerk was built around the end of the fourteenth century, with a large choir, transept, and sacristy. The church also had a tower, although it collapsed in 1512 and was never rebuilt. The current and final architectural form of the church was reached around the first half of the sixteenth century, after the church was expanded during the second half of the fifteenth century and the first half of the sixteenth century (for the appearance of the church in 1675, see Figure 6.1).[9] The Pieterskerk became a Protestant church from 1570s onwards and was a community hub for the Protestant parish, but also had a wider function as a community centre for Leiden citizens and

9 Jan Dröge, with John Veerman, 'Inleiding: De bouwgeschiedenis', in Den Hartog, *De Pieterskerk in Leiden*, pp. 13–71, esp. pp. 53–56.

foreign travellers. Many of the altars, altarpieces, and statues were destroyed during the iconoclasms of 1566 and 1572. This fundamentally changed the interior appearance of the church: it became more subdued and minimalist, although the general outline of the church remained largely unchanged with its pulpit, organ, and stained-glass windows.[10]

The history of the funerary monuments in the Pieterskerk is strongly intertwined with the history of Leiden University. First, the church was located close to the Rapenburg canal (south of the church, see Figure 6.2), where most of the university buildings were located. Secondly, over the decades the Pieterskerk was appropriated as the church of the university, ever since it hosted the university's inauguration. Thirdly, the history of the university was strongly tied to the history of the Reformation and the Dutch Revolt against Spanish rule in the Low Countries. Leiden University was one of the first in Europe to have been founded as a Protestant university. The city of Leiden played an important role in the opposition to the Spanish reign in the Eighty Years' War, in particular the *de facto* governor, the Duke of Alba (1507–1582). In 1573 and 1574 the Spanish army besieged Leiden, unsuccessfully.[11] The resilience of the city and its citizens was widely advertised as an important turning point and symbol of the resistance. As a result of the public revolt against Spanish rule, Reformed Protestantism became the public religion in the Dutch Republic.[12] To honour the sustained persistence of Leiden over the course of the year-long siege, Prince William of Orange proposed to the States General of Holland and Zeeland that a university be established. On 8 February 1575, Leiden University was officially inaugurated in the Pieterskerk.[13] This inauguration also cemented the symbolic relationship between the Pieterskerk and Leiden University. In the first few decades of its existence, Leiden University presented itself as a symbol of how education and learning could be carried out in the reformed tradition. Leiden University was therefore the result and a symbol of Dutch Protestant opposition to Spanish rule and Catholicism.

The Reformation and the Counter-Reformation did not necessarily lead to a radical change in funerary practices in Catholic and Protestant churches.

10 Elizabeth den Hartog and John Veerman, 'Inleiding: Het interieur na de Reformatie', in Den Hartog, *De Pieterskerk in Leiden*, pp. 185–209 (pp. 188–201).

11 Jonathan Israel, *The Dutch Republic: Its Rise, Greatness, and Fall, 1477–1806* (Oxford: Oxford University Press, 1995).

12 Charles H. Parker, 'Reformed Protestantism', in *The Cambridge Companion to the Dutch Golden Age*, ed. Helmer Helmers and Geert H. Janssen, 189–207 (Cambridge: Cambridge University Press, 2018).

13 Willem Otterspeer, *Groepsportret met dame* I: *Het bolwerk van de vrijheid: de Leidse universiteit, 1575–1672* (Amsterdam: Bert Bakker, 2000), pp. 62–64.

Although Lutheran theology and doctrine did suggest that funerals should be held without a procession or singing, congregations continued to value and practice such funeral rites.[14] The remembrance of the dead also remained a central part of Protestant memory cultures, despite theological wishes to make such memorial practices more sober. After the years of religious reform, the church remained an important site of memory in Protestant communities.[15] Roísín Watson argued that Lutheran congregations put the remembrance of the dead front and centre:

> Remembering the dead gave a sense of permanence to the church space that was not only important in the early years of reform, but also in moments of confessional tension, which saw pastors return to cataloguing the monuments of the dead as a means to emphasise ownership and control of the space. The right to remember reflected confessional authority.[16]

In Protestant churches, local communities continued the medieval tradition of placing tombstones for members of noble families in churches, although the emphasis in the remembrance moved from heraldry to the local heritage and community. Watson concludes that the 'church as a site of memory changed as the memory of confessional struggle receded and burial sites transitioned from being a component in the active construction of dynastic memory to an emblem of a territory's heritage.'[17] The monuments in the Pieterskerk generally confirm this trend, since the professors' monuments were commonly framed as part of the city's heritage and university tradition, rather than as dynastic memory, although most epitaphs did mention the deceased's noble descent if there was any. The emphasis in Protestant church design was on the delivery of God's word, giving a central place to the pulpit, but the funerary traditions were generally in place.[18]

14 Craig M. Koslofsky, *The Reformation of the Dead: Death and Ritual in Early Modern Germany, 1450–1700* (Basingstoke; New York: Macmillan, 2000), p. 92.

15 Roísín Watson, 'Funeral Monuments, Ritual and Print: Strategies of Memorialization at the Württemberg Court', *Past & Present* 234, Issue supplement 12, 'Cultures of Lutheranism: Reformation Repertoires in Early Modern Germany' (2017): 139–164 (pp. 154–157, 164).

16 Watson, 'Remembering and Forgetting the Dead in the Churches of Reformation Germany', in *Remembering the Reformation*, ed. Brian Cummings et al., 154–172 (London; New York: Routledge, 2020), p. 169.

17 Ibid., p. 155.

18 C. A. van Swigchem, T. Brouwer, and W. van Os, *Een huis voor het Woord: Het protestantse kerkinterireur in Nederland tot 1900* (The Hague: Staatsuitgeverij; Zeist: Rijksdienst voor de Monumentenzorg, 1984).

In Calvinist and Lutheran theology, the saving of souls through prayer after death was deemed impossible. This rendered meaningless the Catholic practice of burying someone as close to the altar as possible to lessen the time a soul had to spend in purgatory.[19] Calvinist and Lutheran theology denounced the burial of the deceased in the holy ground of the church and did not acknowledge the belief in purgatory. As a consequence, the bodies were increasingly buried outside the walls of the church and the physical bodies were disconnected from the space of the living.[20] From the end of the sixteenth century onwards, burials were devoid of any superstition, sermon, or prayers to save the souls of the dead.[21] On the tombstones in the Pieterskerk the epitaphs rarely appeal to the onlooker to pray for the deceased and be mindful of death (*memento mori*). Often, the monuments are explicitly dedicated 'to the dear memory' ('bonae memoriae') of the deceased, underscoring the desire to monumentalise the deceased with a hanging monument or tombstone.

Despite the rejection of the extensive and ceremonial remembrance and memory of Catholic ritual, the remembrance of the deceased's life and virtues played a prominent role in Protestant theology and in Protestant communities.[22] Funerary monuments continued to be built, both in churches and in print.[23] Protestant communities still had a need for social display and remembrance in their communities. The social status of illustrious members of the community was collectively remembered and cherished. Funeral orations were common for important members of the community or in the case of Leiden, the university community.[24] The case of Leiden exemplifies the visibility of social display and remembrance in university and humanist cultures throughout Europe. Humanist culture stressed the concept of immortality through *fama* (fame or renown), and funerary monuments, epitaphs, and sepulchres ensured that one's *fama* continued to thrive posthumously. Hence, a trope among humanists and scholars was to attain 'eternal life in marble', to which numerous epitaphs also attest.

19 Koslofsky, *Reformation of the Dead*, pp. 17–78.

20 Koslofsky, *Reformation of the Dead*; Gordon D. Raeburn, 'The Reformation of Burial in the Protestant Churches', in *A Companion to Death, Burial, and Remembrance in Late Medieval and Early Modern Europe, c. 1300–1700*, ed. Philip Booth and Elizabeth Tingle, 156–174 (Leiden; Boston: Brill, 2021).

21 Ibid., pp. 160–161.

22 P. den Boer, 'Naar een geschiedenis van de dood', *Tijdschrift voor Geschiedenis* 89, no. 2 (1976): 161–201.

23 Watson, 'Funeral Monuments, Ritual and Print', p. 164.

24 John M. McManamon, *Funeral Oratory and the Cultural Ideals of Italian Humanism* (Chapel Hill; London: University of North Carolina Press, 1989), pp. 123–152.

Universities and academies often left a distinct mark on the memory culture of their cities. In the previous chapter, we discussed the memory cult around the professors in Bologna and Padua in the thirteenth and fourteenth centuries.[25] University scholars often presented themselves as a distinct social group throughout the late medieval and early modern periods, when the university degree was both a distinction of learning as well as a defining feature of the social elite.[26] Even without a university, learned men obtained central positions in society. The examples of Leonardo Bruni and Carlo Marsuppini, extensively discussed in the previous chapter, illustrate the importance of learning and erudition for the position of the secretary of the Florentine Republic in the fourteenth century and beyond (see Chapter 5.3.1).

After the Reformation, the newly founded and reformed Protestant universities helped to create a new idealised scholar, and these universities presented themselves as beacons of the Protestant faith.[27] Leiden University was founded in a similar vein as a new bastion of Protestant learning and education. It was founded as a 'breeding school of the church and of the state'.[28] From its inception, the university and the magistrate of the city of Leiden were closely connected: the university was managed by the senate of the university (containing professors), the *curatores*, who were appointed by the States of Holland, and the burgomasters of Leiden.[29] These burgomasters and *curatores* were often prominent citizens and had the power to appoint new professors.[30] The university would become a flourishing centre of humanist learning and Calvinist theology. As a result, the Leiden memory cultures are characterised by the tension between Leiden University as a transnational, Protestant, and humanist centre of learning on the one hand, and the university as a local, civic academy on the other hand. The entangled relation between religion and education at Leiden is reflected in the many tombstones and monuments found in the Pieterskerk.

25 Renzo Grandi, *I monumenti dei dottori e la scultura a Bologna (1267–1348)* (Bologna: Commune di Bologna / Istituto per la Storia di Bologna, 1982); Jill Emilee Carrington, 'Sculpted Tombs of the Professors of the University of Padua, c. 1358–c. 1557' (PhD diss., Syracuse University, 1996).

26 Richard Kirwan, ed., *Scholarly Self-Fashioning and Community in the Early Modern University* (Farnham; Burlington, VT: Ashgate, 2013); Kirwan, 'Miscellanies of Memory: From Scholarly Biography to Institutional History in the Early German University', in *Memory and Identity in the Learned World: Community Formation in the Early Modern World of Learning and Science*, ed. Koen Scholten et al., 235–254 (Leiden; Boston: Brill, 2022).

27 Kenneth Austin, 'Academic Exchanges: Letters, the Reformation and Scholarly Self-Fashioning', in Kirwan, *Scholarly Self-Fashioning*, pp. 39–58.

28 In Latin: 'Academia Ecclesiae et Reipublicae seminarium'. Otterspeer, *Groepsportret* I, p. 65.

29 Ibid., pp. 67–68.

30 Ibid., pp. 75–80.

Material memory dedicated to scholars and professors in the Pieterskerk takes four forms. First, there are the tombstones on top of graves in the church; secondly, hanging monuments that were usually placed next to the tomb of the scholar to honour his life and work. These wall-mounted epitaphs became popular in the early modern period, especially in the northern parts of Europe, and were considered an apt way to honour the more eminent members of the community.[31] From the sixteenth century onwards, the placement of hanging monuments with a poetic epitaph became increasingly popular, and the epitaph evolved into a literary style that can be characterised as a textual form of memory.[32] Thirdly, there is the memorial tomb with an elaborate sepulchre and epitaph. This particular type is rare in the Pieterskerk and none of these are dedicated to early modern scholars or professors. Finally, there is the memorial without a tomb or sepulchre, such as the monumental urn for Herman Boerhaave and the memorial for Petrus Camper, both placed in the eighteenth century. These expensive monuments aimed to cherish the memory of the deceased (*memoria*) and present an example of virtue (*exemplum virtutis*) for the religious community and for visitors of the church.[33]

The transformation of the Pieterskerk into a coherent site of memory was part of an attempt by Leiden scholars to put Leiden University on the map as a centre of learning in the seventeenth century. The memory culture in the Pieterskerk is exemplary of the ways in which local learned communities appropriated the wider ideals of humanist scholarship and the Republic of Letters. In other words, in the Republic of Letters there was no such thing as a general memory culture that presented scholars and professors as part of a pan-European community of learning—only local interpretations of the ideals of scholarship. The eminent scholar, therefore, was appropriated as the glory and honour of the city and local community, as we have also seen in the previous chapter on the Santa Croce in Florence, where learned men were placed in a Florentine succession of learning. In the Pieterskerk, the scholars were specifically remembered in a tradition of teaching and research at Leiden University.

31 Douglas Brine, *Pious Memories: The Wall-Mounted Memorial in the Burgundian Netherlands* (Leiden; Boston: Brill, 2015).

32 Scott L. Newstok, *Quoting Death in Early Modern England: The Poetics of Epitaphs Beyond the Tomb* (Basingstoke; New York: Palgrave Macmillan, 2009), pp. 28–31.

33 Frits Scholten, *Sumptuous Memories: Studies in Seventeenth-Century Dutch Tomb Sculpture* (Zwolle: Waanders, 2003), pp. 15–22.

## 6.2 The Young University: Honour for Life and Learning, 1575–c. 1600

The problems of undocumented reforms and redesigns in reconstructing the history of the Santa Croce in Florence are also relevant to the history of the Pieterskerk. Fortunately, the monuments that were present in the Pieterskerk in the early seventeenth century have been documented in Buchelius's *Inscriptiones monumentaque*. This meticulous catalogue consists of a general description of the Pieterskerk and its decorations, with particular attention to the heraldic signs on the tombstones, monuments, and insignia.[34]

Around the 1620s, Buchelius visited the Pieterskerk as part of his project to compile a catalogue of the interiors of churches in the Low Countries. Buchelius approached his compilation work as an antiquarian, rather than as a historian.[35] In the early modern period, one could expect the historian to render a vivid and analytical narrative of political history. Based on the example of the historians of antiquity such as Tacitus, the historian was not so much concerned with a meticulous analysis of historical sources, but rather with the production of a literary work. The antiquarian, however, dedicated his time to collecting, studying, and critically assessing the remains from the past, such as texts (philology), coins (numismatics), archaeological finds, and monuments.[36] Buchelius's work was antiquarian, since the goal of much of his work was to describe the many objects that emphasised the role of the Utrecht diocese in its political and cultural development during the medieval period.[37]

The Buchelius manuscript reveals a number of tombstones dedicated to professors and learned men that have previously been overlooked. A few tombstones from the 1580s dedicated to learned men show that from the foundation of Leiden University in 1575 onwards, learned men were buried in the Pieterskerk and their tombstones explicitly mentioned the merit of the deceased's learning. Two tombstones illustrate the elevated status of learned men before the foundation of the university. The first tombstone that was dedicated to a learned man was of a certain Philips Codde († 1480). According to Buchelius's inventory, Codde's tombstone carried one of the few inscriptions about learned men in the Pieterskerk that was written in Dutch:

34 Buchelius, 'Inscriptiones monumentaque'.

35 Sandra Langereis, 'Geschiedenis als ambacht: oudheidkunde in de Gouden Eeuw: Arnoldus Buchelius en Petrus Scriverius' (PhD diss., University of Amsterdam 2001).

36 Ibid., p. 13.

37 Ibid., pp. 265–290. For Buchelius's religious sensibilities and conversion, see Judith Pollmann, 'Another Road to God: The Religious Development of Arnoldus Buchelius (1565–1641)' (PhD diss., University of Amsterdam, 1998).

> Sir Philips Codde doctor is buried here, very renowned in medicine, who died on Saint Bartholomew's eve, in the year of our Lord 1480, pray for the soul.
>
> Hier leijt begraven mr. Ph[i]l[ip]s Cod-
> de doctor seer vermaert inde
> medicinen die sterf [*sic*] op S[int] Barto[-]
> lomeus avent int iaer ons
> heere MCCCCLXXX bidt voor
> de siel.[38]

The tombstone of Philips Codde cannot be found in the Pieterskerk any more and little is known about him.[39] The epitaph is noteworthy, though, because it only refers to Codde as a 'very renowned doctor in medicine'. The monument presents Codde solely as a learned man, without any mention of his descent, official positions, or merit. This indicates that the title of 'doctor' was a signifier of social standing well before the foundation of the university. It suggests that it is important for posterity to remember that Codde was a doctor in medicine.

The second tombstone is dedicated to Quintinus Weytsen (c. 1518–1565), refers to the profession of law, but not to any virtues of learning. According to Buchelius, the inscription on Weytsen's tombstone read:

> To the greatest and almighty God. To the most outstanding lawyer, most skilled in civil affairs, sir Quinten Weytsen, Lord of Brandwijck, in the region of The Hague, councillor of the states [of Holland, Zeeland and West Frisia]. The most beloved wife and children placed [this monument] with tears. He died in the year 1565, on April 13, aged 47.
>
> D[eo] O[ptimo] M[aximo]
> Praestantiss[im]o iurisconsulto, rebus
> civilibus tractandis exercitatiss[im]o

38 Buchelius, 'Inscriptiones monumentaque', fol. 42$^{v}$.

39 The epitaph bears a close resemblance to the inscription on the tombstone of (probably) another Philips Codde († 1471), at the Witte Nonnen chapel in Leiden; see the Lakenhal inventory, 'Grafzerk van Philips Codde, uit de kapel van de Witte Nonnen te Leiden', inventory number 928, www.lakenhal.nl/en/collection/928. The epitaph reads: 'Philips Codde doctor [...] in medicine die sterf op // sinte Katherine avent int // annos heren MCCCCLXXI bid voor die ziel om gods wille[.]' The tombstone Buchelius transcribed in the Pieterskerk is clearly a reference to this tombstone.

viro D. Quintino Weitsen D[omi]no
in Brandwijck regio in Haga
Comitis consillario ord[inari]o charis-
sima coniunx liberiq[ue] cum
lachrimis posuere. Obiit a[nn]o
MDLXV idib[us] Aprilis aetatis
suae XLVII[40]

This tombstone suggests that Weytsen wanted to be remembered as an excellent lawyer and legal advisor (*iurisconsultus*), since it stresses his skill and learning in civil affairs. The epitaph further lays emphasis on his claim to nobility as a Lord of Brandwijck and his role as a counsellor of Prince William.[41] The tombstone praises the merit of a learned *iurisconsultus* who could act as a wise councillor. The foundation of the university in 1575, however, impacted the remembrance of the learned man in Leiden, making the professor the archetypical ideal.

The earliest monument of a professor in the Pieterskerk is dedicated to Rembertus Dodonaeus (Dodoens, 1516–1585). He was appointed as the university's third professor of medicine in 1582, alongside Gerardus Bontius and Johannes Heurnius.[42] Dodonaeus was a famous herbalist, especially renowned for his well-known *Cruydeboeck*, written in the vernacular and first published in 1554. Apart from medicine, the curators of the university decided he would teach physics too. The hanging monument in the Pieterskerk alludes to his exceptional wisdom in all the fields of knowledge in which he taught.

To the greatest and highest God. To Rembertus Dodonaeus of Mechelen, physician and councillor of emperors Maximilian II and Rudolph II, whose erudition in matters of astronomy, plants, and medicine became famous through his writings. Already a grey man, he was public professor of medicine at Leiden University among the Batavians, and died in peace in the year 1585, March 10, his age was 68. The son Rembertus Dodonaeus placed this monument.

D[eo] O[ptimo] M[aximo]
Remberto Dodonaeo Machlin[iae]

40 Buchelius, 'Inscriptiones monumentaque', fol 45$^{v}$.

41 For a cursory reference to Quenten Weytsen, see C. de Waal, 'Haeck, Pieter', *NNBW*, vol. 3 (1914), pp. 522–523.

42 Otterspeer, *Groepsportret* I, p. 142.

D. Maximiliani II et Rudolphi II imp[eratorum]
Medico et consiliario
Cuius in re astronomica, herba[ria], medic[a]
Eruditio scriptis inclaruit
Qui iam senex in acad[emia] Lugdunensi
Apud Batavos, publicus
Medicinae professor feliciter obiit,
Anno MDXXCV ad VI idus Martias
Aetatis suae LXVIII.
Rembertus Dodonaeus
Fil[ius] M[onumentum] P[osuit][43]

This monument shows the desire of illustrious professors to associate themselves with Leiden University. It is a departure in comparison to the two previously mentioned tombstones which are not as extensive nor as rich in references to the fields of knowledge or the state. Dodonaeus's epitaph first offers an image of an outstanding man of learning, who was the personal physician and councillor of Holy Roman Emperors Maximilian II and Rudolph II. The epitaph continues to laud Dodonaeus for his knowledge of astronomy and medicinal plants, two essential fields of knowledge at the young university. In the last few lines, the epitaph leaves no doubt where Dodonaeus, 'already a grey man', spent his last years: 'at Leiden University among the Dutch'.[44] The term *batavi* refers to Hollanders in this instance and the newfound civic identity of Leiden as *Lugdunum Batavorum*, the settlement of the Batavians who resisted Roman occupation.[45] The city of Leiden became known as *Lugdunum Batavorum* and *Lugdunum apud Batavos*, especially after Janus Dousa dubbed Leiden University *Academia Lugduno-Batava*.[46] This reference shows the two overarching themes in the memory culture in the Pieterskerk: the pride and heritage of the university, as well as civic pride. The historical consciousness of the city of Leiden mixed seamlessly with that of the university.

The next monument dedicated to a professor was to Johannes Holmannus (1523–1586), a professor of theology. Holmannus was a student of

43 Buchelius, 'Inscriptiones monumentaque', fol. 38v.

44 '[I]n acad[emia] Lugdunensi apud Batavos'.

45 I. Schöffer, 'The Batavian Myth During the Sixteenth and Seventeenth Centuries', in *Some Political Mythologies: Papers Delivered to the Fifth Anglo-Dutch Historical Conference*, ed. J. S. Bromley and E. H. Kossmann, 78–101 (The Hague: Martinus Nijhoff, 1975).

46 Wim Bonis, 'Lugdunum Batavorum en het Keltische erfgoed', *Leids Jaarboekje* (2005): 45–52.

Melanchthon and allegedly quite popular among students.[47] His tombstone is an exception when compared to all the other tombstones dedicated to professors, since the inscription is written in the vernacular rather than Latin, the common tongue of academia. According to Buchelius's description, the following words were inscribed on the tombstone:

> On the 26th day of December in the year 1586, the very learned man Johannes Holmannus Secundus was buried here, during his life professor of sacred scripture in the school of Leiden.

> Opden 26 dach Decembris A[nn]o
> 1586 is hier begraven de
> eerweerdige hoochgeleerde
> heer. Heer Johannes Holmans
> Secundus by syn leven professor
> der godlicker scrift in de
> schole van Leyden[48]

The inscription on the tombstone is solemn and simple: it characterises Holmannus as a 'very learned man' and a professor of theology. Gotfried Hegenitius's published travelogue of his travel to the Low Countries in 1626–1627, the *Iter Frisio-Hollandicum,* mentions the existence of a hanging monument next to the tombstone as well as a funerary poem on parchment glued onto a wooden panel.[49] The poem invited the visitor to consider Holmannus's innate brilliance, learning, and the bequest of his library to the university.[50] The monument carried the following Latin epitaph:

> To the greatest God and the memory of theologian and professor Johannes Holmannus Secundus, praised for his life and learning. The wife, Beke of Edenbuttels, placed [this monument] in mourning. He lived 63 years, he died in the year 1586, December 26.

> Deo Max[imo]
> Et memoriae Iohannis Holmanni Secundi

47 Otterspeer, *Groepsportret* I, pp. 144–145; F. S. Knipscheer, 'Holmannus, Johannes', *NNBW*, vol. 8 (1930), pp. 798–799.

48 Buchelius, 'Inscriptiones monumentaque', fol. 39$^{r}$.

49 Hegenitius, 'Iter-Frisio-Hollandicum', pp. 116–122.

50 Ibid., 116–117. The Holmannus bequest is also mentioned in relation to the bequest of Joseph Justus Scaliger and Carolus Clusius; see ibid., pp. 101–102.

Theologi et professoris vita doctrinaque
Laudati Beke ab Edenbuttels uxor
Heu moesta[51] posuit
Vixit annos LXIII obiit anno
MDLXXXVI VII Kal[endas] Ian[uarias].[52]

The epitaph on the hanging monument bears a close resemblance to the Dutch inscription on the tombstone, which gives the impression that the hanging monument and tombstone were attuned to different readers in two different languages. The Dutch tombstone inscription appears to address a local Dutch community of churchgoers, while the Latin monument addresses a Latinate, highly educated audience of students, scholars, and visitors. The hanging monument was a type of honour bestowed on the deceased, a practice that reflected the elevated social status of learned men and professors in Leiden from the 1580s onwards. Holmannus Secundus was presented to the reader as a theologian and professor, who should be remembered for his life and learning. Thus, the professor was characterised foremost by his learning and exemplary life.

The Latin hanging monument became a commonplace for eminent professors from the 1590s onwards in the Pieterskerk. The tombstones in the church were dedicated to many notable people from Leiden, but with the hanging monument the academic community quite literally elevated themselves as a distinct social group. The hanging monument dedicated to Jacob Ramsay (Ramsaeus, c. 1565–1593) is a good example of the common epitaph on the hanging monument for the professor. Ramsay was recruited by Justus Lipsius to enrich the philosophy education at Leiden University and was appointed as extraordinary professor in logic in 1588.[53] He died only five years later. His epitaph in the Pieterskerk was extensive and full of praise according to Buchelius.

> To the greatest and highest God. Here lies the young most learned and most renowned Jacob Ramsay of Scotland, Doctor of both Laws [Roman and Canon law] and of Philosophy, professor at Leiden University, exceeding in manifold science, gifted with manifold prudence, moved up to various ranks of honour. Having lived his whole life praiseworthily,

51 The transcription in Hegenitius has 'molesta' (irksome, troublesome), but this is likely a typographical error.

52 Ibid., p. 122.

53 Otterspeer, *Groepsportret* I, p. 169.

he died on May 21, 1593. He lived 28 years. Grieve, you reader and university.

D[eo] O[ptimo] M[aximo]
Hic situs est iuvenis doctiss[imus]
et clariss[imus] M. Iacobus Rasaeus [*sic*]
Scotus I[uris] U[triusque] D[octor] et philosophiae in
Acad[emia] Lugd[unensi] professor multi[i]uga
scientia eximius, varia prudentia
praeditus, ad varios honoris
gradus evectus et omni vita
laudabiliter acta obiit 12
Καλ Iun[io] an[no] MDXCIII
vixit an[nos] XXVIII
lector et academia dolete[54]

The epitaph displays a different array of virtues and characteristics that are praised. First, it mentions his degree as doctor in civil and church law, and philosophy as well as his official position of professor at the university; secondly, the learned virtue of exceeding in many types of knowledge; and finally, the personal virtue of prudence and having lived a praiseworthy life. The death of Ramsay is presented as both a personal loss as well as a loss to the university. The remembrance is thus binary: he is remembered on account of his individual virtues as well as a professor.

More tombstones and monuments were placed in honour of professors in the 1590s. According to Hegenitius's travelogue, the Pieterskerk also had a tombstone for Franciscus Raphelengius (1539–1597), professor of Hebrew, although the inscription on the tomb appears to have been incomplete in the seventeenth century and was never fully transcribed. Hegenitius dryly noted in his description the sole word of the inscription that was visible—'Raphe / lengio / rum'—and added that 'more does not exist'.[55] Hegenitius does mention a few other funerary monuments that are no longer present in the Pieterskerk. In his account, he gives an impression of the customary way to honour a professor. For example, he jotted down the inscription on the hanging monument dedicated to professor and rector magnificus Gerardus Bontius (1538–1599). According to the travelogue, the epitaph read:

54 Buchelius, 'Inscriptiones monumentaque', fol. 38$^{v}$.

55 '[P]lus non exstat'. Hegenitius, 'Iter Frisio-Hollandicum', p. 113.

For the great and almighty God, and for the memory of lord doctor Gerardus Bontius, a man renowned due to the praises of all disciplines and especially from both forms of medicine, who was professor *primarius* of medicine at the most flourishing Leiden University. He has taught publicly for 24 complete years and was also rector magnificus. He has left life on September 15, 1599. He lived 72 years [*sic*],[56] 6 months, and 11 days. The wife and children placed [this monument] according to his merit.

D[eo] O[ptimo] M[aximo]
et
memoriae D[omini] D[octoris] Gerardi
Bonti viri cum o[mn]ium
discipl[inarum] tum praecipue
utriusq[ue] medicinae laudib[us]
ornatiss[imi] qui in florentiss[ima]
Bat[avorum] academia primarius
medic[inae] professor magna
famae celebritate totis
XXIV an[nis] publ[ice] docuit
idemque rector magnif[icus]
rectoratu simul cum
vita defunctus est XV VIIbris an[no] MDIC vix[it] an[nos]
LXXII [*sic*] m[enses] VI d[ies] XI con[iunx] et lib[eri] m[erito] p[osuerunt][57]

Bontius had been a professor since the foundation of the university in 1575 onwards and a principal teaching professor, especially in the early days. The epitaph presented him as a learned man who excelled in his knowledge, but also as a famous and praiseworthy teacher. There is no reference to the city of Leiden, and Bontius is placed solely within the memory culture of the university, 'the most flourishing Leiden University'. His wife and children commissioned the monument, since the senate, *curatores* or burgomasters of the university did not play a role in the raising of monuments to its professors at this point.

56 Hegenitius probably made a mistake when transcribing the number of years Bontius lived, because by all accounts Bontius was born in 1538 and thus lived sixty-two years, not seventy-two. See, e.g., Johann Heinrich Zedlers, ed., *Grosses vollständiges Universal-Lexicon aller Wissenschafften und Künste*, vol. 4 (Halle; Leipzig: Verlegts Johann Heinrich Zedlers, 1733), p. 684.
57 Hegenitius, 'Iter Frisio-Hollandicum', p. 116.

It is important to stress that most graves do not have mottos or a long explanation. This appears to be an honour specifically reserved for professors in the Pieterskerk. Most graves have a common phrase in Dutch such as 'Hier leyt begraven 't huysch van ...' ('here lies the household of ...') and do not contain any reference to official positions, let alone virtues or achievements. The monuments dedicated to Bontius, Ramsay, and others in the 1580s and 1590s *all* refer to the university and the position the deceased held there. The inscription on Bontius's monument even explicitly mentions that he 'publicly taught for 24 complete years', besides identifying him as a *primarius* professor and noting that he died when he was rector of the university. He is characterised as a teacher, with an emphasis on his virtue of teaching and fame, without any reference to learning, erudition, or publications.

Nicolaus Stochius van Recklinghausen (1534–1593), who had been rector of the Latin school since 1556 and was by then a prominent Leiden citizen, was buried in the Pieterskerk and received a hanging monument.[58] This epitaph was executed in bronze according to Buchelius's account.

> To God, most good, most great, to Nicolaus Stochius of Recklinghausen, who following a call to Leiden from the esteemed Leiden City Council as rector of the public gymnasium, after spending 36 consecutive years in laudably managing the schools, piously and calmly left life. In August 1603, his faithful wife and grieving supporter of his work Adeleida van Boot placed this monument. He lived 58 years.

> D[eo] O[ptimo] M[aximo]
> Nicolao Stochio
> Reclinhusio,
> qui
> amplissimi senatus Leydens[is]
> accersitu.
> Lugdunum Batavorum publici
> gymnasii rector perpetuis
> annis XXXVII in scolarum admi-
> nistratione laudabiliter peractis.
> Pie placideque vita functus
> est, anno MDCIII sexti cal[endis]
> uxor Adeleida van Boot
> fidelis et laboris socia moerens

58 'Stochius, Nicolaas', *BWN*, vol. 17, p. 1012.

h[oc] m[onumentum] p[osuit].
Vixit annos LVIII.[59]

Stochius is presented here as an eminent Leiden citizen and senator, as well as a rector of the Latin school. The hanging monument was placed ten years after Stochius's death and is thus a true site of memory to remember his work as a rector. He is praised for his successful management of the school rather than for his teaching. His wife Adeleida van Boot positions herself on the monument as a 'faithful wife' and a 'supporter of his work', which is a rare acknowledgement of the important role played by women in managing a Latin school. Stochius and Van Boot thus wanted to be remembered as the couple who ran the Latin school in Leiden with piety and hard work.

The virtues of learning and teaching were generally reserved for professors, but also rarely associated with other eminent Leiden citizens. Take, for example, the tombstone of merchant Daniel van der Meulen (1554–1600) in the Pieterskerk.[60] Van der Meulen was extraordinary ambassador of the States General for several German princes, and in the last years of his life led many negotiations with eminent politicians. As a diplomat, he talked with archduke Albrecht about a possible peace between Spain and the Dutch Republic. He was a staunch defender of the freedom of religion and a mediator between Spain and the Dutch Republic. According to the *Inscriptiones monumentaque* of Buchelius, the inscription on his tomb read:

> Here lies the most honourable sir Daniel van der Meulen, pious, honest, virtuous, most equipped with learning, loved by the public and king. He was born in Antwerp, 23 October, in the year 1554. He died in Leiden on July 25 in the year 1600.
>
> Hic situs honoratiss[imus]
> D. Daniel vander
> Meulen, pietate pro-
> bitate, virtute, doc-
> trina instructiss[imus]
> reique publi[cae] amatiss[imus]
> natus Antwerpiae
> ante diem 23 Octob[ris]

59 Buchelius, 'Inscriptiones monumentaque', fol. 38$^{r}$.
60 *NNBW*, vol. 4, p. 568; *BWN*, vol. 12.2, pp. 734–735.

Anno 1554. Denatus
Leydae ante diem 25 Iuly A[nn]o 1600.[61]

The virtues and characteristics by which Van der Meulen was to be remembered were his piety, honesty, virtue, learning, and praise from the public and the king. The virtue of learning was thus an important distinguishing feature of the Leiden elite. It was commonly ascribed to ambassadors and political figures who had received higher education.

Learning was valued before the foundation of the university, and professors were honoured as a social and cultural elite from the foundation of the university onwards. In the 1580s and 1590s, the tombstones and hanging monuments dedicated to eminent Leiden citizens also stressed the virtues of knowledge and learning. The epitaphs dedicated to Weytsen, Dodonaeus, Holmannus, Ramsay, and Bontius all associate the deceased with the virtue of learning and a virtuous, exemplary life. From the very start of the university in 1575 the professor was hailed as an exemplary figure whose virtuous life and learning should inspire emulation. The professors were associated with exceptional learning and the piety and dutifulness with which they fulfilled their function. From the 1600s onwards, the cultural symbol of the professor would become ubiquitous and an even clearer marker of scholarly excellence and ideal citizenship.

## 6.3 Glory and Teaching: The Pride and Joy of *Batavia*, c. 1600–1640

From the 1590s onwards, the university wanted to establish itself as *the* Protestant university in the north of Europe. New professors were appointed, and new buildings and sites of knowledge built. The Faliebagijnkerk, for instance, was rebuilt as a library in 1591. Janus Dousa the Younger (1571–1596), the son of the famous curator Janus Dousa (1545–1604), became the new librarian in 1595, and presided over the relatively small collection, most of which had been bequeathed by the theologian Johannes Holmannus.[62] The senate of the university also had a plan to build a botanical garden from the early 1570s onwards. With the arrival of the eminent botanist Carolus Clusius

61 Buchelius, 'Inscriptiones monumentaque', fol. 39$^{v}$.

62 'Vorderinge van de Bibliotheecke', 29 December 1586, in *Bronnen tot de geschiedenis der Leidsche universiteit 1574–1811*, vol. 1, ed. P. C. Molhuysen (The Hague: Martinus Nijhoff, 1913), p. 138.

(1526–1609) in 1592, the plan to build a *hortus botanicus* finally materialised, and in 1594 it was finished.[63] In the same period, Justus Lipsius (1547–1606), one of the central figures in the development of the university in its early years, left. His departure in 1591 was considered an enormous loss for the university, especially by its curators.[64]

Around this time, a distinct Netherlandish historical awareness started to emerge. The fight against a common enemy, that is, the Spanish empire, strengthened patriotic feeling and helped create the idea of the Low Countries as a fatherland.[65] Yet due to the fragmented structure of the new state that emerged from the war against Spain, which gave great power to the individual provinces, the idea of a common fatherland gave way to local forms of identity, based on confessional and civic identities. After the resistance to Spanish and Catholic hegemony, the Dutch Republic emerged in 1588 with a desire to present itself as a flourishing state, continuing in the tradition of the Batavians who resisted Roman rule before them.[66] There was no such thing as a general Netherlandish identity in the early seventeenth century, but rather an amalgamation of identities often based on regional or civic histories of dissent and resistance.[67] In this context, the city of Leiden began to develop a distinct memory culture grounded on the importance of the siege of Leiden in the war against Spanish rule, as well as a newly discovered civic history as the centre of a Batavian settlement. In the formation of the Leiden civil historical consciousness, texts from antiquity, such as Tacitus's *Germania*, were appropriated to mould a new identity.[68] The Leiden historian Petrus Scriverius (1576–1660), for example, wanted to demonstrate in his work that the powerful position of the Dutch

63 Otterspeer, *Groepsportret* I, pp. 172–175.

64 Ibid., pp. 176–179.

65 Alastair Duke, 'In Defence of the Common Fatherland: Patriotism and Liberty in the Low Countries, 1555–1576', in *Networks, Regions and Nations: Shaping Identities in the Low Countries, 1300–1650*, ed. Robert Stein and Judith Pollmann, 217–240 (Leiden; Boston: Brill, 2010), p. 227.

66 Langereis, 'Geschiedenis als ambacht'; N. C. F. van Sas, ed., *Vaderland: Een geschiedenis van de vijftiende eeuw tot 1940* (Amsterdam: Amsterdam University Press, 1999).

67 Judith Pollmann, 'No Man's Land: Reinventing Netherlandish Identities, 1585–1621', in *Networks, Regions and Nations*, ed. Stein and Pollmann, pp. 241–261.

68 Two citations from Tacitus are the core of this so-called Batavian myth: *Germania*, 29, and *Historiae* IV, 14. For the development of the so-called Batavian myth, see Schöffer, 'The Batavian Myth'; H. Kampinga, *De opvattingen over onze oudere vaderlandsche geschiedenis bij de Hollandsche historici der XVIe en XVIIe eeuw* (Utrecht: Hes, 1980), esp. 56–86; Isidore Leonard Leeb, *The Ideological Origins of the Batavian Revolution: History and Politics in the Dutch Republic, 1747–1800* (The Hague: Martinus Nijhoff, 1973). For the appropriation of antiquity in the early modern period, see Karl A. E. Enenkel and Koen Ottenheym, *Oudheid als ambitie: De zoektocht naar een passend verleden, 1400–1700* (Nijmegen: Uitgeverij Vantilt, 2017); and Enenkel and

Republic dated back to influential people and cities in the medieval period and antiquity in 'Batavia', understood as the territory of the province of Holland.[69] From the 1600s onwards, Leiden University was thus a driving force behind the emergence of a strong sense of civic pride and identity.

The improving military and economic situation in the 1590s boosted the financial and intellectual situation of Leiden University, releasing funds for professors and facilities, and stimulating a steady increase in the student population.[70] In particular, Leiden University attracted many students of medicine owing to its botanical garden and its eminent professors, such as Carolus Clusius.[71]

Amid this cultural and economic growth, Leiden developed into a distinct university town. In turn, the university created an avalanche of memorial objects in print that were meant to visualise, represent, and memorialise the thriving university. Funerary monuments gained popularity and increased visibility in the Dutch Republic and Leiden was no exception.[72] Additionally, Leiden University had amassed a rich collection of portraits of illustrious scholars and professors which first appeared in the library in the first half of the seventeenth century.[73] This started with the gift of a portrait of Desiderius Erasmus by Richard Walius from Dordrecht to the university in 1597: from that moment onwards, the library of the university was arrayed with portraits.[74] The library's portraits can also be seen in the Woudanus prints, a set of four engravings from 1610 (see Figures 6.3–6.6). In the print of the library (see Figure 6.6), one can see the case of books that Joseph Justus Scaliger bequeathed to the university library after his death in 1609 on the right. Scaliger's collection of books, often termed the *Arca Scaligerana* (Scaliger's Ark), was one of the university library's prized possessions, as well as a tangible reminder that the great Scaliger had once adorned the university with the learning that he had gathered from the onslaught of history and safeguarded for scholarship. Scaliger did not necessarily see

Ottenheym, eds., *The Quest for an Appropriate Past in Literature, Art and Architecture* (Leiden; Boston: Brill, 2018).

69 Sandra Langereis, 'Geschiedenis als ambacht', pp. 203–264.

70 Israel, *The Dutch Republic*, pp. 569–575.

71 Ole Peter Grell, 'The Attraction of Leiden University for English Students of Medicine and Theology, 1590–1642', in *The Great Emporium: The Low Countries as a Cultural Crossroads in the Renaissance and the Eighteenth Century*, ed. C. C. Barfoot and R. Todd, 83–104 (Amsterdam; Atlanta, GA: Rodopi, 1992).

72 Scholten, *Sumptuous Memories*.

73 R. E. O. Ekkart, ed., *Icones Leidenses: De portretverzameling van de Rijksuniversiteit te Leiden* (Leiden: Universitaire Pers, 1973), esp. pp. 12–17.

74 Ibid., p. 9.

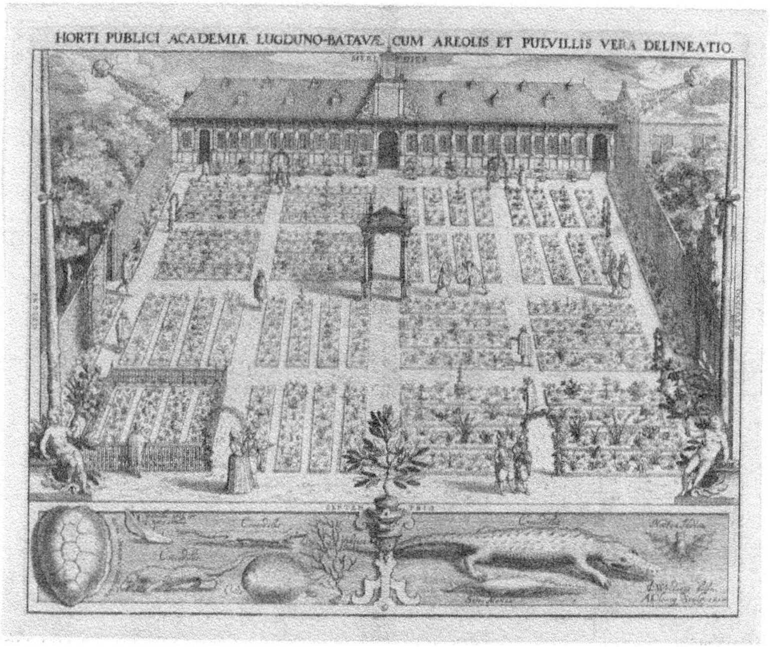


Figure 6.3. Willem Isaacz. van Swanenburg, *Hortus Botanicus of Leiden University*, after a drawing by Jan Cornelisz. van Woudt. Etch and engraving, 328 × 404mm. Leiden: Andries Clouck, 1610. Rijksmuseum: RP-P-1893-A-18089

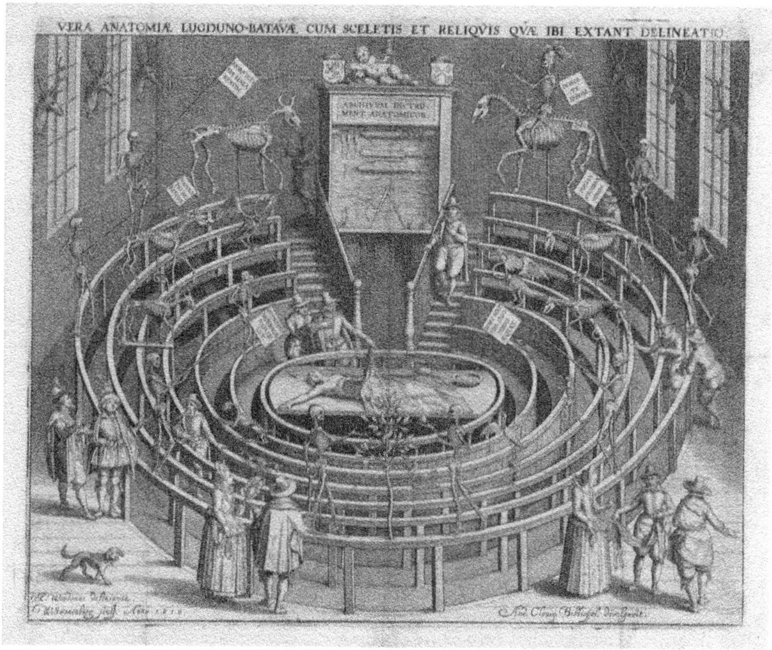


Figure 6.4. Willem Isaacz. van Swanenburg, *Anatomical Theatre of Leiden University*, after a drawing by Jan Cornelisz. van Woudt. Etch and engraving, 330 × 394mm. Leiden: Andries Clouck, 1610. Rijksmuseum: RP-P-1893-A-18090

Figure 6.5. Willem Isaacz. van Swanenburg, *Fencing School of Leiden University*, after a drawing by Jan Cornelisz. van Woudt. Etch and engraving, 333 × 402mm. Leiden: Andries Clouck, 1610. Rijksmuseum: RP-P-1893-A-18091

Figure 6.6. Willem Isaacz. van Swanenburg, *Library of Leiden University*, after a drawing by Jan Cornelisz. van Woudt. Etch and engraving, 330 × 403mm. Leiden: Andries Clouck, 1610. Rijksmuseum: RP-P-1893-A-18092

his bequest as an essential addition to the library collection, but rather one that underlined his own noble descent and his strong ties to Leiden University with the bequest.[75]

A clear example of the emergence of an academic memory culture at Leiden is Johannes Meursius's *Illustris Academia Lugd[uno]-Batava* (The illustrious Leiden University), a collection of printed portraits and eulogies of eminent professors of Leiden University published in 1613.[76] The *Illustris Academia* is a reflection of the standing of the university in the early seventeenth century, as well as the desire to put Leiden on the map as an institution with a rich history and illustrious professors. It showed the overlapping institutional and civic identities that Leiden University wanted to present to the learned world. The booklet was republished and expanded in 1625 with the title *Athenae Batavae* (Athens of Batavia), with the same frontispiece (see Figure 6.7). The four engravings that show parts of the university were also part of Meursius's *Illustris Academia* and *Athenae Batavae*. These four engravings portrayed the four most important and unique sites of the university: the botanical garden, the anatomical theatre, the fencing school, and the library (see Figures 6.3–6.6).

Meursius's booklet and the engravings were part of a campaign to put Leiden University on the map as an outstanding, modern university. Anthony Grafton emphasised the research-driven education that made Leiden unique in the first two decades of the seventeenth century, due to the extraordinary professors Clusius and Scaliger, who were not required to teach 'normal' seminars but did inspire advanced students to carry out their own research.[77] Meursius's *Illustris Academia* and *Athenae Batavae* stress the importance of publication, writing, and research for the fame of a professor, as well as the university at large. This novel way of thinking about research can clearly be seen in the monuments built for Joseph Scaliger and Carolus Clusius, which will be closely analysed further on in this section.

The Pieterskerk was the main church for the burial and remembrance of eminent professors from the 1600s onwards. The funerary monuments and tomb stones from this period reflect the reconfiguration of the institutional identity of the university. Important professors obtained prominent hanging

75 Kasper van Ommen, '"Tous mes livres de langues estrangeres": het oosterse legaat van Josephus Justus Scaliger in de Universiteitsbibliotheek Leiden' (PhD diss., Leiden University, 2020), pp. 251–252.

76 Johannes Meursius, *Illustris Academia Lugd[uno]-Batava: id est Virorum Clarissimorum Icones, Elogia ac vitae, qui eam scriptis suis illustrarunt* (Leiden: Andries Clouck, 1613).

77 Anthony Grafton, *Athenae Batavae: The Research Imperative at Leiden, 1575–1650* (Leiden: Primavera Pers, 2003), esp. 14–31.

RP-P-1885-A-9084

Figure 6.7. Willem Isaacsz. van Swanenburg, title page of Joannes Meursius, *Athenae Batavae. Sive, De Urbe Leidensi, & Academia, Virisque claris; qui utramque ingenio suo, atque scriptis illustrarunt: libri duo*, 2 vols. (Leiden: Andries Clouck; Elsevier, 1625). Rijksmuseum: RP-P-1885-A-9084. The same frontispiece was also used for Meursius, *Illustris Academia Lugd[uno]-Batava* […] (Leiden: Andries Clouck, 1613)

monuments and the texts presented them as adornments of the city and university. A prime example is Johannes Heurnius, professor of medicine from 1581 until his death in 1601. He had also been *rector magnificus* six times.[78] The hanging monument of Johannes Heurnius (1543–1601) related

78 E. D. Baumann, 'Heurnius, Johannes', *NNBW*, vol. 4 (1918), pp. 745–746.

his greatness to both the city of Leiden and the university and promotes the concept of a city or state that owes something to scholars. The inscription reads:

> Heurnius-scion, the sun of Batavian physicians, has passed
> While he lived, Hygeia always lived
> As Cos owes much to Hippocrates, and Pergamon to Galen
> So much, great Heurnius, the land of Batavia owes to you
>
> Occidit Heurniades medicorum Bataviae sol,
> quo vivo semper vivida Hygeia[79] fuit.
> Cos[80] quantum Hippocrati, Galeno Pergama[81] debent,
> Tantum, magne Heurni, terra Batava tibi

Heurnius is represented in this epitaph as the centre of the community of Leiden physicians. The comparison with Hippocrates and Galen positions him in a tradition of great healers and scholars of medicine. What is more, these men are presented as adornments of the cities where they lived: Cos and Hippocrates; Pergamon and Galen; Leiden and Heurnius. Heurnius is confidently presented to the onlookers as one of the greatest healers the world has ever known. After Hippocrates and Galen, the epitome of medicinal knowledge could now be found in the Athens of the North.

It had also become common practice by that time to honour deceased professors in funerary poems. Daniel Heinsius (1580–1655) became the official occasional orator of Leiden University from 1600 onwards and was a devoted writer of funerary poetry.[82] The monuments in the Pieterskerk and the funerary poetry have a different focus. In the funerary poem dedicated to Heurnius, Heinsius wrote: 'Here that Heurnius-scion lies, to whom so many great names are indebted, to whom medicine owes herself'.[83] He stresses Heurnius's position in the field of medicine, as well as his marriage

79 Hygeia is the Ancient Greek goddess of health, cleanliness, and sanitation.

80 Reference to the city of Cos, the birthplace of Hippocrates.

81 Reference to the city of Pergamon, the birthplace of Galen.

82 Barbara Becker-Cantarino, *Daniel Heinsius* (Boston, MA: Twayne, 1978); D. J. H. ter Horst, 'Daniel Heinsius (1580–1655)' (PhD diss., Utrecht University, 1934); Sonja Fortunette Witstein, 'Funeraire poëzie in de Nederlandse renaissance: enkele funeraire gedichten van Heinsius, Hooft, Huygens en Vondel, bezien tegen de achtergrond van de theorie betreffende het genre' (PhD diss., Utrecht University 1969); Eckard Lefèvre, 'Daniel Heinsius' *Manes Lipsiani* nebst einer Erklärung von Joseph Scaligers *Epicedium* auf Lipsius', in *Daniel Heinsius: Klassischer Philologe und Poet*, ed. Eckard Lefèvre and Eckart Schäfer, 203–248 (Tübingen: Gunter Narr Verlag, 2008).

83 'Hic ille Heurniades, cui se tot maxima debent / Nomina, cui debet se medicina, iacet'.

to his 'most faithful wife'.[84] Of this, Heinsius writes, 'Just as they were once conjoined in marriage, now joined in the grave, to continue the faithful pact'.[85] The tone of the funerary poem is different from the epitaph in the Pieterskerk. Heinsius's poem does not address the fame of the city of Leiden or even the university, but rather portrays Heurnius as a modest, pious, and skilful man of medicine. But the epitaph is more noteworthy due to the comparison with the death of Erasmus: 'The cruel executioner of the great Erasmus took him away; which he himself took away from others so often with the skills of his art'. Heurnius and Erasmus died of bladderstones and the epitaph claims that Heurnius cured many others from that torturous disease. The reference suggests that Heurnius was seen as a great man of learning as much as 'the great Erasmus'. In the poem, Heurnius is not praised by the city of Leiden or the university, but his death is seen as a loss for the entire discipline of medicine. This suggests a more intimate circle of readers, rather than the general onlooker and student in the Pieterskerk.

This display of restraint and piety is a common theme in the epitaphs and funerary poems. The epitaph dedicated to theology professor Franciscus Junius the Elder (1545–1602) is an example of such restraint and simplicity. The original tombstone has been lost, but both Buchelius and Hegenitius made transcriptions of the complex, original chronogram in 1630, in which the capitalised letters add up to 1602 in Roman numerals.

> Thrice the sixth plus fifth day of October [i.e., 23 October] shines bright, and the sixth hour resounds twice on Leiden soil, in the world where the Titan Francis Junius, touched by the black plague, reaches out to the blessed kingdom of heaven.
>
> OCtobrIs ter sena DIes & qVIna refVLget
> SenaqVe LeyDensI bIs sonat hora soLo
> OrbIs VbI tItan franCIsCVs IVnIVs atra
> TaCtVs peste, aethrae regna beata petIt.[86]

84 '[F]idissima coniunx'.

85 'Sic olim coniuncta thoro, nunc iuncta sepulcro, / Continuat pactam, nondum aliena, fidem'. The entire epitaph reads: 'Hic ille Heurniades, cui se tot maxima debent / Nomina, cui debet se medicina, iacet. / Sustulit hunc magni tortor crudelis Erasmi: / Quem toties aliis sustulit artis ope. / Ne tumulo quid desit, adest fidissima coniunx; / Et vitae comitem nunc quoque mortis habet. / Sic olim coniuncta thoro, nunc iuncta sepulcro, / Continuat pactam, nondum aliena, fidem.' Printed in Adam, *Vitae Germanorum medicorum* [...], *Vitae II* (Heidelberg: Jona Rosa, 1620), p. 374.

86 Hegenitius, 'Iter Frisio-Hollandicum', p. 114; Buchelius, 'Inscriptiones monumentaque', fol. 37$^{v}$.

The tombstone simply declared Junius's death, without any praise or reference to his position as a theology professor, and with very complex references to the date and hour he died. In contrast, the poetry and lamentations that followed framed his death as a loss to the university and the learned world. The written funerary culture cultivated and expressed a stronger academic identity compared to the funerary monuments in the Pieterskerk. An example of this difference can be seen in the lamentation on Junius written by Joseph Justus Scaliger:

> Junius, mighty in learning, surrounded by the docile circle of youth who hangs on the lips of the speaker in a crowded meeting at the Leiden Gymnasium [...] A mourning school weeps for you, their teacher; the bereaved church [weeps] for you, her parent; and the whole world laments you, a learned man.
>
> Iuni, quem modo literis potentem,
> Pleni Gymnasii frequente coetu,
> Cingebat docilis corona pubis
> Docto pendula disserentis ore:
> [...]
> Te moerens scola flet suum magistrum,
> Orba ecclesia te suum parentem,
> Doctorem gemit orbis universus.[87]

In this lamentation, Scaliger remembered Junius foremost as a professor at Leiden University, in the middle of a crowded classroom. Scaliger identified three groups that mourned Junius in a different way: the university missed an exquisite teacher, the church missed a parent, and the 'whole world' missed a learned man. These three different audiences clarify the different aspects of the memory culture we see in the Pieterskerk. In the church and on the monuments, the professors were remembered less as teachers than as exquisite pinnacles of knowledge that enriched the university. There is, however, rarely a reference to a wider learned world, such as a *respublica literaria*. Universal acclaim in the learned world is often phrased as a variation of 'praised by everyone' ('ab omnibus laudatus'). The move away from the characterisation of professors as teachers to outstanding

87 Joseph Justus Scaliger, *Poemata omnia: ex museio Petri Scriverii* ([Leiden]: ex Officina Plantiniana Raphelengii, 1615), p. 106; also found in Gerardus Johannes Vossius, *De historicis latinis libri tres* (Leiden: Johannes Maire, 1627), [sig. **3$^{r-v}$].

men of learning who bring fame to Leiden reaches its culmination after the death of two extraordinary professors in 1609.

The most prominent remembrance of learned men was on display in 1609 after the death of two honorary professors, Carolus Clusius and Joseph Justus Scaliger (1540–1609). They had no teaching obligations, and the sole aim of their tenure was to elevate the status and fame of Leiden University. Both men died in 1609 and received an epitaph and a grave, financed from university funds. Honorary professors were officially not professors since they did not attend senate meetings, nor was their name on the list of professors. The honorary professor in Leiden can best be understood as a public *exemplum* of learning and virtue whose excellence was meant to cement the reputation of the university as the most erudite centre in Europe. The text below the portrait of Scaliger in the university collection likewise does not describe him as a professor, but as 'the ornament of the academy' ('decus academiae'), attesting to the contemporary perception of the honorary professor as one who bestows fame on the university.[88]

Clusius was appointed professor at Leiden University in October 1593, and also became the prefect of the botanical garden, one of the earliest and best-known university-owned botanical gardens at that time in Europe, which was also displayed (and boasted about) in Meursius's *Illustris Academia*. The arrival of Clusius in Leiden had sparked the study of natural history and botany there and consequently attracted many students of medicine. Clusius brought the court fashion of botanical knowledge and *naturalia* into an academic discipline and thereby put Leiden on the map as a centre of botanical knowledge.[89] After his death in 1609, Clusius received a hanging monument in the Walloon church, which was later moved to the Pieterskerk (see Figure 6.8). The inscription read:

> Placed in loving memory of Carolus Clusius of Arras, who was appointed at the court of the Holy Roman Emperor, because of the fame of his name, acquired on account of his honesty, erudition, and in particular his explanation of herbal matter; after various travels he was finally summoned to this city by the most noble and highest curators and consuls to help adorn the university, and was honoured with a public stipend for 16 years. In the 84th year of his life, he died unmarried on April 4, 1609.

88 Otterspeer, *Groepsportret* I, pp. 86–87.

89 Florike Egmond, *The World of Carolus Clusius: Natural History in the Making, 1550–1610* (London: Pickering & Chatto, 2010), pp. 157–173; Egmond and P. G. M. Hoftijzer, eds., *Carolus Clusius: Towards a Cultural History of a Renaissance Naturalist* (Amsterdam: Koninklijke Akademie van Wetenschappen, 2007).

Figure 6.8. Hanging monument for Carolus Clusius, c. 1609. Photograph by author, courtesy of Pieterskerk, Leiden

Bonae memoriae
Caroli Clusii Atrebatis
pos[uit]
Qui ob nominis celebritatem probitate, eruditione, tum rei inprimis Herbariae illustratione
partam, inter aulae Caes[aris] familiares allectus, et post varias peregrinationes a nob[ili]b[us] demum et ampliss[imis] D[ominis] cur[ato]r[ibus] et co[n]s[ulibu]s, in hanc urbem condecorandae Academiae evocatus, et

stipendio publico per annos XVI honoratus. XXCIV ae[tatis] s[uae] annum ingressus, obiit caelebs IV. apr. MDCIX.

The epitaph dedicated to Clusius showcases the ideal of the learned man in the context of Leiden University. Since he was no ordinary professor, but rather an ornament of the university, the epitaph helps us understand the contemporary criteria for being perceived as such an outstanding man of learning. First, the epitaph lists the virtues and successes of Clusius, and by extension of the ideal learned man: he was famous 'on account of his honesty, erudition, and in particular his explanation of herbal matter'. The combination of the virtues honesty (*probitas*) and erudition (*eruditio*) is striking, because it refers to the Protestant ideal of a humble, honest scholar.

The university paid for the cost of the hanging monument and half of the costs of the funeral. The 'most noble and highest curators and consuls' prominently positioned themselves on the epitaph.[90] They, after all, honoured Clusius with a stipend and enticed him to leave the court of Maximilian II in Vienna. The honorary professor deserved a monument that reflected his standing, affirming Clusius's presence at Leiden University to heighten its fame.

Joseph Justus Scaliger, the second honorary professor who died in 1609, received similar treatment. Scaliger was closely connected with the prime curator of the university, Janus Dousa, who was a well-respected man of learning with good connections in French Protestant circles. The two met each other in Paris when Dousa studied there, and Scaliger was omnipresent in Dousa's scholarly work. He called Scaliger 'the most learned of all', 'divine', and 'our great ruler'.[91] According to his funeral oration, Dousa personally took credit for bringing both Lipsius and Scaliger to Leiden: 'He used to attribute to himself the glory of two most famous men: Lipsius and Scaliger.'[92] Dousa was tasked with finding a replacement for Justus Lipsius when he left the university, and was responsible for Scaliger's arrival in Leiden in 1593.[93]

90 '[N]obilissimi et amplissimi domini curatores et consules'.

91 '[O]mnium doctissimus'; 'divinus'; 'Imperator ille noster, quem dico, Scaliger'. C. L. Heesakkers, 'Scaliger en de Dousa's', in *Adelaar in de wolken: De Leidse jaren van Josephus Justus Scaliger, 1593–1609*, ed. P. G. Hoftijzer, 36–57 (Leiden: Scaliger Instituut; Universiteitsbibliotheek Leiden, 2005), p. 39.

92 'Duorum tamen virorum celeberrimorum gloriam sibi uni solitus asserere: Lipsii et Scaligeri.' Petrus Bertius, *Een welbesteed leven en een vroom verscheiden: Lijkrede op Janus Dousa (1604)*, ed. C. L. Heesakkers and W. G. Heesakkers-Kamerbeek (Amersfoort: Florivallis, 2004), pp. 38–39; cited in Heesakkers, 'Scaliger en de Dousa's', p. 43, n. 7.

93 P. C. Molhuysen, *De komst van Scaliger in Leiden* (Leiden: A. W. Sijthoff, 1913).

Figure 6.9. Hanging monument for Joseph Justus Scaliger, c. 1609. Photograph by author, courtesy of Pieterskerk, Leiden

Both the States General and the political elite of Leiden were elated when Scaliger accepted their offer to come to Leiden. The burgomasters of Leiden and the curators of the university welcomed him with open arms, since they had successfully tied one of the most prominent scholars to their city and university.[94] Scaliger brought considerable allure to Leiden University: the number of French students greatly increased and the university's European appeal became stronger.[95]

Scaliger's death in 1609 offered an occasion to solidify the reputation of Scaliger, the city of Leiden, and the university. The university was closed for an entire week of mourning. Scaliger's star pupil, Daniel Heinsius, wrote an extensive eulogy and bemoaned Scaliger's death as a loss for the learned world and the city of Leiden.[96] The curators of the university ordered a hanging monument to be put above his tombstone. Quite ironically, the placement of the monument was strictly against Scaliger's own last will and testament, which stipulated he wanted a modest tombstone without opulence.[97] Gotfried Hegenitius commended Scaliger's modesty in his travelogue, especially because of the short, yet striking, inscription on his tombstone: 'Joseph Justus Scaliger, son of Julius Caesar, here I await my resurrection'.[98] The hanging monument placed above the tombstone, however, was a far cry from a display of moderation (see Figure 6.9).[99] The Scaliger tomb and hanging monument were originally placed in the Walloon church, the Vrouwekerk, but were moved to the Pieterskerk in 1819.[100] When the hanging monument was placed in 1609, Daniel Heinsius once again gave an oration to inaugurate the monument.[101] The hanging monument with epitaph can still be found in the Pieterskerk and reads:

> Shrine to the greatest and highest God and to the eternal memory of Joseph Justus Scaliger, son of Julius Caesar from Burden, descendant of the princes of Verona.

94 Ibid., pp. 11–12.

95 H. J. de Jonge, 'Scaliger in Leiden', *Jaarboekje voor Geschiedenis en Oudheidkunde van Leiden en omstreken* 71 (1979): 71–94.

96 Daniel Heinsius, *In obitum v[iri] illustr[i] Iosephi Scaligeri Iul[ii] Caes[ari] a Burden. f[ilii] eruditorum Princeps Orationes Duae* […] (Leiden: Ex Officina Plantiniana Raphelengii, 1609).

97 Henk Jan de Jonge, 'The Latin Testament of Joseph Scaliger, 1607', *Lias* 2, no. 2 (1975): 249–263.

98 'Iosephus Iustus Scaliger / Iul[ii] Caes[aris] Fil[ius] / Hic expecto resurrectionem'. The full passage reads: 'Ipse alias, qua era modestia, non aliud monumento incidi voluit, nisi hoc breve oraculum, quod etiamnum illic legitur: Iosephus Iustus Scaliger / Iul[ii] Caes[aris] Fil[ius] / Hic expecto resurrectionem. In extrema parte lapidis sepulchralis: *Terra haec ab Ecclesia, em[p]ta est, / Nemini cadaver huc inferre licet.*' Hegenitius, 'Iter Frisio-Hollandicum', pp. 99–100.

99 Henk Jan de Jonge, 'Grafsteen en graf van Scaliger', *Jaarboekje voor geschiedenis en oudheidkunde van Leiden en omstreken* 70 (1978): 91–96; R. E. O. Ekkart, 'Het grafmonument van Scaliger', *Jaarboekje voor geschiedenis en oudheidkunde van Leiden en omstreken* 70 (1978): 81–90.

100 Molhuysen, *De komst van Scaliger in Leiden*, p. 13; De Jonge, 'Scaliger in Leiden', pp. 87–88.

101 Daniel Heinsius, 'Oratio II, Ad Amplissimos Academiae Curatores, & Urbis Consules, de monumento Iosepho Scaligero ab iis decreto', in Heinsius, *In obitum Iosephi Scaligeri*, pp. 33–39.

For the man, who with undefeated spirit, together with his father, a mighty hero, rose against fate and claimed his right to an imperium, taken from his forefather. He reclaimed [it], almost fatally, with his excellent brilliance, his indefatigable work and an erudition that was extraordinary in the *respublica literaria*.
Yet, [this monument is] above all for his modesty. The monument that he forbade to be made for himself, the *curatores* of the academy and the burgomasters of the city, who called him to this city, willingly made sure to be placed here.
He himself left a monument to himself in the minds of men.

Deo Opt[imo] Max[imo] sacrum et
aeternae memoriae Iosephi Iusti
Scaligeri I[ulii] Caes[aris] a Burden F[ilii]
principum Veronensium nepotis
viri qui invicto animo una cum
parente heroe maximo contra for-
tunam adsurgens ac ius suum sibi
persequens imperium maioribus
ereptum ingenio excelso labore
indefesso eruditione inusitata in
literaria rep[ublica] quasi fataliter recu-
peravit sed praesertim eiusdem
modestiae quod sibi fieri vetuit
iidem qui in urbem hanc vocarunt
curatores academiae ac urb[is] co[n]s[ule]s
hoc in loco monumentum p[onendum] e[rigendum] l[ibentes] c[uraverunt]
ipse sibi aeternum in animis hominum reliquit[102]

How, then, was Scaliger memorialised and eulogised, and on what accounts was he presented as a model of the Leiden university scholar? First, the epitaph tries to position Scaliger within a lineage of Veronese princes. Scaliger himself had written a defence of his ancestry, and (erroneously) fashioned himself as a descendant of the noble Della Scala family.[103]

102 Transcription from the epitaph in the Pieterskerk. For a design of the epitaph with a copy of the text, see Heinsius, *In obitum Iosephi Scaligeri*, p. 30.
103 Karl A. E. Enenkel, *Die Erfindung des Menschen: Die Autobiographik des Frühneuzeitlichen Humanismus von Petrarca bis Lipsius* (Berlin; New York: Walter de Gruyter, 2008), ch. 24, 'Autobiographie als Genealogie: Joseph Scaligers *Epistola de vetustate et splendore gentis Scaligerae* (1594)', pp. 728–755.

Figure 6.10. Hanging monument for Willebrord Snellius, c. 1627. Photograph by author, courtesy of Pieterskerk, Leiden

Noble descent, after all, was considered an essential characteristic of an eminent man, even in learned circles.[104] Second, the epitaph presents Scaliger as a man of distinct scholarly virtue: who can be characterised

104 Marian Füssel, 'A Struggle for Nobility: 'Nobilitas literaria' as Academic Self-Fashioning in Early Modern Germany', in Kirwan, *Scholarly Self-Fashioning*, pp. 103–120.

by his 'excellent brilliance', 'indefatigable work', and 'erudition that was extraordinary in the *respublica literaria*'. Thirdly and finally, the religious and Protestant virtue of modesty is on display, since the pompous language of the hanging monument contrasts strongly with the modest tombstone Scaliger wanted for himself, dryly noting he is awaiting his resurrection in the tomb beneath the inscription ('hic expecto resurrectionem'). The tombstone and its short inscription were seen as a clear example of his modesty by Hegenitius, among others, and so were Scaliger's protestations at being honoured in Leiden at all, although the sincerity of these claims is dubious at best. Regardless, the university would not miss a chance to elevate their 'jewel' and frame him as an example of outstanding learning at Leiden University.

Clusius and Scaliger's monuments were examples of how to honour extraordinary men of learning in Leiden. The monuments that followed adopted the same hyperbole and pomp. An example of this is the monument designed for Rudolph Snellius (1546–1613), professor of mathematics. It is not known whether the monument really existed or if the extant drawing is only a detailed sketch of a monument that never materialised.[105] Despite the uncertain status of the monument, the epitaph does allow us to see how Rudolph's son, Willebrord Snellius (1580–1626), wanted to remember his father. The design of the hanging monument shows the family emblem on the top, with two pillars on the side. The inscription explains that Rudolphus Snellius taught at Leiden University, 'with great care of the faith and for the common good' and that he was *rector magnificus* twice.[106] Surprisingly, the epitaph does not refer to Rudolph's status as a professor or even his discipline of mathematics. The son, Willebrord Snellius, himself became an influential professor of mathematics and astronomy in Leiden. In turn, his children placed a monument—which the epitaph refers to as a 'shrine' ('sacrum')—in the Pieterskerk for both Willebrord and his wife Maria de Lange (see Figure 6.10), inscribed with an epitaph to remember Willebrord as a famous professor:

> A shrine, to the greatest and almighty God and to posterity. To the most famous and learned man, Willebrord Snellius from Royen, the mathematicians' eyes among the Batavians, and professor of mathematics at the university here, most famous wherever he turned, most skilled, most

105 For details on Rudolph Snellius's monument, see Knöll, 'Creating Academic Communities', pp. 418–419.

106 '[C]um cara fide et bono publico exegit'.

esteemed, most worthy; and to his sweetest, most chaste, most beloved wife Maria de Lange[.]

D[eo] O[ptimo] M[aximo]
et
Posteritati sacrum.
Clarissimo doctissimoque viro d[omi]no
Willebrordo Snellio a Royen
Mathematicorum in Batavis
Ocello et in academia quae hic
Celeberrima mathematum pro-
fessori qua qua versum celeber-
rimo, sollertissimo, dignissimo
Meritissimo
Nec non
Lectissimae Castissimaeque matronae
Maria de Lange coniugi charissimae [.][107]

The epitaph hails Willebrord Snellius as an outstanding professor. The emphasis on teaching that was present in earlier monuments has made way for an emphasis on personal virtue and fame. Snellius was skilled, esteemed, and worthy, and famous everywhere he turned. Maria de Lange, in turn, is characterised here as the wife of Snellius and by the female virtue of chasteness, reminding us that women were often only honoured in relation to a highly esteemed professor husband.

The epitaph on the hanging monument dedicated to Petrus Bockenberg (1548–1617), a Leiden historiographer, represents him as a political figure without references to virtues of knowledge or learning (see Figure 6.11).[108] Bockenberg was married to the daughter of Johannes Wijkersloot, rector of the Latin School in Woerden and then lived in Leiden from 1589 onwards. He dedicated himself to historical studies, financially supported by the States of Holland and Zeeland. Oldenbarnevelt appointed him historian of

107 The full inscription reads: D[eo] O[ptimo] M[aximo] / et / Posteritati sacrum. / Clarissimo doctissimoque viro d[omi]no / Willebrordo Snellio a Royen / Mathematicorum in Batavis / Ocello et in academia quae hic / Celeberrima mathematum pro: / fessori qua qua versum celeber: / rimo, sollertissimo, dicnissimo / Meritissimo / Nec non / Lectissimae Castissimaeque matronae / Maria de Lange coniugi charissi: / mae hoc qualecunque / MNHMOSYNON / Debitae erga parentes observantiae / TEKMHRION sive indubitatum si / gnum: liberi moesti posuere. / Denatus XXX Octob: / MDCXXVI / Denata XI Novemb: MDCXXVII.'

108 H. Brugmans, 'Bockenberg, Pieter Corneliszn.', *NNBW*, vol. 6, pp. 123–124.

Figure 6.11. Hanging monument for Petrus Bockenberg, c. 1617.
Photograph by author, courtesy of Pieterskerk, Leiden

Holland in 1591. Janus Dousa and professor Dominicus Baudius (1561–1613) thought that this was a mistake and provoked and criticised Bockenberg, prompting an extensive polemic.[109] Despite all his work as a historiographer, the epitaph Arnoldus Buchelius encountered in the Pieterskerk did not elaborate on this at all and shows that the careers of scholars were told in terms of the positions that they held.

109 See, e.g., the polemical work of Bockenberg, *Ad nonnullas Jani Dousae aspera scripta responsio illustribus Hollandiae et Westfrisiae ordinibus reddita* (Delft: Ex officina Brunonis H. Schinckelii, 1601).

To Petrus Bockenberg, historiographer of Holland and Zeeland, and to the dutiful grief of his pious wife, who placed this monument for her husband. He was born in Gouda in the year 1548; died in Leiden in the year 1617, on January 16th; having exceeded the 68th year of his life.

Petro
Bockenbergio
Historiographo
Hollandiae Zeelandiae
et piae coniugis officio-
sus in maritum dolor
monum[entum] pos[uit]
Natus Goudae A[nn]o MDXLVIII
Denatus Leydae A[nn]o MDCXVII
Ianuarii die XVI.
Egressus annum vitae LXVIII[110]

Bockenberg had no connection to Leiden University and is rendered a 'historiographer of Holland and Zeeland'. This shows that the memory culture in the Pieterskerk after the foundation of the university was not exclusively reserved for professors but encompassed the remembrance of Leiden learned men more generally, as could also be seen on the monument dedicated to Quintinus Weytsen.

After Clusius and Scaliger, most Leiden professors were honoured with elaborate hanging monuments. They often recounted the life story of the professors and their numerous virtues and qualities. Reinier Bontius (1576–1623), professor of medicine, received a hanging monument that characterised him as a 'doctor and professor of medicine in this university' and the son of Gerardus Bontius, whose hanging monument has been discussed above.[111] The epitaph stresses Bontius's modesty, erudition, and humanity, as well as

110 Buchelius, 'Inscriptiones monumentaque', fol. 41$^{r}$. The last line in Buchelius reads 'Annu[m] vitae LXVIII egressus', but the monument clearly reads 'Egressus annum vitae LXVIII' (see Fig. 6.11).

111 'Medicinae doctor ac professor in hac academia'. The full epitaph reads: D[eo] O[ptimo] M[aximo] / et felici memoriae Reineri Bontii / Medici eximii Gerardi Medicinae doctoris / ac professoris in hac academia praeclari / Filio praeter eruditionem comitate / ac humanitate generis humani delicio / Modestiae ac suavitatis exemplo qui / Annos aliquot medicinam ac philosophiam / Ibidem professus ac magnifici rectoris / Functus munere summi insuper excelsique / Herois Mauritii D.G. Principis Auriaci / Comitis Nassauiae Castrensis medicus / Domesticusque inaffectata morum / Dexteritate ita invictum armis expugnavit / Animum ut arcanis quibusdam rebusque / ab eo adhibitus gravissimis invito ac / cum ingenti eius luctu quem

Figure 6.12. Hanging monument for Thomas Erpenius, c. 1624. Photograph by author, courtesy of Pieterskerk, Leiden

his function as *rector magnificus* and personal physician of Prince Maurice of Orange.

The hanging monument and tombstone of Thomas Erpenius (1584–1624), who became the first professor of oriental languages (Arabic and Hebrew) in 1613, is similar to Scaliger's tombstone and monument (see Figure 6.12). The inscription on Erpenius's tombstone was very simple and short, clearly

mirifice testatus / est ereptus fuerit haeredes liberique / hoc in loco monumentum publicum / L[ubentes] P[ondendum] D[edicandumque] C[uraverunt]'.

modelled after Scaliger's tomb. According to Hegenitius's travelogue, the tombstone read: 'Thomas Erpenius awaits the resurrection of the faithful here'.[112] In contrast with the frugal tombstone, the hanging monument gives a long account of Erpenius's life and achievements.[113] Its epitaph characterised him as a 'man exceptional in all types learning' and emphasised his tenure as a Leiden professor.[114] Erpenius was hailed for his 'distinguished life of integrity, piety, probity, modesty, and the pleasantness of his manners'.[115] Learning was the first qualification, but the distinct Protestant virtues of piety and modesty were crucial characteristics of a professor at Leiden University as well, reflected in both the restraint on Erpenius's tombstone and the enumeration of virtues on the hanging monument.

The monuments from the period between roughly 1600 and 1640 represent the ideal man of learning as a well-known and distinguished professor, who is not so much concerned with teaching, but rather with bringing fame to the university and city, while living a virtuous and pious life. This is in accord with Anthony Grafton's observation that Leiden University became a centre for research-driven education.[116] This development can also be seen in the representations on the monuments: the monuments dedicated to professors in the period before 1600 stressed the virtues of learning and teaching, but after the examples set by the non-teaching extraordinary professors Clusius and Scaliger, the focus shifted to a representation of erudition and renown. Additionally, the professors were increasingly represented as the glory of Leiden and the university. After all, the curators of the university and the burgomasters of Leiden paid for the monuments of Scaliger and Clusius, to safeguard their memory as the pride of the city. There are cursory references to nobility and descent, such as the presentation of Scaliger as a

112 'Thomas Erpenius hic fidelium resurrectionem expectat'. Hegenitius, 'Iter Frisio-Hollandicum', p. 72.

113 The full inscription reads: 'Thomae Erpenio, / Viro in omni genere literarum eximio, / orientalium linguarum peritissimo, / earumdemque in alma Batavorum Academia / professori longe praestantissimo, / librisque suis et Arabicis typis claro, / vitae integritate, pietate, probitate, / modestia, morumque suavitate insigni, / An[no] MDLXXXIV, III eidus Septemb., / Gorcomii in Batavis bono egregii publici nato / eiusque incomparabili damno Lugduni Batavorum / in flore (heu!) aetatis suae eidib[us] Novembr[is] denato, / An[no] MDCXXIV / Iacoba Buyesia / ampliss[imi] et integerrimi Hollandicae curiae senatoris / Iohannis Buyesii Monachodamensis Ic[ti] filia, optimo et benemerenti coniugi, cum quo octennium / sine querela vixit, trium ex eo liberum parens / contra votum moestissima posuit / *vale vale dulcissime—sit tibi terra levis.* / Homo bulla—Ruit hora.'

114 'Vir in omni genere literarum eximius'.

115 '[V]ita integritate, pietate, probitate, modestia, morumque suavitate insignis'.

116 Grafton, *Athenae Batavae*.

descendant of a noble Veronese family. As we will see in the next section, this representation of learned men as originating from noble families became more prominent in the second half of the seventeenth century.

## 6.4 Family, Nobility, and Personal Greatness, c. 1640–c. 1750

The monumentalisation of professors in the Pieterskerk continued in the second half of the seventeenth century, although the emphasis on research and excellent scholarship gave way to remembrance in a personal and familial context. Hanging monuments slowly lost their popularity and made place for larger monuments, such as ornamental tombs, busts, and sculptures.

The professor of theology Johannes Polyander van Kerckhoven (1568–1646) received one of the last hanging monuments in the Pieterskerk. It was made by the sculptor Rombout Verhulst (1624–1698) and placed by the professor's son, Johannes Polyander Jnr. Verhulst was a famed sculptor who embodied the rising interest in decorative tombs and busts in the Low Countries halfway through the seventeenth century. He made the funerary monuments in honour of Dutch sea heroes, such as those of Maarten Tromp in Delft (1657) and Michiel de Ruyter in Amsterdam (1681).[117] Polyander was appointed professor of theology in 1611 and served as *rector magnificus* of the university.[118] The epitaph on the hanging monument stresses his noble descent, as well as his roles as a religious leader, professor *primarius*, and *rector magnificus*.

> Dedicated to the greatest and highest God, and to the eternal memory of the well-respected and most noble sir Johannes Polyander from Kerkhoven, from the ancient and most noble family of the Ghentian Kerkhovens, minister in the Walloon church of Dordrecht for 20 years, doctor and *primarius* professor at the university in Leiden for 35 years, eight times rector magnificus, and president of 23 synods of the Walloon Church[.]

117 See, e.g., Scholten, *Sumptuous Memories*, pp. 58–60, 64–66, and 198–207; Elisabeth Neurdenburg, *De zeventiende eewsche beeldhouwkunst in de noordelijke Nederlanden: Hendrick de Keyser, Artus Quellinus, Rombout Verhulst* (Amsterdam: Meulenhoff, 1948). For monuments by Verhulst in the province of Groningen, see Frits Scholten, *Rombout Verhulst in Groningen: Zeventiende eeuwse praalgraven in Midwolde en Stedum* (Utrecht: Stichting Matrijs, 1983).

118 A. J. Lamping, *Johannes Polyander: Een dienaar van kerk en universiteit* (Leiden: E. J. Brill, 1980).

> Deo Opt[imo] Max[imo] Sacrum, Aeternae memoriae reverendi et nobilissimi viri D[omin]i Johannis Polyandri a Kerkhoven, ex antiquissima et nobilissima Kerkhovium Gandavensium familia, in ecclesia Gallo Belgica Dordrechtana annos XX pastoris, in academia Lugduni Batavorum doctoris et professoris primarii annos XXXV, octies rectoris magnifici, et XXIII Synodi Gallo Belgici praesidis[.][119]

First, Johannes Polyander Jnr. wanted to honour his father and strengthen the fame of the 'ancient and most noble family of the Ghentian Kerkhovens', one of the first references in the epitaph on the tomb monument. At the end of the epitaph, it notes that 'The only son Johannes Polyander, Lord of Heenvliet, prefect of Holland, placed [this monument] in mourning.' Polyander Jnr. also praised the virtues of his father in the epitaph, who was celebrated by everyone because of his 'piety, prudence, honesty, his well-mannered, affable, modest soul, [and] love of peace and concord'. The second stakeholder was the university community, who wanted to honour and publicly display the professor who brought them fame. The epitaph goes to great lengths to explain Polyander's merit to the university. He taught for thirty-five years and was *rector magnificus* eight times. He was 'clear in teaching, sharp in disputation, eloquent in preaching'.[120] Thirdly, the religious communities wanted to remember him as a president of many synods. Polyander thus served as an example for a variety of communities. The funeral was a ceremonial, public funeral attended by major political figures such as Prince William II of Orange, and many funerary poems and lamentations were written and published shortly after Polyander's death. His burial in the Pieterskerk attested to the contemporary appeal of the church as a site of memory. Due to this burial, he was placed among fellow professors, rather than fellow members of his religious community.

119 The full text on the hanging monument reads: 'Deo / Opt[imo] Max[imo] Sacrum / Aeternae memoriae / Reverendi et nobilissimi viri / D[omin]i Johannis Polyandri a Kerkhove / Ex antiquissima et nobilissima / Kerkhovium Gandavensium familia / in ecclesia Gallo Belgica Dordrechtana / annos XX pastoris / in academia Lugd[uni] / Bat[avorum] doctoris et professoris / primarii annos XXXV / octies rectoris magnifici / et XXIII Synodi Gallo-Belg. praesidis. Hic / pietate, prudentia, probitate, morum sua- / vitate, liberalitate animi moderatione, pacis et concordiae amore, nemini secundus, omnibus gratus et charus, nulli gravis, in docendo perspicuus, in disputando / promptus, in concionando facundus, / memoriae et iudicii incomparabilis. / Eadem qua vixit tranquillitate placide / supremum diem clausit, in fine rectoratus / sui octavi MDCXLVI. IV. Febr. / Aetatis suae LXXVIII / Iohannes a Kerkhove, Heenvlietae dominus / saltuum Hollandiae praefectus / filius unicus moerens posuit.' Also published in Lamping, *Johannes Polyander*, p. 126.

120 '[I]n docendo perspicuus, in disputando promptus, in concionando facundus'.

His son, Johannes Polyander Jnr., an advisor of Frederick Henry, Prince of Orange (1584–1647), died twenty years later and was placed in a monumental marble *gisant* tomb in the Pieterskerk, also made by Rombout Verhulst (1624–1698). Thus, a family tradition of burial in the Pieterskerk began, underlining the importance of heraldry, nobility, and family lines. His tomb monument is clearly identified by the family emblem, yet there is no reference to learning.[121] It is ironic that one of the biggest ornamental tombs in the Pieterskerk is dedicated to the son of a professor, not an eminent professor or curator of Leiden University. The Polyander Jnr. ornamental tomb shows that eminent men were increasingly remembered in the context of their family and the nobility they had attained, rather than the context of learning, the city of Leiden, or the university.

The last hanging monuments dedicated to professors in the second half of the seventeenth century were relatively standard. The epitaphs followed the example set by earlier epitaphs and contain the same four elements: first, the name of the deceased and possible titles; second, the tenure at the university; third, the virtues of learning and piety; and finally, the name of the person who raised the monument. For example, the monument of Ewaldus Screvelius (1575–1647), professor of medicine in Leiden since 1624, follows this pattern. The virtues associated with Screvelius are 'learning' ('doctrina') and 'virtue' ('virtus').[122] The monument stresses his tenure as a professor *primarius* in medicine as well as his virtuous life. Furthermore, Maria van Swaenswijck, Screvelius's wife, presents herself as the mourning benefactor of the monument.

The last hanging monument in the Pieterskerk was dedicated to Johannes Cocceius (1603–1669), professor of theology at Leiden University from 1650 until his death (see Figure 6.13).[123] The monument was ordered in 1712—almost four decades after Cocceius's death—by Maria Heinsius, widow of Cocceius's

121 For more information, see Bieke van der Mark, 'Een praalgraf voor een *schoolmeesters soon*, het grafmonument voor Johannes Polyander van Kerckhoven door Rombout Verhulst', in *Pieterskerk*, ed. Den Hertog, pp. 386–394.

122 The full inscription of the hanging monument reads: 'D[eo] O[ptimo] M[aximo] / et / Ewaldo Screvelio, / Adriani, trigesimo Hagae quae Batavorum / Aula est, consulatu gesto insignis, Filio, An[no] MD / LXXV ibidem nato, Senatori et Medico, dein / Medicinae in Leidensi Academia Professori / Primario, et Rectori Magnifico, singulari doctrina, / virtute, et prolixa in omnes comitate clarissimo / cui in vita nihil charius, quam aliis eam velut / dare[.] Nihil in morte iucundius fuit, quam ad / meliorem et immortalem transire, An[no] MD / CXLVII denato, / Maria Van Swaenswijck / uxor marito, et Liberi Parenti dulcissimo, / desideratissimo moestissimi / H[oc] M[onumentum] P[osuerunt] / Sit tibi, qui nemini gravis vixisti, / terra levis.' See Kneppelhout van Sterkenburg, *Gedenkteekenen in de Pieters-kerk*, p. 50.

123 H. E. Knappert, 'Coccejus, Johannes', *NNBW*, vol. 1 (1911), pp. 616–618.

Figure 6.13. Hanging monument for Johannes Cocceius, c. 1669. Photograph by author, courtesy of Pieterskerk, Leiden

only son, Johannes Henricus Cocceius. Johannes Henricus Cocceius was the broker and defender of the memory of his father during his life and published Cocceius's complete works in eight volumes.[124] Maria Heinsius

124 Edward Grasman, 'Het omstreden gedenkteken voor Coccejus', in Den Hartog, *De Pieterskerk in Leiden*, pp. 396–402.

thought it was appropriate to place a hanging monument for father and son. In the early eighteenth century, a scholarly skirmish broke out concerning this monument: in essence, this was a continuation of the theological disagreements Cocceius had once had with Utrecht professor Gisbertus Voetius (1589–1676), which divided Reformed Protestants throughout the Dutch Republic. The placement of the monument in the Pieterskerk rubbed Voetian Protestants the wrong way, since for them Cocceius had become a symbol of an incorrect interpretation of Scripture.[125] The monument came nevertheless and stressed Cocceius's superior reading of the Bible:

> The whole world knows his skill, sagacity, and zeal in scrutinising the Sacred Scriptures and explaining, collating, and considering them; and also the work that he fulfilled at the universities of Franeker and Leiden.
>
> [...]
> In scrutandis sacr[is] litt[eris] iisque
> exponendis, conferendis et meditandis
> solertiam, sagacitatem et industriam,
> exhaustosque in academiis
> Franek[erensi] et Leyd[ensi] labores
> totus orbis novit.[126]

Cocceius is presented as a universal man of skill and wisdom whom the whole world knows about (which is not the same as agreement with his work and teachings). Even after four decades, the polemic surrounding this monument clearly reflected the importance and honour attached to placing a monument in the Pieterskerk. Cocceius's detractors, after all, argued against

125 The polemic about the monument in the Pieterskerk is chronicled in the following book: [N.N.], *Poetische schermutseling met grote en kleine flitsjes over het oprichten van het marmeren gedenkteken en sierlijk grafschrift van den Christiaanschen en Pauliaansche Coccejus, in de Pieterskerk te Leiden* (Flitszenburg: Gedrukt by de Erfgenamen der waarheit inde gelasterde afbeelding, 1715).

126 The full epitaph reads: 'Exuvias / Insignis theologi / Johannis Coccei / Christo redituro sacras / hoc loco composuerunt / vidua et liberi / anno MDCLXIX aet[atis] LXVI / cuius / In scrutandis sacr[is] litt[eris] iisque / expondendis, conferendis et meditandis / solertiam, sagacitatem et industriam, / exhaustosque in academiis / Franek[erensi] et Leyd[ensi] / totus orbis novit. / et filii eius unici / Joh[annis] Henrici Coccei S[acro]S[anctae] Th[eologiae] D[octoris] / Feud[orum] Holland[iae] Graphiarii / et rational[ium] Ord[inum] General[ium] a secret[is] / hic cum parentibus condi et suas addi / nurus et vidua / Maria Heinsius / soceroque et coniugi optimis / h[oc] m[onumentum] supr[a?] tab[ulas] p[onendum] v[oluit] / MDCCXII.' See Kneppelhout van Sterkenburg, *Gedenkteekenen in de Pieters-kerk*, p. 38.

having him in the hall of honour of the university and cementing his memory as a worthy and momentous professor. Clearly, his son, alongside his own spouse, wanted to defend him and associate themselves with the fame and honour of Cocceius by commissioning a monument for him, as well as for themselves. The curators of the university, however, were not involved in this dispute, underlining that the university was no longer involved in cementing the posthumous honour of a professor by commissioning a monument, as it had been in the first half of the seventeenth century.

Most of the signs of remembrance of professors in the Pieterskerk in the second half of the seventeenth century are in the form of tombstones, such as those of Johannes Valckenier (1617–1670), professor of theology; Franciscus Sylvius (1614–1672), professor of medicine; Johannes Friedrich Böckelmannus (1632–1681), professor of law; and Antonius Hulsius (1615–1685), professor of theology.[127] The professors of theology are remembered not just as professors, but also as ministers of the church. The tombstone of Antonius Hulsius, for example, mentions that 'the church had him as a pastor, the school as a rector, the college as a regent, the university as a professor, [his] children as a pious parent'.[128] Hulsius's tombstone is reminiscent of Scaliger's characterisation of Franciscus Junius as a teacher, parent of the church, and learned man. The epitaph shows the central role of theology professors in Leiden society, as well as the overlapping array of identities they represented: a minister, rector, regent, professor, and pious family man. The strong focus on teaching or excellent research seems to have waned at this point and made way for an enumeration of the functions the deceased fulfilled. Once again, there is no reference to a wider learned world, and the context of the deceased professors' memory is the local church and university. The professors are not remembered in a university context or for the virtues of learning and teaching, but rather through attestations of personal virtue and brilliance.

In the eighteenth century, monuments were dedicated to two illustrious professors amidst a new wave of memorialisations in the Pieterskerk. These

127 For the monuments of Johannes Valckenier, Franciscus Sylvius, Johannes Friedrich Böckelmannus, and Antonius Hulsius, see Kneppelhout van Sterkenburg, *De gedenkteekenen in de Pieters-kerk*, p. 17, no. 25; p. 71, no. 326; p. 72, no. 335; p. 53, no. 248; respectively.

128 'Quem habuerunt / ecclesia pastorem / schola rectorem / collegium regentem / academia professorem / filii parentem pium'. The full inscription on the tombstone read: 'Antonius Hulsius / Theologus / natus Hildae Montium / anno salutis MDCXV / denatus Leidae in Batavis / MDCLXXXV / Quem habuerunt / ecclesia pastorem / schola rectorem / collegium regentem / academia professorem / filii parentem pium / Anima deo placide reddita / optato hic fruitur somno'.

monuments were dedicated to Herman Boerhaave (1668–1738) and Petrus Camper (1722–1789), both eminent physicians and professors of medicine at Leiden University. The two monuments represent a unique and final phase of the monumentalisation of learned men in the Pieterskerk. Additionally, the two monuments represent a departure from the more traditional tombstones or hanging monuments in form and style, and testify to the popularity of unique monuments in the eighteenth century.[129] The Boerhaave monument is an urn on a pedestal, which carries the inscription 'Dedicated to the health-bringing brilliance of Boerhaave' (see Figure 6.14).[130] There is no long epitaph, no explanation, not even a reference to Boerhaave's life, his field of knowledge, or his association with Leiden University. All these things are assumed to be common knowledge shared by the onlookers. The monument's unique form is in stark contrast with previous hanging monuments and tombstones. The monument does not present Boerhaave in the context of his professorship, the university, or the city, but rather as a singular instantiation of genius, detached from tradition or context. The placement in the Pieterskerk, however, does suggest a desire to position Boerhaave in a wider tradition of eminent professors.

The monument built for Petrus Camper is an even sharper departure from tradition, since he never was a professor at Leiden University, only a student. A professor at the Athenaeum in Amsterdam and the university in Groningen, his sole relationship with Leiden consisted of his birth and family ties. The Camper monument is a bust on a pedestal, which carries a simple, matter-of-fact Dutch inscription without any mention of merit or achievements (see Figure 6.15). The inscription only mentions his name, place and year of birth, and place and time of death.[131] Much like Boerhaave's monument, the onlooker is assumed to know who Camper is. Moreover, the placement of the monument in the Pieterskerk suggested a link with Camper and Leiden University, although none is apparent. Neither of the monuments' commissioners deemed it necessary to characterise either man as a professor or even a man of learning. The monuments suggest that it was self-evident that any monument in the Pieterskerk must be dedicated to a man of learning or a professor.

129 See, e.g., the development of monument in Westminster Abbey: Matthew Craske, 'Westminster Abbey 1720–70: a Public Pantheon Built Upon Private Interest', in *Pantheons: Transformations of a Monumental Idea*, ed. Richard Wrigley and Matthew Craske, 57–79 (Aldershot; Burlington, VT: Ashgate, 2004).

130 'Salutifero / Boerhaavii / Genio / Sacrum'.

131 The inscription reads: 'Petrus Camper / geboren te Leyden / den XI May / MDCCXXII / Overleden / in s'Gravenhage / den VII April / MDCCLXXXIX [1789].'

Figure 6.14. Abraham Delfos, *Monument for Herman Boerhaave*, c. 1808–1820. Etch and pencil, 295 × 171mm. Rijksmuseum:RP-P-BI-6938

Figure 6.15. Daniël Vrijdag, *Funerary Monument for Petrus Camper*, after Anthony Ziesenis, 1792. Etch and engraving, 138 × 81mm. Rijksmuseum: RP-P-1905-4522

The monuments and tombstones dedicated to professors in the second half of the seventeenth century and into the eighteenth show that the Pieterskerk had become the standard burial site for professors. Exemplary and outstanding professors continued to receive a hanging monument in the seventeenth century, with the odd exception who received a funerary monument such as the one dedicated to Polyander Jnr. In the eighteenth century, the hanging monument went out of fashion, and instead, exemplary professors such as Boerhaave and Camper received a standing monument. Overall, fewer monuments were placed in this period and the number of monuments from the seventeenth century greatly outnumber those from the eighteenth century, during which Boerhaave and Camper's were the only monuments dedicated to learned men. The literary flourish that characterised the seventeenth-century hanging monument and epitaph had made way for symbolic monuments with short inscriptions.

## 6.5 Conclusion

The first conclusion we can draw is that the remembrance and monumentalisation of learned men was shaped by local conventions in Leiden, as reflected in the numerous attestations to the university and the city. The memory culture in the Pieterskerk was not a curated and planned, but a familial and individual affair. With some exceptions that were paid for by the university, the majority of the monuments were ordered by the family. In both Leiden and Florence (see Chapter 5), the remembrance was strongly influenced by the local context. In Florence, the prevailing narrative of civic humanism characterised leading political figures as learned men and rendered learned men examples of civic success and pride. In Leiden, the learned man was seen first as a teaching professor, and later as an adornment of the university and city.

In her study *Creating Academic Communities*, Stefanie Knöll concluded that the funerary monuments of professors reflected both moral and academic capital:[132] moral capital, since the epitaphs all reminded the reader of the virtuous and exemplary life of the deceased professor; and academic capital, referring to scholarly virtues such as learning and erudition. The relationship between the deceased professors and the university described in the funerary inscriptions in the Pieterskerk appears to be somewhat different from those in Oxford and Tübingen, although not in the way Knöll described. According to Knöll, the epitaphs in the Pieterskerk presented the professors in terms of scholarly virtues, while the professors' monuments in Oxford and Tübingen 'stress the loyalty they owe their university/college, their State and their Church.'[133] When we take into consideration Buchelius and Hegenitius's accounts, we see numerous monuments that do construct a relation between the university and the city. The professor–university relationship, however, was not presented as one of loyalty, but rather of glory and prosperity.

The learned man of the Republic of Letters is thus an abstraction that was appropriated by and amended in local memory cultures. In comparison with the ideal learned man that emerged from the *vitae* and *elogia* (see Chapter 1), the funerary monuments do not express an oppositional confessional identity. The monuments in Leiden do stress the piety and solemnity of the dead, yet there were no references to the restoration of learning after its destruction at the end of the sixteenth century. In Florence, only Giovanni

132 Knöll, 'Creating Academic Communities', p. 250.

133 Ibid., p. 215.

Lami's epitaph referred to his scholarly defence of the Catholic faith 'against the impiety of anti-trinitarians'.[134] Taking all of this into account, it can be suggested that the collective remembrance in a collection of *vitae* was a better platform to imagine a confessional learned community, while the remembrance of learned men in churches was more often adjusted as befit the individual men who were remembered.

It has become evident that a grave or funeral monument is tied to a certain place, whereas a collective biography in print is far less static, although the example of Meusius's *Illustris Academia* shows how print was used to advertise a single university.[135] And funerary culture itself reverberated beyond the borders of the town: both Kool and Hegenitius copied out epitaphs, and they were not exceptional cases. Funerary monuments were *lieux de savoir*, where members of the community confirmed their identity by stressing the excellence of their kin, and where travellers sought out to familiarise themselves with local, regional, or even 'national' heroes of learning. The only way to take these home, of course, was to draw them in notebooks or describe them in travelogues. When such accounts were printed, as in the case of Hegenitius's *Iter*, the material pantheon was communicated to other local groups or regional pockets, showing at home that elsewhere, in Leiden or Florence, there were other heroes of learning who were likewise part of the larger choir of the Republic of Letters. The funerary monuments in the Pieterskerk create the sense of a microcosm of the Republic of Letters in Leiden. The Pieterskerk is a site of memory that visualises the collective history of the university and helps to imagine a transhistorical learned community. The monuments invite visitors to be aware of the fact that they are standing in the middle of a tradition of virtuous scholarship at Leiden University.

134 The two relevant sentences on Giovanni Lami's epitaph read: 'De catholica religione quam eruditis eruditissimis commentariis adversus / Arianorum impietatem defendit optime merito morum candore veritatis'.

135 Meursius, *Illustris Academia*; Meursius, *Athenae Batavae*; both discussed in Grafton, *Athenae Batavae*, pp. 14–31.

# Conclusion

**Abstract:** The conclusion argues that the Republic of Letters should be understood as an imagined community shaped by scholarly virtues, collective memory, and intellectual exchange. It emphasises that scholars' letters, life-writing, travel accounts, and funerary monuments were central to creating and perpetuating the idea of a transnational commonwealth of learning. The conclusion also argues that the Republic of Letters was not monolithic but rather a collection of diverse, interconnected learned communities. It argues for the importance of studying media representations of scholars to better understand how intellectuals constructed their identities and how these identities helped sustain the ideals of the Republic of Letters.

**Keywords:** Republic of Letters, scholarly identity, scholarly memory, transnational learned communities

In an article on the Republic of Letters and Erasmus's position in it, Peter Burke argued that the Republic of Letters should be understood as an imagined community:

> But what did humanists mean by speaking of a *respublica*? The republic was what we might call an invisible or 'imagined community'. In this respect it was like the *respublica christiana* (another common phrase of the time), or the 'invisible church', a phrase used in the seventeenth century in particular to describe the community which—it was hoped—underlay the all-too-visible divisions and conflicts between Latin, Orthodox, Catholic and Protestant [...] An important element in the idea of the republic was that of consensus, what the Spanish humanist Vives once called the *consensus hominum doctorum pariter et bonorum* [...] Another was the idea of intellectual equality, at least in principle. Another important element was the ideal of scholarly co-operation, of sharing information.[1]

1 Peter Burke, 'Erasmus and the Republic of Letters', *European Review* 7, no. 1 (1999): 5–17 (p. 8).

Scholten, Koen. *Rethinking the Republic of Letters: Memory and Identity in Early Modern Learned Communities.* Taylor & Francis Group, 2025
DOI 10.5117/9789048559855_CON

Burke pointed to seven ways the Republic of Letters was brought into practice and that illustrated its historical importance: intellectual mobility; the high regard for libraries; extensive Latin epistolary exchanges; scholarly visits and travels; scholarly *alba amicorum*; the rise of learned academies; and the emergence of learned journals. These are all facets of the early modern, pan-European community dedicated to the pursuit of learning. Still, the question remains how they helped form an imagined community. This book has built on Burke's observations to offer a more profound understanding of the Republic of Letters as an imagined community by looking at its media representations in early modern Europe.

Despite efforts to define the Republic of Letters as an imagined community, historians have used the term in a variety of ways without a clear definition or demarcation between the actors' category and the historiographical category. In other words, historians have used the term 'Republic of Letters' as a metaphor for the entire learned world, as a concept to describe the early modern moral economy of knowledge, and epistolary networks (see Introduction). In this monograph, I have taken the concept of an imagined community as a starting point for my historical analysis.

Studying the Republic of Letters as an imagined community has methodological consequences, since imagined communities are generally studied through their representations. Benedict Anderson argued that the nation could be understood as a political imagined community:

> It is *imagined* because the members of even the smallest nation will never know most of their fellow-members, meet them, or even hear of them, yet in the minds of each lives the image of their communion. [...] Communities are to be distinguished, not by their falsity/genuineness, but by the style in which they are imagined.[2]

If we apply Anderson's concept of an imagined community to the Republic of Letters, then, we should not ask whether it was real, genuine, or merely an ideal, but rather how the community was imagined in the early modern period. In this monograph I have argued that to understand the Republic of Letters as an imagined community it is necessary to look at the representations of learned communities, with a particular attention to identity formation and memory. In so doing, my aim has been to deepen the conception of the Republic of Letters as an imagined community by looking at the early

2 Benedict Anderson, *Imagined Communities: Reflections on the Origin and Spread of Nationalism*, rev. ed. (London: Verso, 2006), pp. 5–6.

modern media and experiences that were vital to the formation of learned communities: collections of lives and eulogies, correspondence, travels, memorials, and funerary monuments.

To study the collective aspect of these representations of learned communities, quantitative, I have used digital methods. Digital textual analysis—or 'distant reading'—offered a quantitative approach to the early modern collections of *vitae* and *elogia* in Chapters 1 and 2, as well as to the correspondence of Hugo Grotius in Chapter 3. The vital requirement of the implementation of digital textual analysis is the availability of clean, high-quality data. Fortunately, the full-text versions of the compendia of *vitae* and *elogia* were available in the *Thesaurus Eruditionis*.[3] The digital Grotius correspondence was made available by the Huygens Institute both through *The Correspondence of Hugo Grotius* as well as the *ePistolarium* project.[4] Because the quality of the digitised sources was high (without, for instance, rampant Optical Character Recognition errors), the Latin texts could be fruitfully analysed using AntConc software.[5] The digital textual analysis helped to identify common tropes and virtues associated with learned men. These findings then pointed to passages in the texts that could be subjected to close reading to properly contextualise the usage of the tropes and virtues. This method offers a fresh perspective on bulk sources, such as compendia of *vitae* and correspondence collections. In the writing of this monograph, it helped identify the main scholarly virtues (see 1.2, esp. Figure 1.1) that were appropriated in the *vitae*, but also the local interpretations of the learned man as a Reformed scholar (see 2.2) and the pride of a region or state (see 2.3). In the Grotius correspondence, it helped identify shared citation patterns (see 3.2) as well as the recurrence of scholarly virtues and values (see 3.3). Other digitised correspondences could be used in future research to expand the study of scholarly identity and virtues to include more parts of Europe (for example, Spain, Poland, or Hungary), or even other continents if high-quality data is available. None of this research was, or would be, possible without operable, clean data.

3 The *Thesaurus Eruditionis* was compiled by the Deutsche Forschungsgemeinschaft for the CAMENA and TERMINI projects between 1999 and 2008: see http://mateo.uni-mannheim.de/camenahtdocs/ camenaref_e.html. See also Appendix 1.

4 *The Correspondence of Hugo Grotius* is a digital version of *BW*: see http://grotius.huygens.knaw.nl. The *ePistolarium* was part of the *Circulation of Knowledge and Learned Practices in the 17th-Century Dutch Republic* project: see http://ckcc.huygens.knaw.nl.

5 In this project I used AntConc 3.5.7 (Macintosh OC X), developed by Laurence Anthony, Faculty of Science and Engineering, Waseda University, Japan.

The diverse case studies in this book have challenged the implicit historiographical assumption that the Republic of Letters was commonly imagined as a pan-European, transconfessional community of learning. The Republic of Letters as an imagined community is a paradox: learned men were often framed as *exempla* of their local learned communities, while the phenomenon in which learned men were praised in their local communities was widespread throughout Europe in the early modern period. The studies of *vitae* in Chapters 1 and 2 clearly show that there was no pan-European history of the Republic of Letters in place with recurring *exempla* throughout Europe. The *vitae* did indeed appeal to a European audience, but the scholarly identity of the learned man was rarely represented as a universal, pan-European figure over the course of the early modern period. 'Cosmopolitanism', 'internationalism', or 'transconfessionality' are never found as scholarly virtues. Admittedly, these are anachronistic categories, but the Erasmian desire to be a *civis mundi*, a citizen of the world, is never appealed to in the praise of learned men. Even a non-polemical, peace-loving standpoint in religious debates is hardly ever praised, let alone a truly ecumenical standpoint. Rather, the compilers of many collections wanted to inscribe themselves and their learned community into a pan-European community that valued the accumulation, production, and diffusion of knowledge. Additionally, the studies of the learned sites of memory in Florence and Leiden both show that the learned man was remembered and framed in local contexts, rather than within a pan-European narrative of learning.

These observations have three implications for our understanding of the Republic of Letters. First, based on these case studies, it becomes clear that in the sixteenth and seventeenth centuries, there was no master narrative of the Republic of Letters in terms of a shared history or recurring *exempla*. Johannes Fichard's collection of lives from 1536 came closest to such a master narrative,[6] although this collection is the exception rather than the rule. Second, local communities appropriated the general ideal of the pursuit of knowledge and learning, as well as the scholarly virtues that were generally associated with the man of learning—erudition, learning, and innate brilliance—to fit their own sense of history and identity. In other words, the Republic of Letters as a pan-European, imagined community of learning was kept alive by local representations that remembered scholars as part of their own learned community. Third, the sustained

6 Johannes Fichard, *Virorum qui superiori nostroque seculo eruditione et doctrina illustres atque memorabiles fuerunt, Vitae* [...] (Frankfurt a.M.: Christianus Egenolphus, 1536).

travels, correspondence, and publication of collections of *vitae* show that these learned communities imagined that similar communities existed throughout Europe, or, to quote Benedict Anderson, that 'in the minds of each lives the image of their communion.'[7]

This Republic of Letters was tacitly and implicitly imagined as a male community and the archetypical scholar as a virile learned man throughout the early modern period. Male scholars were commonly referred to as 'learned men' (*viri docti*), while female scholars were rarely called 'learned women' (*doctae mulieres*) and were more generally included in the broad category of 'bright women' (*clarae mulieres*), which also included martyrs, saints, and spiritual examples. The ideal female scholar was not constructed in relation with or opposition to the male scholar and considered part of the Republic of Letters. The achievements of female scholars were placed in a succession of female *exempla*. This thesis underscores Carol Pal's conclusion that, in the seventeenth century, 'the work of moral, scientific, and intellectual reform networks in Europe was in fact a dually gendered enterprise.'[8] The male and female scholarly imagined communities were effectively two different spheres with their own set of virtues, *exempla*, and senses of history. To Pal's argument, I would add that the Latin-speaking Republic of Letters was a community of *virtus*—male virtue—and that the archetypical scholar was praised on account of male virtues that were unattainable for female scholars (see 1.3). While there were many female scholars and learned women in the early modern period, the Latin-speaking Republic of Letters was a masculine imagined community that excluded female scholars, because it almost exclusively extolled male scholarly virtues and *exempla*.

The digital textual analysis and the conceptual approach of the imagined community are two factors that help us reassess the Republic of Letters from the perspective of collective identity formation. This framework goes beyond individual (self)representation and focuses on the value and importance of collective scholarly identity and memory in the construction of learned communities. That is why, in this monograph, I have moved away from studying the ideas, practices, and self-fashioning of well-known and well-established early modern scholars such as Petrarch, Galileo, or Erasmus.[9] By contrast, this book has shown how lesser-known figures

7 Anderson, *Imagined Communities*, p. 5.

8 Carol Pal, *Republic of Women: Rethinking the Republic of Letters in the Seventeenth Century* (Cambridge: Cambridge University Press, 2012), p. 6.

9 See, e.g., Karl A. E. Enenkel, Betsy de Jong-Crane, and Peter Liebregts, eds., *Modelling the Individual: Biography and Portrait in the Renaissance. With a Critical Edition of Petrarch's Letter to Posterity* (Amsterdam; Atlanta, GA: Rodopi, 1998); Peter Burke, 'Representations of the Self

such as Melchior Adam, Johannes Fichard, Nicolaus Reusner, and Gian Vittorio Rossi appropriated history and formed collective scholarly identities. Chapter 3 showed how scholarly connections were made and maintained on the basis of a shared history and shared values, most notably that of friendship (*amicitia*). These connections would then be carefully maintained and inherited by later generations, as my close examination of the early modern scholarly journey—the *peregrinatio literaria*—has shown (see Chapter 4). The scholarly traveller could rely on the shared history of travels of yore and would continue the exchange of knowledge based on shared values by travelling himself. A continued sense of an existing and evolving community dedicated to the pursuit of knowledge all over Europe was crucially kept alive through correspondences, continued travels, and memory cultures with funerary monuments in churches.

This focus on scholarly memory — remembrance in *vitae*, correspondence, and funerary monuments — has shed new light on the representation of scholarly identity. Few studies approached memory and remembrance in scholarship and science from this perspective.[10] Throughout this study, exemplary figures have reappeared as the embodiment of scholarly virtues and as essential conveyors of a sense of the past. The Republic of Letters was an imagined community because learned communities throughout Europe imagined themselves as part of a larger European scholarly community dedicated to the pursuit of learning. The local learned communities praised and appreciated diverse scholarly identities and senses of history, which underscores the importance of regional and temporal differences and contexts for the representation of learned communities. Learned communities represented themselves according to different virtues and *exempla*. This divergence can be seen, for example, in the different *exempla* portrayed in the *vitae* in the seventeenth century, as well as the different memory cultures in Florence and Leiden that framed learned men as part of their local community rather than a pan-European community of learning. The praise of learned men was universal, but the specific narrative and memory culture in which they were placed differed.

from Petrarch to Descartes', in *Rewriting the Self, Histories from the Renaissance to the Present*, ed. Roy Porter, 17–28 (London; New York: Routledge, 1997); Mario Biagioli, *Galileo, Courtier: The Practice of Science in the Culture of Absolutism* (Chicago; London: University of Chicago Press, 1993); Lisa Jardine, *Erasmus, Man of Letters: The Construction of Charisma in Print* (Princeton, NJ: Princeton University Press, 1993).

10 See, e.g., Pnina G. Abir-Am and Clark A. Elliot, eds., *Commemorative Practices in Science: Historical Perspectives on the Politics of Collective Memory* (Chicago: University of Chicago Press, 1999).

This study has argued for an understanding of the Republic of Letters as an imagined community realised through distinct memory cultures and the representation and identification with collective identities. The early modern *vitae*, eulogies, letters, travelogues, and funerary monuments all elevate the cultural symbol of the learned man as a vital part of culture and society. The Republic of Letters was no monolithic community, but rather a set of interrelated communities that each appropriated the ideal of learning in a local history, context, and culture. It was not a community that one could point to and clearly identify its members, but rather an imagined community that was *made real* because of the many different representations of learning and learned communities throughout Europe. Every tome filled with *vitae* and every church that honoured *viri docti* with tomb monuments contributed to the idea that learning was universally valued, which made learning into a common good. Every act of remembering a scholar for his learning asserted *res literaria* as a *bonum commune. Viri docti* throughout Europe imagined the existence of similarly minded colleague scholars in other European cities. As we saw in Chapter 2, Melchior Adam, in the preface of his five-volume *Vitae*, visualised this virtuous imagined community as a learned theatre (*doctum theatrum*) in these Ovidian lines:

*Those whom a first generation has admired, a second generation will also admire*
*The learned theatres resound in applause for them*

*Quos prior est mirata, sequens mirabitur aetas:*
*In quorum plausus docta theatra sonant.*[11]

This is the Republic of Letters as an imagined community: men of learning who admired their predecessors in the theatre of knowledge. They elevated themselves by their virtues—erudition, learning, piety, benefiting the public good, and friendship—and received praise and applause in the memory cultures of local learned communities. There were many learned theatres, to return to Adam's imagery, each with their own heroes and *exempla*. Yet the applause for their outstanding virtues reverberated throughout early

11 Melchior Adam, *Vitae Germanorum Philosophorum* [...], *Vitae* I (Frankfurt a.M.; Heidelberg: Jona Rosa 1615), 'Epistola dedicatoria', sig. *4v; cited and discussed in Chapter 2, p. 127. Adam changed Ovid's 'tota theatra' (all the theatres) to 'docta theatra' (the learned theatres). The original lines in Ovid, *Ex Ponto*, II, VI, 27–28 read: 'quos prior est mirata, sequens mirabitur aetas, in quorum plausus tota theatra sonant.' My translation is based on Ovid, *Tristia; Ex Ponto*, trans. Arthur Leslie Wheeler (Cambridge, MA: Harvard University Press, 1975), pp. 348–349.

modern Europe. The learned communities were local and regional; the scholarly virtues they radiated were meant to transcend borders and reach other confessions. The pan-European applause for scholarly virtue and the pursuit of learning could inspire many a young aspiring scholar to blow the dust from manuscripts, page through ancient tomes, and polish his gift of innate brilliance, in the hopes of becoming a man of learning himself, of gaining the status of erudite man, and perhaps, of being admired for years to come as an *exemplum* of his learned community, who would perpetuate the pan-European imagined community of learning, the Republic of Letters.

# Bibliography

## List of Abbreviations

| | |
|---|---|
| AE | Archief Eemland, Amersfoort |
| ASD | Desiderius Erasmus, Opera Omnia, ed. Jan Waszink et al., 9 parts, 55 vols. (Amsterdam: North-Holland Publishing Company; 1969–1994; Amsterdam and Boston: Elsevier, 1997–2008; Leiden and Boston: Brill, 2009–2022) |
| ASFi | Archivio di Stato di Firenze, Florence |
| *ADB* | *Allgemeine Deutsche Biographie*, 56 vols. (Leipzig: Duncker & Humblot, 1875–1912) |
| BiASA | Biblioteca di Archeologia e Storia dell'Arte, Rome |
| BNCF | Biblioteca Nazionale Centrale di Firenze, Florence |
| *BW* | Hugo Grotius, *Briefwisseling van Hugo Grotius*, ed. P.C. Molhuysen et al., 17 vols. (The Hague: Martinus Nijhoff, 1928–2001) |
| *BWN* | *Biographisch woordenboek der Nederlanden*, 21 vols., ed. A.J. van der Aa (Haarlem: J.J. Brederode, 1852–1872) |
| *DBI* | *Dizionario Biografico degli Italiani*, 100 vols. (Rome: Istituto della Enciclopedia italiana, 1960–2020) |
| HUA | Het Utrechts Archief, Utrecht |
| *ILE* | Justus Lipsius, *Iusti Lipsi epistolae*, ed. Aloïs Gerlo et al., 13 vols. (Brussel; Leuven; Paris: Peeters, 1978–) |
| KB | Koninklijke Bibliotheek (Royal Library), The Hague |
| NA | Nationaal Archief (National Archive), The Hague |
| *NDB* | *Neue Deutsche Biographie*, 28 vols. (Berlin: Duncker & Humblot, 1953–) |
| *NBW* | *Nationaal Biografisch Woordenboek*, ed. J. Duverger et al., 10 vols. (Brussel: Paleis der Academiën, 1964–1983) |
| *NNBW* | *Nieuw Nederlandsch Biografisch Woordenboek*, ed. P.C. Molhuysen and P.J. Blok, 10 vols. (Leiden: A.W. Sijthoff, 1911–1937) |
| *ODNB* | *Oxford Dictionary of National Biography*, 60 vols. (Oxford: Oxford University Press, 2004), accessed online: oxforddnb.com |
| SCArch | L'Archivio Storico di Santa Croce, Florence |
| UBA | University Library, University of Amsterdam |
| UBL | University Library, Leiden University |
| UBU | University Library, Utrecht University |

## Manuscript Sources

'Annuncio della morte di Giovanni Targioni Tozzetti'. 8 January 1783. BNCF: Arch. Magl., filza XI, fols. 3r–4r.

Buchelius, Arnoldus. 'Commentarius rerum quotidianarum, in quo, praeter itinera diversarum regionum, urbium, oppidorumque situs, antiquitates, principes, instituta, mores, multa eorum quae tam inter publicos quam privatos contingere solent, occurrent exempla'. c. 1593–1600. Utrecht, UL, Hs. 798 (Hs 6 E 15).

Buchelius, Arnoldus. 'Inscriptiones monumentaque in templis et monasteriis Belgicis inventa'. c. 1609–1632. UBU: hs. 1648 (Hs 4 L 16).

Dussen, Jacob van der, and Pieter van der Dussen. 'Dagboek van een reis van Den Haag door Duitsland en Oostenrijk naar Italië en Zwitserland'. 1699–1700. The Hague, KB, 128 D 34.

Gronovius, Jacob. 'Daghverhaal eener Reis naar Spanje en Italien (1672 en 1673) door Jac. Gronovius'. 1672–1673. UBL, Ltk 859, 860.

Gronovius, Laurentius Theodorus. 'Itinerarium Laurentii Theodori Gronovii qui A.C. MDCLXXIX die III maii Haga Comitum profectus iter suscepit in Italiam et Lugdunum Batavorum rediit die XXI aprilis A.C. MDCLXXXIII'. 1679–1682. The Hague, KB, 76 H 27 I.

Gronovius, Laurentius Theodorus. 'Reize van Gronovius na en door Italie enz.'. 1693–1695. The Hague, KB, 76 H 27 II.

Kool, Joannes. 'Journaal van een rijse gedaan door Italien &c in den Jaare Anno 1698 door Mr. Jan Guilliam Kool'. 1698–1699. 3 vols. Rome, BiASA, Fondo Lanciani, Mss 34 A 1–2–3.

*Opera di Santa Croce*. Florence, Santa Croce, L'Archivio Storico.

Rosselli, Stefano. 'Sepultuario Fiorentino, ovvero descrizione delle chiese, capelle, e sepolture loro armi et incrizioni della citta di Firenze e suoi contorni'. 1657. Florence, Archivio privato famiglia Rosselli del Turco, ms. 262.

Rosselli, Stefano. 'Sepultuario Fiorentino ovvero descrizione delle chiese capelle e sepolture loro armi & inscrizioni della citta di Firenze e suoi contorni'. 1657. Florence, ASFi, mss. 624–625. For a digital edition of this manuscript, see Di Stasi, *Stefano di Francesco Rosselli*.

Ruysch, Coenraad. 'Journaal van een reis naar Genève, Italië en Frankrijk van Coenraad Ruysch met zijn neef Dirck van Hoogeveen'. 1674–1677. The Hague, NA, FA Teding van Berkhout, 1408.

Vossius, Isaac. 'Variae adnotationes de Ignatio, latine'. 1642. Leiden, UBL, VGQ 72.

## Printed Sources, Before 1800

Adam, Melchior. *Vitae Germanorum Philosophorum: qui seculo superiori, et quod excurrit, philosophis ac humanioribus Literis clari floruerunt. Vitae* I. Frankfurt a.M.; Heidelberg: Jona Rosa, 1615.

Adam, Melchior. *Vitae Germanorum Medicorum: qui seculo superiori, et quod excurrit, claruerunt. Vitae* II. Heidelberg: Jona Rosa, 1620.

Adam, Melchior. *Vitae Germanorum iureconsultorum et politicorum: qui superiori seculo, et quod excurrit, floruerunt. Vitae* III. Heidelberg: Jona Rosa, 1620.

Adam, Melchior. *Vitae Germanorum Theologorum, qui superiori seculo ecclesiam Christi voce scriptisque propagarunt et propugnarunt. Vitae* IV. Heidelberg: Jona Rosa, 1620.

Adam, Melchior. *Decades duae continentes Vitas Theologorum exterorum principum, qui ecclesiam Christi superiori seculo propagarunt et propugnarunt. Vitae* V. Frankfurt a.M.: Jona Rosa's widow, 1623.

Allegrini, Giuseppe, ed. *Serie di ritratti d'uomini illustri toscani con gli elogi istorici dei medesimi.* 4 vols. Florence: Allegrini, 1766–1773.

Allegrini, Giuseppe, ed. *Serie di ritratti d'uomini illustri toscani con gli elogi istorici dei medesimi.* 4 vols. Lucca: [n.p.], 1771–1774.

Ayloffe, Joseph, and John Nichols. *An account of some ancient monuments in Westminster Abbey.* London: J. Nichols, 1780.

Barbaro, Francesco, et al. *Francisci Barbari et aliorum ad ipsum Epistolae. Ab anno Chr. MCCCCXXV. ad Annum MCCCCLIII. Nunc primum editae ex duplici MS. Cod. Brixiano, & Vaticano uno. Accedit ampla earundem mantissa ex Foroiuliensibus Biblioth. Guernerianae MSS.* Brescia: Joannes-Maria Rizzardi, 1743.

Bayle, Pierre. 'Article II: *Les Eloges des hommes Sçavans tirez de l'Histoire de M. de Thou, avec des additions contenant l'Abregé de leur vie, le jugement & le Catalogue de leurs Ouvrages, par Antoine Teissier Avocat au Présidial de Nimes.* A Geneve chez Jean Herman Widerhold 1683. 2 vol. in 12.' *Nouvelle de la République des Lettres*, October 1684, pp. 772–773. Amsterdam: Henry Desbordes, 1686.

Bayle, Pierre. *Dictionaire historique et critique.* Rotterdam: Reinier Leers, 1697.

Bemmel, Abraham van. *Beschryving van de Stadt Amersfoort, dezelver Regeeringsvorm, Burgereye en Gilden, Jaarmarkten, Handel en Negotie, Vryheid en Grondgebied.* 2 vols. Utrecht: Henricus Spruyt, 1760.

Beza, Theodore. *Icones, id est Verae imagines virorum doctrina simul et pietate illustrium, quorum praecipue ministerio partim bonarum literarum studia sunt restituta, partim vera Religio in variis orbis Christiani regionibus, nostra patrumque memoria fuit instaurata: additis eorundem vitae & operae descriptionibus, quibus adiectae sunt nonnullae picturae quas Emblemata vocant.* [Geneva]: Jean de Laon, 1580.

*Bibliotheca Kooliana Instructa elegantissima ac nitida supellectile Librorum, Omnis generis ac facultatis praestantissimorum: inter quos reperiuntur Theologi insignes, JCti de meliore nota, &, quod caput rei est, Miscellanei, constantes praecipue, ex Antiquariis & Numismaticis rarissimis, Poëtis Graecis & Latinis, Lexicographis aliisque Litteratoribus, in quibus multae eminent antiquae Aldi, Stephani &c. editiones. His accedunt diversorum Auctorum Codices manu exarati, non spernendi generis & antiquitatis. Deinde prisci operis studiosi invenient aliquot sictilia & aerea opera, veluti statuas, urnas diversae magnitudinis, lucernas &c. Tandem politissimum ac pretiosissimum sacrarum aureis, argenteis & aereis numismatibus repletum, quorum index brevi prodibit. Thesaurum hunc Musis dicatum reliquit vir Doctissimus D. Joannes Kool, J.U.D. Senator Urbis Amisfurtensis. Publica licitatione distractio fiet Ultrajecti (op de oude gragt by de Viebrug.) Ad diem Lunae 9. Maji 1712.* Utrecht: Willem van de Water, 1712.

Blaeu, Johannes Willemszoon. *Toonneel der steden van de Vereenighde Nederlanden.* 2 vols. Amsterdam: Johannes Willemszoon Blaeu, 1652.

Bocchi, Francesco. *Elogiorum; quibus Viri Doctissimi nati Florentiae decorantur, Liber Primus.* Florence: apud Iuntas, 1609.

Bockenberg, Petrus. *Ad nonnullas Jani Dousae aspera scripta responsio illustribus Hollandiae et Westfrisiae ordinibus reddita.* Delft: Ex officina Brunonis H. Schinckelii, 1601.

Burman, Pieter, ed. *Sylloges epistolarum a viris illustribus scriptarum tomi quinque.* 5 vols. Leiden: Samuel Luchtmans, 1724–1727.

Collins, John, et al. *Commercium epistolicum D. Johannis Collins, et aliorum, de analysi promota.* London: Ex officina J. Johnson & J. Watts, 1722.

De' Nelli, Giovanni Battista Clemente, ed. *Grati animi monumenta Vincentii Viviani.* Florence: Franciscus Moücke, 1791.

De' Nelli, Giovanni Battista Clemente. *Vita e commercio litterario di Galileo Galilei nobile e patrizio fiorentino mattematico e filosofo sopraordinario de' Gran Duchi di Toscana Cosimo e Ferdinando II.* 2 vols. Lausanne: Stamperia Moücke, 1793.

Donati, Alessandro. *Roma vetus ac recens utriusque aedificiis ad eruditam cognitionem expositis.* Amsterdam: Janssons van Waesberge & Johannes Watters, 1695.

Dubos, Jean Baptiste. *Histoire des quatre Gordiens, prouvée et illustrée par les medailles.* Paris: Florentin & Piere Delaulne, 1695.

Earle, John. *Micro-cosmographie: Or, a Piece of the World Discovered; in essays, and characters.* 8th ed. London: R. D. for P. C., 1664.

Erasmus, Desiderius. *Libellus de conscribendis epistolis.* Cambridge: Siberch, 1521.

Erasmus, Desiderius. *Opus de conscribendis epistolis recognitum denuo ab autore & locupletatum.* Cologne: apud Heronem Alopecium, 1523.

Erasmus, Desiderius. *De civilitate morum puerilum per Des. Erasmum Roterodamum libellus nunc primum conditus & aeditus.* Basel: Froben, 1530.

Erpenius, Thomas. *De peregrinatione Gallica Utiliter instuenda Tractatus. Item Brevis admodum totius Galliae descriptio: et Iusti Lipsii V.C. Epistola de Peregrinatione Italica.* Leiden: Franciscus Hegerus, 1631.

Fichard, Johannes. *Virorum qui superiori nostroque seculo eruditione et doctrina illustres atque memorabiles fuerunt, Vitae. Iamprimum in hoc Volumen collectae.* Frankfurt a.M.: Christianus Egenolphus excudebat, 1536.

Fontani, Francesco, ed. *Elogio di Carlo Roberto Dati recitato nella reale Accademia Fiorentina nell'adunanza del dì 30. di settembre 1790.* Florence: Gaetano Cambiagi Stampatore Granducale, 1790.

Foresti, Giacomo Filippo. *De plurimis claris selectisque mulieribus.* Ferrara: Lorenzo de' Rossi, 1497.

Freher, Paul. *Theatrum Virorum Eruditione Clarorum. In quo Vitae & Scripta Theologorum, Iureconsultorum, Medicorum & Philosophorum, Tam in Germania Superiore & Inferiore, quam in aliis Europae Regionibus, Graecia nempe, Hispania, Italia, Gallia, Polonia, Hungaria, Bohemia, Dania & Suecia A Seculis Aliquot, Ad Haec Usque Tempora, Florentium, Secundum Annorum Emortalium Seriem, Tanquam Variis in Scenis Repraesentantur. Opus omnibus Eruditis Lectu iucundissimum in quatuor partes divisum, quarum I. Theologos varios II. Magnates, Iurisconsultos & Politicos. III. Medicos, Chymicos, Botanicos, Anatomicos &c. & IV. Philosophos, Philologos, Historicos, Mathematicos, Poetas &c. complectitur.* 2 vols. Nuremberg: Johannes Hofmann and Andrea Knorz's widow, 1688.

Giovio, Paolo. *Elogia veris clarorum virorum imaginibus apposita: Quae in musaeo Ioviano Comi spectantur.* Venice: apud Michaelem Tramezinum, 1546.

Giovio, Paolo. *Elogia virorum bellica virtute illustrium veris imaginibus supposita, quae apud Musaeum spectantur, in libros septem digesta.* 7 vols. Florence: Laurentius Torrentinus, 1551.

Giovio, Paolo. *Elogia doctorum virorum ab avorum memoria publicatis ingenii monumentis illustrium. Authore Paulo Iovio Novocomense, episcopo Nucerino. Praeter nova Ioan. Latomi Bergani in singulos Epigrammata: adiecimus ad priora Italice editionis, illustrium aliquot Poetarum alia.* Antwerp: Jean Bellère, 1557.

Giovio, Paolo. *Elogia Virorum literis illustrium: quotquot vel nostra vel avorum memoria vixere. Ex eiusdem Musaeo (cuius descriptionem una exhibemus) ad vivum expressis imaginibus exornata.* Basel: Petrus Perna, 1577.

Graevius, Johann Georg, ed. *Thesaurus antiquitatum romanarum.* 12 vols. Utrecht: François Halma; Leiden: Petrus vander Aa, 1694–1699.

Graevius, Johann Georg. *Callimachi hymni, epigrammata, et fragmenta.* Utrecht: François Halma & Willem vande Water, 1697.

Gronovius, Jacob, ed. *Thesaurus Graecarum antiquitatum*. 13 vols. Leiden: Petrus and Baldvinus vander Aa, 1697.

Gronovius, Jacob. *Manethonis apotelesmaticorum libri sex: Nunc primum ex Bibliotheca Mediceae editi*. Leiden: Fredericus Haaring, 1698.

Grotius, Hugo. *Batavi Syntagma Arateorum: Opus Antiquitatis et Astronomiae Studiosis utilissimum: Quo quae contineantur versa pagella indicabit*. Leiden: Officina Plantiniana, apud Christophorum Raphelengium, 1600.

Grotius, Hugo. *Sacra in quibus Adamus exul tragoedia aliorumque eiusdem generis carminum cumulus consecrate Franci principi*. The Hague: Albertus Henrici, 1601.

Grotius, Hugo. *Mare liberum sive de iure quod Batavis competit ad indicana commercia dissertatio*. Leiden: Lodwijk Elzevir, 1609.

Grotius, Hugo. *Liber de Antiquitate Reipublicae Batavicae*. Leiden: Officina Plantiniana Raphelengii, 1610.

Grotius, Hugo. *Ordinum Hollandiae ac Westfrisiae pietas. Ab Improbissimis Multorum Calumniis, simulque a nuper a Sibrandi Lubberti Epistola, quam ad Reverendissimum Archiepiscopum Cantuariensem scripsit, vindicata*. Leiden: Ioannes Patius, 1613.

Grotius, Hugo. *De iure belli ac pacis libri tres. In quibus ius naturae & Gentium: item iuris publici praecipua explicantur*. Paris: Nicolaus Buon, 1625.

Grotius, Hugo. *Of the Government and Rites of the Ancient Church, Conciliation of Grace and Free-will, Certainty and Assurance of Salvation, Government of the Highest Powers in Church-affairs. In a letter To the States Embassador*. London: Printed for the translator, 1675.

Gude, Marquard. *Antiquae inscriptiones quum graecae tum latinae, olim a Marquardo Gudio collectae; nuper a Ioanne Koolio digestae hortatu consilioque Ioannis Georgii Graevii; nunc a Francisco Hesselio editae cum adnotationibus eorum*. Leeuwarden: François Hesselius, heir of François Halma, 1731.

Hegenitius, Gotfried. 'Iter Frisio-Hollandicum'. In Gotfried Hegenitius and Abraham Ortelius, *Itinerarium Frisio-Hollandicum, et Itinerarium Gallo Brabanticum. In quibus quae visu quae lectu digna*, pp. 11–117. Leiden: Elzevier, 1630.

Heinsius, Daniel. *In obitum v[iri] illustr[i] Iosephi Scaligeri Iul[ii] Caes[ari] a Burden. f[ilii] eruditorum Princeps Orationes Duae. Accedunt* Epicedia *eiusdem & aliorum: effigies item ac monumentum Scaligeri, & Principum Veronensium aeri incisa*. Leiden: Ex Officina Plantiniana Raphelengii, 1609.

Heinsius, Nicolaas. *Opera P. Ovidii Nasonis editio nova, accurante Nic. Heinsio. Dan. fil.* 3 vols. Amsterdam: Elsevier, 1664.

Heinsius, Nicolaas. *Poematum nova editio, prioribus longe auctior*. 2 vols. Amsterdam: Elsevier, 1666.

Holstenius, Lucas. *Notae & castigationes in Stephanum Byzantium de urbibus. Accedunt Scymni Chii fragmenta Graeca, cum versione Latina Lucae Holstenii, Nec non eiusdem Commentariolus in veterem picturam Nymphaei. Item Dissertationes III. de Pila Staffilari, de Milliario Aureo, atque in Laudem Boreae. Omnia ex recensione Theodori Rykii*. Leiden: Pieter van der Aa, 1692.

Imperiali, Giovanni. *Museaum Historicum et Physicum. In primo illustrium literis virorum imagines ad vivum expressae continentur. Additis elogiis eorundem vitas, et mores notantibus. In secundo animorum imagines, sive ingeniorum naturae, differentiae, causae, ac signa physicae perpenduntur. Adeo ut artis exactissimae loco possit esse, dignoscendi ad quam quisquam artem, ac scientiam sit habitis*. 2 vols. Venice: Giunti, 1640.

Kool, Joannes. *Oratio de artibus quibus Augustus callide Rempublicam R. invasit. Habita publice In illustri Academia Trajectina IV. Non. Decemb.* Utrecht: François Halma, 1696.

Kool, Joannes. *Disputatio juridica inauguralis ad legem Ateriam Tarpejam de multis quam, annuente summo numine Ex Auctoritate Magnifici D. Rectoris, D. Hermanni Witsii, S.S. Theol. Doctoris, ejusdemque in Inclyta Academia Ultrajectina Professoris Ordinarii, & Ecclesiae ibidem Pastoris, nec non Amplissimi Senatus Academici Consensu, & Nobilissimae Facultatis Juridicae Decreto, pro gradu doctoratus, Summique in utroque jure Honoribus & Privilegiis rite ac legitime More Majorum consequendis, Publico Examini submittit Joannes Kool, Ultrajectinus. Ad diem 17. Februarii horis locoque solitis.* Doctoral diss., Utrecht University. Utrecht: François Halma, 1698.

Lipsius, Justus. *Epistolarum centuriae duae: Quarum prior innovata, altera nova.* Leiden: Ex officina Plantiana, apud Franciscum Raphelengium, 1590.

Lipsius, Justus. *Epistola de peregrinatione Italica.* Leiden: Franciscus Haegerus, 1592.

Lipsius, Justus. *Dissertatiuncula apud Principes: item C. Plini Panegyricus liber Traiano dictus, cum eiusdem Lipsii perpetuo commentario.* Antwerp: Plantiniana apud Ioannem Moretum, 1600.

Lipsius, Justus. 'De Ratione cum fructu peregrinandi, & praesertim in Italia'. In Thomas Erpenius, *De peregrinatione Gallica Utiliter instuenda Tractatus. Item Brevis admodum totius Galliae descriptio: et Iusti Lipsii V.C. Epistola de Peregrinatione Italica*, pp. 105–128. Leiden: Franciscus Hegerus, 1631.

Lipsius, Justus. 'Een groot oordeel van dien grooten en uytsteeckenden Justus Lipsius, over het reysen'. In N. van Ravesteyn, *Wegh-Wyser, Vertoonende de besonderste vremde vermaecklijckheden die in t Reysen door Vranckryck en een aengrensende Landen te sien zijn.* Amsterdam: Nicolaes van Ravesteyn, 1647.

Meursius, Johannes. *Illustris Academia Lugd*[uno]*-Batava: id est Virorum Clarissimorum Icones, Elogia ac vitae, qui eam scriptis suis illustrarunt.* Leiden: Andries Clouck, 1613.

Meursius, Johannes. *Illustrium Hollandiae & Westfrisiae ordinum alma Academia Leidensis: contenta proxima pagina docebit.* Leiden: Jacob Marcus & Joost van Colster, 1614.

Meursius, Johannes. *Athenae Batavae. Sive, De Urbe Leidensi, & Academia, Virisque claris; qui utramque ingenio suo, atque scriptis illustrarunt: libri duo.* 2 vols. Leiden: Andries Clouck; Elsevier, 1625.

Meursius, Johannes. *Opera.* Edited by Giovanni Lami. Florence: apud Tartinium et Franchium, 1772.

[N. N.]. *Poetische schermutseling met grote en kleine flitsjes over het oprichten van het marmeren gedenkteken en sierlijk grafschrift van den Christiaanschen en Pauliaansche Coccejus, in de Pieterskerk te Leiden.* Flitszenburg: Gedrukt by de Erfgenamen der waarheit inde gelasterde afbeelding, 1715.

Outhof, Gerardus. *Levensbeschryvinge van bekende en geleerde mannen met hedendaagsche sterfgevallen en andere nieuwigheden.* 6 vols. Amsterdam: A Wor & Erve G. onder de Linden, 1730–1733.

Ravisius, Johannes. *De memorabilibus et claris mulieribus: aliquot diversorum scriptorum opera.* Paris: Simon de Colines, 1521.

Reusner, Nicolaus. *Imagines virorum literis illustrium quorum fide et doctrina religionis & bonarum literarum studia, nostra patrumque memoria, in Germania praesertim, in integrum sunt restituta. Additis eorundem elogiis diversorum auctorum.* Strasbourg: Bernard Jobin, 1587.

Reusner, Nicolaus. *Icones sive imagines vivae, literis Cl*[arorum] *virorum, Italiae, Graeciae, Germaniae, Galliae, Angliae, Ungariae.* Basel: Conrad Waldkirch, 1589.

Reusner, Nicolaus. *Icones sive Imagines virorum literis illustrium, quorum fide et doctrina religionis et bonarum literarum studia, nostra patrumque memoria, in Germania praesertim, in integrum sunt restituta. Additis eorundem elogiis diversorum auctorum.* Strasbourg: Bernhard Jobin, 1590; 1593.

Reusner, Nicolaus. *Icones sive Imagines virorum literis illustrium: qui Seculo XV. praesertim doctrina Religionis aliarumque bonarum scientiarum tanquam lumina in Germania nostra claruere. Olim a Tobia Stimmero pictore suae aetatis perfectissimo ad vivum expressae & Nic. Reusnero IC. cum brevi descriptione eorum vitarum & operum in lucem editae, nunc Reipublicae literariae bono iterum recusae & ab interitu vindicatae cura & impensis C.A.Z.J.C.A.* Johann Balthasar Diehl, 1714; 1719.

Richa, Giuseppe. *Notizie istoriche delle chiese fiorentine.* 10 vols. Florence: Pietro Gaetano Viviani, 1754–1762.

Rossi, Gian Vittorio. *Pinacotheca imaginum illustrium, doctrinae vel ingenii laude, Virorum, qui, auctore superstite, diem suum obierunt.* 3 vols. [Amsterdam: Johannes Blaeu], 1642–1645 — Leipzig: Thomas Fritsch, 1712 — Wolfenbüttel: J. C. Meisner, 1729.

Scaliger, Joseph Justus. *Poemata omnia: ex museio Petri Scriverii.* [Leiden]: Ex Officina Plantiniana Raphelengii, 1615.

Spitzel, Gottlieb. *Templum honoris reseratum. Sive illustrium aevi huius theologorum et philologorum imagines ac elogia. In quo L. illustrium aevi huius, orthodoxorum, ac beate defunctorum theologorum philologorumque imagines exhibentur, et quibus sive in sacram, sive literariam rem meritis, quibus item monumentis librisque editis vel mss. inclaruerint diserte ostenditur.* Augsburg: Koppmayer, 1673.

Stepling, Joseph, et al. *Litterarum Commercium.* Bratislava: Guil. Theoph. Kornius, 1782.

Targioni Tozzetti, Giovanni, ed. *Clarorum Belgarum ad Ant. Magliabechium nonnullosque alios epistolae. Ex Autographis in Biblioth. Magliabechiana, quae nunc Publica Florentinorum est, adservatis descriptae.* 2 vols. Florence: Ex Typographia ad Insigne Apollinis in Platea Magni Ducis, 1745.

*Ter bruiloft van den heere Joan Kool, Raad en Schepen der Stadt Amersfoort, en juffrouw Elizabeth Looten: In den Echt vereenigt binnen Amersfoort den 6 van Sprokkelmaand.* Amsterdam: François Halma, 1703.

Teissier, Antoine, ed. *Les Eloges des Hommes Sçavans Tirez de l'Histoire de M. de Thou. Avec des additions Contenans l'Abrégé de leur vie, le jugement, et le Catalogue de leurs Ouvrages.* Geneva: Wilderhold, 1683.

Teissier, Antoine, ed. *Les Eloges des Hommes Sçavans Tirez de Histoire de M. de Thou.* 2 vols. Utrecht: François Halma, 1696.

Tomasini, Giacomo Filippo. *Illustrium Virorum Elogia iconibus exornata. Illustriss. et reverendiss. DD. Io. Baptistae Agucchiae Archiepiscopo Amasiensi, & Apostolico Nuncio Venetiis pro Santiss. D. Urbano VIII. D.* Padua: Donatus Pasquardus, 1630.

Tomasini, Giacomo Filippo. *Elogia Virorum Literis & Sapientia Illustrium ad vivum expressis imaginibus exornata.* Patavia: Ex Typographia Sebastiani Sardi, 1644.

Thou, Jacques Auguste de. *Historiarum sui temporis pars prima.* Paris: Ambrose and Jerome Drouart, 1606.

Thou, Jacques Auguste de. *Historiarum sui temporis libri.* Edited by Thomas Carte. 7 vols. London: Samuel Buckley, 1733.

Trithemius, Johannes. *Cathalogus illustrium virorum Germaniam suis ingeniis et lucubrationibus omnifariam exornantium: domini Iohannis Tritemii abbatis Spanhemensis ordinis sancti Benedicti: ad Jacobum Wimpfelingium Sletstatinum theologum.* [Mainz: Peter von Friedberg], c. 1495.

Valla, Lorenzo. *Laurentii Vallae Opera, nunc primo non mediocribus vigiliis & iudicio quorundam eruditiss[imorum] virorum in unum volumen collecta, &, exemplaribus variis collatis, emendata. Ludimagistris, aut alias bonas literas profitentibus incredibiliter utilia adeoque necessaria. Quam ob rem rectissime a doctioribus fere omnibus iudicantur neque docti neque vere studiosi, qui non omnes huius autoris libros habent, idque praecipuo loco.* Basel: Heinrick Petri, 1540.

Vilani, Filippo. *Le vite d'uomini illustri florentini*. Venice: Giambatista Pasquali, 1747.

Viviani, Vincenzo. *De locis solidis secunda divinatio geometrica in quinque libros iniuria temporum amissos Aristaei Senioris geometrae autore Vincentio Viviani magni Galilaei novissimo discipulo regiae celsitud. Cosmi III. M.D. Etruriae mathematico primario a Ludovico Magno inter octo exteros regiae academ. scientiar. socios adscripso et regalis societatis Londini sodali*. Florence: Hyppolyto Navesi, 1701.

Voorst, Lucas van. *Disputatio juridica inauguralis de hereditatis petitione*. Doctoral diss., Utrecht University. Utrecht: François Halma, 1693.

Vossius, Gerardus Johannes. *De historicis latinis libri tres*. Leiden: Johannes Maire, 1627.

Vossius, Isaac, ed. *Epistolae genuinae S. Ignatii martyris; quae nunc primum lucem vident ex bibliotheca Florentina. Adduntur S. Ignatii epistolae, qules vulgo circumferuntur. Adhaec S. Barnabae epistola. Accessit universis translatio vetus. Edidit, & Notas addidit, Isaacus Vossius*. Amsterdam: Joan Blaeu, 1646.

Vossius, Isaac. 'Viro Cl. Carolo Dato Isaacus Vossius S.'. In *Elogio di Carlo Roberto Dati recitato nella reale Accademia Fiorentina nell'adunanza del dì 30. di settembre 1790*, edited by Francesco Fontani, 66–67. Florence: Gaetano Cambiagi Stampatore Granducale, 1790.

Zedlers, Johann Heinrich, ed. *Grosses vollständiges Universal-Lexicon aller Wissenschafften und Künste*. 64 vols. Halle; Leipzig: Verlegts Johann Heinrich Zedlers, 1733.

## Printed Sources, Modern

Alberti, Leon Battista. *L'Architettura* [*De re aedificatoria*]. Edited and translated by Giovanni Orlandi. Milan: Edizioni il Polifilo, 1966.

Bertius, Petrus. *Een welbesteed leven en een vroom verscheiden: Lijkrede op Janus Dousa (1604)*. Edited by C. L. Heesakkers and W. G. Heesakkers-Kamerbeek. Amersfoort: Florivallis, 2004.

Billi, Antonio. *Il libro di Antonio Billi esistente in due copie nella Bibliotheca Nazionale di Firenze*. Edited by Carl Frey. Berlin: G. Grote'sche Verlagsbuchhandlung, 1892.

Blok, F. F., and C. S. M. Rademaker. 'Isaac Vossius' Grand Tour, 1641–1644: The Correspondence between Isaac and his Parents'. *Lias* 33, no. 2 (2006): 150–216.

Blok, F. F., and C. S. M. Rademaker. 'Isaac Vossius' Grand Tour, 1641–1644: The Correspondence between Isaac and his Parents. Part II: Isaac Vossius in Italy (with 3 ills.)'. *Lias* 35, nos. 1–2 (2008): 209–279.

Blok, F. F., and C. S. M. Rademaker. 'Isaac Vossius' Grand Tour, 1641–1644: The Correspondence between Isaac and his Parents. Part III: Isaac in Paris'. *Lias* 36, no. 2 (2009): 295–385.

Boccaccio, Giovanni. *Famous Women*. Edited and translated by Virginia Brown. Cambridge, MA: Harvard University Press, 2001.

Bracciolini, Poggius, and Nicolaus de Niccolis. *Two Renaissance Book Hunters: The Letters of Poggius Bracciolini to Nicolaus de Niccolis*. Translated and annotated by Phylis Walter Goodhart Gordan. New York: Columbia University Press, 1974.

Bruni, Leonardi. *Laudatio Florentinae urbis*. Edited by Stefano U. Baldassarri. Florence: SISMEL edizioni del Galluzzo, 2000.

Bruni, Leonardi. *Panegirico della città di Firenze: testo italiano a fronte di frate Lazaro di Padova*. Edited by Giuseppe de Toffol. Florence: Nuova Italia Editrice, 1974.

Cereta, Laura. *Collected Letters of a Renaissance Feminist*. Edited and translated by Diana Robin. The Other Voice in Early Modern Europe. Chicago; London: University of Chicago Press, 1997.

Chapelain, Jean. *Les lettres authentiques à Nicolas Heinsius (1649–1672): Une amitié érudite entre France et Hollande*. Bibliothèque des correspondances, mémoires et journaux 22. Edited, annotated, and introduction by Bernard Bray. Paris: Honoré Champion Éditeur, 2005.

Cicero. *De finibus bonorum et malorum* [*On Ends*]. Translated by H. Rackham. Loeb Classical Library 40. 2nd ed., repr. Cambridge, MA: Harvard University Press; London: Heinemann, 1971.

Cicero. *On the Republic; On the Laws*. Translated by Clinton Walker Keyes. Loeb Classical Library 213. Cambridge, MA; London: Harvard University Press, 1928.

Dante Alighieri. *The Divine Comedy*, vol. 1: *Inferno*. Translated and commentary by John D. Sinclair. New York: Oxford University Press, 1961.

Diogenes Laertius. *Lives of the Eminent Philosophers*. Translated by Pamela Mensch, edited by James Miller. Oxford; New York: Oxford University Press, 2018.

Erasmus, Desiderius. 'On the Writing of Letters / De conscribendis epistolis'. Translated and annotated by Charles Fantazzi. In *Collected Works of Erasmus: Literary and Educational Writings*, vol. 25, edited by J. K. Sowards, 1–254. Toronto; Buffalo; London: University of Toronto Press, 1985.

Dupuy, Jacques, and Nicolaas Heinsius. *Correspondance de Jacques Dupuy et Nicolas Heinsius (1646–1656)*. Edited by Hans Bots. The Hague: Martinus Nijhoff, 1971.

Galilei, Galileo. *Opere di Galileo Galilei*. Edited by Antonio Favaro. 20 vols. Florence: Tipografia Barbèra, 1890–1909.

Gattei, Stefano. 'Galileo's Legacy: A Critical Edition and Translation of the Manuscript of Vincenzo Viviani's *Grati Animi Monumenta*'. *The British Journal for the History of Science* 50, no. 2 (2017): 181–228.

Giovio, Paolo. *An Italian Portrait Gallery; Being Brief Biographies of Scholars Illustrious Within the Memory of Our Grandfathers for the Published Monuments of Their Genius*. Translated by Florence Alden Gragg. Boston: Chapman & Grimes, 1935.

Grotius, Hugo. *The Antiquity of the Batavian Republic: With the Notes by Petrus Scriverius*. Edited and translated by Jan Waszink et al. Assen: Van Gorcum, 2000.

Grotius, Hugo. *Briefwisseling van Hugo Grotius*. Edited by P. C. Molhuysen, B. L. Meulenbroek, Paula P. Witkam, Henk J. M. Nellen, and Cornelia M. Ridderikhoff. 17 vols. Rijks geschiedkundige publicaties. Grote ser. 64, 82, 105, 113, 119, 124, 130, 136, 142, 154, 179, 197, 213, 222, 238, 246, 248. The Hague: Martinus Nijhoff, 1928–2001.

Grotius, Hugo. *Commentary on the Law of Prize and Booty*. Edited and introduction by Martine Julia van Ittesum. Natural Law and Enlightenment Classics. Indianapolis: Liberty Fund, 2006.

Grotius, Hugo. *Ordinum Hollandiae ac Westfrisiae Pietas (1613)*. Edited and translated by E. Rabbie. Studies in the History of Christian Traditions 66. Leiden; Boston: Brill, 1995.

Horace. *Odes and Epodes*. Edited and translated by Niall Rudd. Loeb Classical Library 33. Cambridge, MA; London: Harvard University Press, 2004.

Jerome, Saint. *On Illustrious Men*. Translated by Thomas P. Halton. Fathers of the Church, A New Translation 100. Washington, DC: Catholic University of America Press, 1999.

Lipsius, Justus. *Iusti Lipsi epistolae*. Edited Aloïs Gerlo et al. 13 vols. Brussels; Louvain; Paris: Peeters, 1978–.

Lipsius, Justus. *Principles of Letter-Writing: A Bilingual text of Justi Lipsi Epistolica institutio*. Edited and translated by R. V. Young and M. Thomas Hester. Carbondale: South Illinois University Press, 1996.

Mayer, Thomas F., ed. *The Trial of Galileo, 1612–1633*. Toronto: University of Toronto Press, 2012.

Molhuysen, P. C., ed. *Bronnen tot de geschiedenis der Leidsche universiteit 1574–1811*. Rijks Geschiedkundige Publicatiën 20. The Hague: Martinus Nijhoff, 1913.

Ovid. *Tristia: Ex Ponto*. Translated by Arthur Leslie Wheeler. Ovid VI. Loeb Classical Library 151. Cambridge, MA: Harvard University Press, 1975.

Pagano, Sergio, ed. *I documenti vaticani del processo di Galileo Galilei, 1611–1741*. New ed. Vatican City: Archivio segreto Vaticano, 2009.

Ploos, Jacob Ploos van. *Een tempel voor Hugo de Groot: een rede van Jacob Ploos van Amstel uit 1774*. Edited by Jan Bloemendal and Hans Krist Ploos van Amstel. Amersfoort: Florivallis, 2010.

Ploos van Amstel, Henrick, and Nicolaas Heinsius. *Brieven, 1635–1637*. Edited with introduction and annotations by Hans Kristian Ploos van Amstel. Translated by Frank Tichelman. Woubrugge: Avalon Pers, 2007.

Scaliger, Joseph Justus. *The Correspondence of Joseph Justus Scaliger*. 8 vols. Edited by Paul Botley and Dirk van Miert. Travaux d'Humanisme et Renaissance 507. Geneva: Librairie Droz, 2012.

Seneca the Younger. *Moral Essays, Volume III: De Beneficiis*. Translated by John W. Basore. Loeb Classical Library 310. Cambridge, MA: Harvard University Press, 1935.

Shelley, Mary. *Frankenstein; Or, the Modern Prometheus*. Edited by Maurice Hindle. London: Penguin Books, 2012. Originally published in 3 vols. London: Hughes Harding, Mayor and Jones, 1818.

Trithemius, Johannes. 'Johannes Trithemius: *Catalogue of the Illustrious Men Who Have Adorned Germany in Every Way with Their Talents and Tireless Studies* (selections)'. Edited by Johannes Helmrath, translated by Patrick Baker and Johannes Helmrath. In Baker, *Biography, Historiography, and Modes of Philosophizing*, 209–239.

Vasari, Giorgio. *The Lives of the Artists*. Translated, introduction, and notes by Julia Conaway Bondanella and Peter Bondanella. Oxford World's Classics. Oxford; New York: Oxford University Press, 1991.

Veronese, Guarino. *Carmina*. Edited by Aldo Manetti. Bergamo: Istituto universitario di Bergamo, 1985.

Vossius, Gerardus. *Geschiedenis als wetenschap*. Edited, introduction, and annotations by Cor Rademaker. Geschiedenis van de wijsbegeerte in Nederland 9. Baarn: Ambo, 1990.

Wagenaar, Lodewijk. *Een Toscaanse prins bezoekt Nederland: De twee reizen van Cosimo de' Medici, 1667–1669*. Translated by Berti Eringa. Amsterdam: Uitgeverij Bas Lubberhuizen, 2014.

## Secondary Literature

Abir-Am, Pnina G., and Clark A. Elliot, eds. *Commemorative Practices in Science: Historical Perspectives on the Politics of Collective Memory*. Osiris 14. Chicago: University of Chicago Press, 1999.

Abou-El-Haj, Barbara. *The Medieval Cult of Saints: Formations and Transformations*. Cambridge: Cambridge University Press, 1994.

Adams, Alison. 'The *Emblemata* of Théodore de Bèze (1580)'. In *Mundus Emblematicus: Studies in Neo-Latin Emblem Books*, edited by Karl A. E. Enenkel and Arnoud S.Q. Visser, 71–99, Imago Figurata Studies 4. Turnhout: Brepols, 2003.

Adams, Alison. *Web of Allusion: French Protestant Emblem Books of the Sixteenth Century*. Travaux d'Humanisme et Renaissance 378. Geneva: Librairie Droz, 2003.

al-Musawi, Muhsin J. *The Medieval Islamic Republic of Letters: Arabic Knowledge Construction*. Notre Dame, IN: University of Notre Dame Press, 2015.

Anderson, Benedict. *Imagined Communities: Reflections on the Origin and Spread of Nationalism*. Rev. ed. London; New York: Verso, 2006. Originally published in 1983.

Anderson, Judith H. *Biographical Truth: The Representation of Historical Persons in Tudor-Stuart Writing*. New Haven, CT; London: Yale University Press, 1984.

Apfelbaum, Erika. 'Halbwachs and the Social Properties of Memory'. In Radstone and Schwarz, *Memory*, 77–92.

Austin, Kenneth. 'Academic Exchanges: Letters, the Reformation and Scholarly Self-Fashioning'. In Kirwan, *Scholarly Self-Fashioning*, 39–58.

Azzolini, Monica. 'In Praise of Art: Text and Context of Leonardo's *Paragone* and Its Critique of the Arts and Sciences'. *Renaissance Studies* 19, no. 4 (2005): 487–510.

Baar, Mirjam de, and Brita Rang. 'Minerva of savante, heilige of dweepster: Receptiegeschiedenis van Anna Maria van Schurman sedert de zeventiende eeuw'. In *Anna Maria van Schurman (1607–1678): Een uitzonderlijk geleerde vrouw*, edited by Mirjam de Baar, 9–28. Zutphen: Walburg Pers, 1992.

Babel, R., ed. *Grand Tour: Adeliges Reisen und europäische Kultur vom 14. bis zum 18. Jahrhundert*. Beihefte der Francia 60. Ostfildern: Jan Thorbecke Verlag, 2005.

Backus, Irena. *Life Writing in Reformation Europe: Lives of Reformers by Friends, Disciples and Foes*. St Andrews Studies in Reformation History. Aldershot; Burlington, VT: Ashgate, 2008.

Backus, Irena, ed. *Théodore de Bèze (1519–1605): Actes du Colloque de Genève (septembre 2005)*. Travaux d'Humanisme et Renaissance 424. Geneva: Librairie Droz, 2007.

Backus, Irena. 'What is a Historical Account? Religious Biography and the Reformation's Break with the Middle Ages'. *Archiv für Reformationsgeschichte* 101, no. 1 (2013): 289–304.

Baker, Patrick, ed. *Biography, Historiography, and Modes of Philosophizing: The Tradition of Collective Biography in Early Modern Europe*. Renaissance Society of America 7. Leiden; Boston: Brill, 2017.

Baker, Patrick. *Italian Renaissance Humanism in the Mirror*. Ideas in Context. Cambridge: Cambridge University Press, 2015.

Baker Patrick, Ronny Kaiser, Maike Priesterjahn, and Johannes Helmrath, eds. *Portraying the Prince in the Renaissance: The Humanist Depiction of Rulers in Historiographical and Biographical Texts*. Transformationen der Antike 44. Berlin; Boston: Walter de Gruyter, 2016.

Baldassarri, Stefano U. 'Poggio Bracciolini and Coluccio Salutati: The Epitaph and the 1405–1406 Letters'. In *Poggio Bracciolini and the Re(dis)covery of Antiquity: Textual and Material Traditions*, edited by Roberta Ricci, 71–87, Atti 38. Florence: Firenze University Press, 2020.

Baldini, Umberto, and Bruno Nardini, eds. *Il complesso monumentale di Santa Croce: la basilica, le capelle, i chiostri, il museo*. Florence: Nardini Editore, 1983.

Baldissin Molli, Giovanna, ed. *Frate Francesco Sansone 'de Brixi' Ministro generale OFMConv (1414–1499): Un mecenate francescano del Rinascimento*. Quaderni Museo Antoniano. Padua: Centro Studi Antoniani, 2000.

Balsem, Astrid Cecilia. 'Een biografie van de Bibliotheca Vossiana'. PhD diss., Leiden University, 2020.

Barash, Jeffrey Andrew. *Collective Memory and the Historical Past*. Chicago: University of Chicago Press, 2016.

Barnes, Annie. *Jean le Clerc (1657–1736) et la République des Lettres*. Paris: E. Droz, 1938.

Baron, Hans. *The Crisis of the Early Italian Renaissance: Civic Humanism and Republican Liberty in an Age of Classicism and Tyranny*. 2 vols. Princeton, NJ: Princeton University Press, 1955.

Barrow, Julia. *The Clergy in the Medieval World: Secular Clerics, Their Families and Careers in North-Western Europe, c. 800–c. 1200*. New York: Cambridge University Press, 2016.

Bartoli, Lando. 'La Basilica di Santa Croce al di là spazio e del tempo'. In Baldini and Nardini, *Il complesso monumentale di Santa Croce*, 33–49.

Battersby, Christine. *Gender and Genius: Towards a Feminist Aesthetics*. Bloomington: Indiana University Press, 1989.

Bauer, Stefan. *The Invention of Papal History: Onofrio Panvinio between Renaissance and Catholic Reform*. Oxford–Warburg Studies. Oxford: Oxford University Press, 2020.

Becker, Jochen. *Henrick de Keyser: Standbeeld van Desiderius Erasmus in Rotterdam*. Bloemendaal: H. J. W. Becht, 1993.

Becker-Cantarino, Barbar. *Daniel Heinsius*. Boston, MA: Twayne, 1978.

Bender, Eva. *Die Prinzenreise: Bildungsaufenthalt und Kavalierstour im höfischen Kontext gegen Ende des 17. Jahrhunderts*. Schriften zur Residenzkultur 6. Berlin: Lukas, 2011.

Bensaude-Vincent, Bernadette, and Christine Blondel, eds. *Science and Spectacle in the European Enlightenment*. Aldershot: Ashgate, 2008.

Benson, Pamela Joseph. *The Invention of the Renaissance Woman: The Challenge of Female Independence in the Literature and Thought of Italy and England*. University Park: Pennsylvania State University Press, 1992.

Bernays, Jacob. *Joseph Justus Scaliger*. Berlin: W. Hertz, 1855.

Berti, Luciano, ed. *Il pantheon di Santa Croce a Firenze*. Florence: Cassa di Risparmio di Firenze, 1993.

Bertram, Martin. 'Mittelalterliche Gelehrtengräber in Bologna: Anmerkingen zu dem neuen Werk von Renzo Grandi'. *Quellen und Forschungen aus italienischen Bibliotheken und Archiven* 65 (1985): 427–435.

Beyer, Jürgen, Albrecht Burkardt, and Fred van Lieburg, eds. *Confessional Sanctity (c. 1500–c. 1800)*. Mainz: Philipp von Zabern, 2004.

Biagioli, Mario. *Galileo, Courtier: The Practice of Science in the Culture of Absolutism*. Chicago; London: University of Chicago Press, 1993.

Biagioli, Mario. '"Playing with Evidence"'. *Early Science and Medicine* 1, no. 1 (1996): 70–105.

Bianchi, Eric. 'Bad Latin, Bad Manners: Giovanni Battista Doni, Marin Mersenne, and Literary Style in Seventeenth-Century Music Theory'. *Music & Letters* 96, no. 2 (2015): 167–184.

Biow, Douglas. *Vasari's Words: The* Lives of the Artists *as a History of Ideas in the Italian Renaissance*. Cambridge; New York; Melbourne; New Delphi: Cambridge University Press, 2018.

Black, Jeremy. *The British Abroad: The Grand Tour in the Eighteenth Century*. Stroud: Alan Sutton, 1992.

Black, Jeremy. *France and the Grand Tour*. London: Palgrave Macmillan, 2003.

Black, Jeremy. *Italy and the Grand Tour*. New Haven, CT; London: Yale University Press, 2003.

Blair, Ann. *Too Much To Know: Managing Scholarly Information Before the Modern Age*. New Haven, CT: Yale University Press, 2010.

Blok, F. F. *Isaac Vossius and His Circle: His Life Until His Farewell to Queen Christina of Sweden, 1618–1655*. Translated by Cis van Heertum. Groningen: Egbert Forsten, 2000.

Blok, F. F. *Isaac Vossius en zijn kring: Zijn leven tot zijn afscheid van koningin Christina van Zweden, 1618–1655*. Groningen: Egbert Forsten, 1999.

Blok, F. F. 'Nicolaas Heinsius in dienst van koningin Christina van Zweden'. PhD diss., Leiden University, 1949.

Blok, F. F. *Nicolaas Heinsius in Napels (april–juli 1647)*. Verhandeling der Koninklijke Akademie van Wetenschappen, Afdeling Letterkunde, Nieuwe Reeks 125. Amsterdam; Oxford; New York: Noord-Hollandsche Uitgevers Maatschappij, 1984.

Blom, Hans. 'Hugo Grotius on Trust, Its Causes and Effects'. In *Trust and Happiness in the History of European Political Thought*, edited by László Kontler and Mark Somos, 76–98. Leiden; Boston: Brill, 2018.

Boer, P. den. 'Naar een geschiedenis van de dood'. *Tijdschrift voor Geschiedenis* 89, no. 2 (1976): 161–201.

Boisen, Camilla. 'Predestination'. In Lesaffer and Nijman, *The Cambridge Companion to Hugo Grotius*, 221–242.

Bonis, Wim. 'Lugdunum Batavorum en het Keltische erfgoed'. *Leids Jaarboekje* (2005): 45–52.

Booth, Philip, and Elizabeth Tingle, eds. *A Companion to Death, Burial, and Remembrance in Late Medieval and Early Modern Europe, c. 1300–1700*. Brill's Companions to the Christian Tradition 94. Leiden; Boston: Brill, 2021.

Booth, Philip, and Elizabeth Tingle. 'Introduction: Dying, Death, and Commemoration, 1350–1700'. In Booth and Tingle, *A Companion to Death, Burial, and Remembrance*, 1–30.

Bots, Hans. *Republiek der Letteren: De Europese intellectuele wereld, 1500–1760*. Nijmegen: Uitgeverij Vantilt, 2018.

Bots, Hans. *Republiek der Letteren: Ideaal en Werkelijkheid*. Amsterdam: APA-Holland Universiteits Pers, 1977.

Bots, Hans, and Françoise Waquet, eds. *Commercium Litterarium: La Communication dans la République des Lettres / Forms of Communications in the Republic of Letters, 1600–1750*. Amsterdam; Maarssen: APA-Holland University Press, 1994.

Bots, Hans, and Françoise Waquet. *La République des Lettres*. Europe & Histoire. Paris: Belin, 1997.

Bots, Hans, and Willem Frijhoff. 'Academiereis of educatiereis? Noordbrabantse studenten in het buitenland, 1550–1650'. *Batavia Academica* 1 (1983): 13–30.

Boutier, Jean, Maria Pia Paoli, and Corrado Viola, eds. *Antonio Magliabechi nell'Europea dei saperi*. Pisa: Edizioni della Normale, 2017.

Bouwers, Eveline G. *Public Pantheons in Revolutionary Europe: Comparing Cultures of Remembrance, c. 1790–1840*. Basingstoke; New York: Palgrave Macmillan, 2012.

Brann, Noel L. *The Abbot Trithemius (1462–1516): The Renaissance of Monastic Humanism*. Studies in the History of Christian Thought 24. Leiden; Boston: Brill, 1981.

Bray, Alan. *The Friend*. Chicago; London: University of Chicago Press, 2003.

Bray, John S. 'The Value of Works in the Theology of Calvin and Beza'. *The Sixteenth Century Journal* 4, no. 2 (1973): 77–86.

Breitenberg, Mark. *Anxious Masculinity in Early Modern England*. Cambridge: Cambridge University Press, 1996.

Brennan, Michael G., ed. *The Origins of the Grand Tour: The Travels of Robert Montagu, Lord Mandeville (1649–1654), William Hammond (1655–1658), Banaster Maynard (1660–1663)*. London: Hakluyt Society, 2004.

Brine, Douglas. *Pious Memories: The Wall-Mounted Memorial in the Burgundian Netherlands*. Leiden; Boston: Brill, 2015.

Brockliss, Laurence. *Calvet's Web: Enlightenment and the Republic of Letters in Eighteenth-Century France*. Oxford: Oxford University Press, 2002.

Brown, Peter. *The Cult of the Saints: Its Rise and Function in Late Christianity*. Haskell Lectures on History of Religions, New Series 2. Chicago; London: University of Chicago Press, 1981.

Brucker, Gene. *The Civic World of Early Renaissance Florence*. Princeton, NJ: Princeton University Press, 1977.

Bruhn, John G. *The Sociology of Community Connections*. 2nd ed. Dordrecht; Heidelberg; London; New York: Springer, 2011.

Bryson, Anna. *From Courtesy to Civility: Changing Codes of Conduct in Early Modern England*. Oxford; New York: Oxford University Press, 1998.

Buck, August, ed. *Biographie und Autobiographie in der Renaissance*. Wolfenbütteler Abhandlungen zur Renaissanceforschung 4. Wiesbaden: Otto Harrassowitz, 1983.

Burke, Peter. 'Erasmus and the Republic of Letters'. *European Review* 7, no. 1 (1999): 5–17.

Burke, Peter. 'Representations of the Self from Petrarch to Descartes'. In *Rewriting the Self, Histories from the Renaissance to the Present*, edited by Roy Porter, 17–28. London; New York: Routledge, 1997.

Butterfield, Andrew. 'Social Structure and the Typology of Funerary Monuments in Early Renaissance Florence'. *RES: Anthropology and Aesthetics* 26 (1994): 47–67.

Caldwell, Dorigen. 'The *Paragone* Between Word and Image in *Impresa* Literature'. *Journal of the Warburg and Courtauld Institutes* 63 (2000): 277–286.

Callard, Caroline. 'Diogène au service des princes: Antonio Magliabechi à la cour de Toscane (1633–1714)'. *Histoire, Économie et Société* 19, no. 1 (2000): 85–103.

Cameron, Evan. 'Medieval Heretics as Protestant Martyrs'. *Studies in Church History* 30: Martyrs and Martyrologies (1993): 185–207.

Campbell, Julie D., and Anne R. Larsen, eds. *Early Modern Women and Transnational Communities of Letters*. Women and Gender in the Early Modern World. Farnham; Burlington, VT: Ashgate, 2009.

Cappelli, Adriano. *Lexicon Abbreviaturarum: Wörterbuch lateinischer und italienischer Abkürzungen wie sie in Urkunden und Handschriften besonders des Mittelalters gebräuchlich sind, dargestellt in über 1400 Holzschnittzeichen*. 2nd ed. Leipzig: Verslagsbuchhandlung von J. J. Weber, 1928.

Carbonnier-Burkard, Marianne. '*L'Histoire ecclésiastique des Eglises réformées* ...: la construction bézienne d'un "corps d'histoire" '. In Backus, *Théodore de Bèze*, 145–161.

Carrara, Elena. '*Exequiae, funera, sepolcra*: Rituale e iconografia nei monumenti dei professori dello Studio tra Quattro e Cinquecento'. In Sicca, *Scultura a Pisa nell'età moderna*, 21–73.

Carrington, Jill Emilee. 'Sculpted Tombs of the Professors of the University of Padua, c. 1358–c. 1557'. PhD diss., Syracuse University, 1996.

Casini, Tommaso. *Ritratti parlanti: Collezionismo e biografie illustrate nei secoli XVI e XVII*. Collana di Museologia e Museografia. Florence: Edifir, 2004.

Ceccarelli, Paola, Lutz Doering, Thorsten Fögen, and Ingo Gildenhard, eds. *Letters and Communities: Studies in the Socio-Political Dimensions of Ancient Epistolography*. Oxford: Oxford University Press, 2018.

Cecchi, Alessandro. 'L'estremo omaggio al "padre e maestro di tutte le arti": Il monumento funebre di Michelangelo'. In Berti, *Il pantheon di Santa Croce*, 57–82.

Chambers, David Wade. 'Stereotypic Images of the Scientist: The Draw-A-Scientist Test'. *Science Education* 67, no. 2 (1983): 255–265.

Chazalon, Christophe. 'Les *Icones* de Théodore de Bèze (1580): Entre mémoire et propagande'. *Bibliothèque d'Humanisme et Renaissance* 66, no. 2 (2004): 359–376.

Cochrane, Eric W. *Florence in the Forgotten Centuries, 1527–1800: A History of Florence and the Florentines in the Age of the Grand Dukes*. Chicago; London: University of Chicago Press, 1973.

Cohen, Bernard. 'Commemorations & Memorials: Isaac Newton, 1727–1977'. *Vistae in Astronomy* 22 (1979): 381–394.

Coleman, Patrick, Jayne Elizabeth Lewis, and Jill Anne Kowalik, eds. *Representations of the Self From the Renaissance to Romanticism*. Cambridge; New York; Melbourne: Cambridge University Press, 2000.

Confino, Alon. 'Collective Memory and Cultural History: Problems of Method'. *The American Historical Review* 102, no. 5 (1997): 1386–1403.

Consodine, John. *Dictionaries in Early Modern Europe: Lexicography and the Making of Heritage*. Cambridge; New York: Cambridge University Press, 2008.

Cooper, Donal. 'Experiencing Dominican and Franciscan Churches in Renaissance Italy'. In *Sanctity Pictured: The Art of the Dominican and Franciscan Orders in Renaissance Italy*, edited by Trinita Kennedy, 47–61. Nashville; London: Frist Center for the Visual Arts and Philip Wilson Publishers, 2014.

Covi, Dario A. 'An Unnoticed Verrocchio?'. *The Burlington Magazine* 110, no. 778 (1968): 4–9, 11.

Cox, Patricia. *Biography in Late Antiquity: A Quest for the Holy Man.* The Transformation of the Classical Heritage 5. Berkeley; Los Angeles; London: University of California Press, 1983.

Craske, Matthew. 'Westminster Abbey 1720–70: A Public Pantheon Built Upon Private Interest'. In Wrigley and Craske, *Pantheons*, 57–79.

Cristin, Claude. 'Aux origins de L'histoire Littéraire Française: "Les éloges des Hommes sçavans Tirez de l'histoire de M. de Thou par Antoine Teissier" (1683–1715)'. *Revue d'Histoire littéraire de la France* 72, no. 2 (1972): 238–246.

Cubitt, Geoffrey, ed. *Imagining Nations.* York Studies in Cultural History. Manchester: Manchester University Press, 1998.

Cubitt, Geoffrey. *History and Memory.* Historical Approaches. Manchester; New York: Manchester University Press, 2007.

Cummings, Brian, Ceri Law, Karis Riley, and Alexandra Walsham, eds. *Remembering the Reformation.* Remembering the Medieval and Early Modern Worlds. London; New York: Routledge, 2020.

Damerow, Julia, and Dirk Wintergrün. 'The Hitchhiker's Guide to Data in the History of Science'. *Isis* 110, no. 3 (2019): 513–521.

Daston, Lorraine. 'The Ideal and Reality of the Republic of Letters in the Enlightenment'. *Science in Context* 4, no. 2 (1991): 367–386.

Daston, Lorraine. 'The Moral Economy of Science'. In *Constructing Knowledge in the History of Science*, edited by Arnold Thackray, 3–24. Osiris 10. Chicago: University of Chicago Press, 1995.

Daybell, James. 'Letters'. In *The Cambridge Companion to Early Modern Women's Writing*, edited by Laura Lunger Knoppers, 181–193. Cambridge: Cambridge University Press, 2009.

Daybell, James. *The Material Letter in Early Modern England: Manuscript Letters and the Culture and Practices of Letter-Writing, 1512–1635.* Early Modern Literature in History. Basingstoke; New York: Palgrave Macmillan, 2012.

Deinsen, Lieke van. *The Panpoëticon Batavûm: The Portrait of the Author as a Celebrity.* Rijksmuseum Studies in History 1. Amsterdam: Rijksmuseum, 2016.

De Cesari, Chiara, and Ann Rigney, eds. *Transnational Memory: Circulation, Articulation, Scales.* Media and Cultural Memory 19. Berlin: De Gruyter, 2014.

De Landtsheer, Jeanine. '*Pius Lipsius* or *Lipsius Proteus*'. In De Landtsheer and Nellen, *Between Scylla and Charybdis*, 303–349.

De Landtsheer, Jeanine, and Henk Nellen, eds. *Between Scylla and Charybdis: Learned Letter Writers Navigating the Reefs of Religious and Political Controversy in Early Modern Europe.* Brill's Studies in Intellectual History 192. Leiden; Boston: Brill, 2011.

De Marchi, Andrea. 'Relitti di un naufragio: affreschi di Giotto, Taddeo Gaddi e Maso di Banco nelle navate di Santa Croce'. In *Santa Croce: Oltre le apparenze*, edited by Andrea De Marchi and Giacomo Piraz, 33–71. Pistoia: Gli Ori, 2011.

Dekker, Rudolf. 'Egodocuments in the Netherlands from the Sixteenth to the Nineteenth Century'. In *Envisioning Self and Status: Self-Representation in the Low Countries, 1400–1700*, edited by Eric Griffey, 255–285. Hull: Association for Low Countries Studies in Great Britain and Ireland (ALCS), 1999.

Di Stasi, Michelina. *Stefano di Francesco Rosselli: antiquario fiorentino del XVII sec. e il suo sepoltuario.* Florence: Edizione Polistampa, 2014.

Dibon, Paul, and Françoise Waquet. *Johannes Fredericus Gronovius, pèlerin de la république des lettres: recherches sur le voyage savant au xxiie siècle.* Geneva: Librairie Droz, 1984.

Dilon, Anne. *The Construction of Martyrdom in the English Catholic Community.* St Andrews Studies in Reformation History. London; New York: Routledge, 2016.

Dionisotti, Carlo. 'La galleria degli uomini illustri'. In *Cultura e società nel Rinascimento tra riforme e manierismi*, edited by Vittore Branca and Carlo Ossola, 449–461. Civiltà Veneziana 32. Florence: Olski, 1984.

Dixhoorn, Arjan van, and Susie Speakman Sutch, eds. *The Reach of the Republic of Letters: Literary and Learned Societies in Late Medieval and Early Modern Europe*. 2 vols. Brill's Studies in Intellectual History 168. Leiden; Boston: Brill, 2008.

Domínguez, Freddy C. 'Relics and Saints: Commemoration and Memoralization of the Holy Dead'. In Booth and Tingle, *A Companion to Death, Burial, and Remembrance*, 393–417.

Dongen, Jeroen van, and Herman Paul, eds. *Epistemic Virtues in the Sciences and the Humanities*. Boston Studies in the Philosophy and History of Science 321. Cham: Springer International, 2017.

Dorsman, L. J., and P. J. Knegtmans, eds. *De menselijke maat in de wetenschap: de geleerden(auto) biografie als bron voor wetenschaps- en universiteitsgeschiedenis*. Universiteit en samenleving 11. Hilversum: Uitgeverij Verloren, 2013.

Dröge, Jan, with John Veerman. 'Inleiding: De bouwgeschiedenis'. In Den Hartog, *De Pieterskerk in Leiden*, 13–71.

Duff, Timothy E. *Plutrarch's* Lives*: Exploring Virtues and Vice*. Oxford: Clarendon Press, 1999.

Duke, Alastair. 'In Defence of the Common Fatherland: Patriotism and Liberty in the Low Countries, 1555–1576'. In *Networks, Regions and Nations: Shaping Identities in the Low Countries, 1300–1650*, edited by Robert Stein and Judith Pollmann, 217–240. Studies in Medieval and Reformation Traditions 149. Leiden; Boston: Brill, 2010.

Eden, Kathy. *Friends Hold All Things in Common: Tradition, Intellectual Property, and the* Adages *of Erasmus*. New Haven, CT: Yale University Press, 2001.

Eden, Kathy. *The Renaissance Rediscovery of Intimacy*. Chicago; London: University of Chicago Press, 2012.

Egmond, Florike. *The World of Carolus Clusius: Natural History in the Making, 1550–1610*. Perspectives in Economic and Social History 6. London: Pickering & Chatto, 2010.

Egmond, Florike, and P. G. Hoftijzer, eds. *Carolus Clusius: Towards a Cultural History of a Renaissance Naturalist*. History of Science and Scholarship in the Netherlands 8. Amsterdam: Koninklijke Akademie van Wetenschappen, 2007.

Eichel-Lojkine, Patricia. *Le siècle des grand hommes: Les recueils de vies d'hommes illustres avec portraits du XVIème siècle*. La Republique des Lettres 5. Leuven; Paris; Sterling, VA: Éditions Peeters, 2001.

Eijnatten, Joris van, and Ruben Ros. 'The Eurocentric Fallacy: A Digital-Historical Approach to the Concept of "Modernity", "Civilization" and "Europe" (1840–1990)'. *International Journal for History, Culture and Modernity* 7 (2019): 686–736.

Eisenstein, Elizabeth L. *The Printing Press as an Agent of Change: Communications and Cultural Transformations in Early-Modern Europe*. 2 vols. New York: Cambridge University Press, 1980.

Ekkart, R. E. O. 'Het grafmonument van Scaliger'. *Jaarboekje voor geschiedenis en oudheidkunde van Leiden en omstreken* 70 (1978): 81–90.

Ekkart, R. E. O., ed. *Icones Leidenses: De portretverzameling van de Rijksuniversiteit te Leiden*. Leiden: Universitaire Pers, 1973.

Elsner, Jás and Joan-Pau Rubiés, eds. *Voyages and Visions: Towards a Cultural History of Travel*. London: Reaktion Books, 1999.

Enenkel, Karl A. E. *Die Erfindung des Menschen: Die Autobiographik des Frühneuzeitlichen Humanismus von Petrarca bis Lipsius*. Berlin; New York: Walter de Gruyter, 2008.

Enenkel, Karl A. E. ' "Identities" in Humanist Autobiographies and Related Self-Perception'. In Scholten, Van Miert, and Enenkel, *Memory and Identity in the Learned World*, 31–88.

Enenkel, Karl A. E. *Die Stiftung von Autorschaft in der neulateinischen Literatur (ca. 1350–ca. 1650): Zur autorisierenden und wissenvermittelnden Funktion von Widmungen, Vorworttexten, Autorporträts und Dedikationsbildern.* Mittellateinische Studien und Texte 48. Leiden; Boston: Brill, 2015.

Enenkel, Karl A. E., Betsy de Jong-Crane, and Peter Liebregts, eds. *Modelling the Individual: Biography and Portrait in the Renaissance. With a Critical Edition of Petrarch's Letter to Posterity.* DQR Studies in Literature 23. Amsterdam; Atlanta, GA: Rodopi, 1998.

Enenkel, Karl A. E., and Jan L. de Jong, eds. *Artes Apodemicae and Early Modern Travel Culture, 1550–1700.* Leiden; Boston: Brill, 2019.

Enenkel, Karl A. E., and Koen Ottenheym. *Oudheid als ambitie: De zoektocht naar een passend verleden, 1400–1700.* Nijmegen: Uitgeverij Vantilt, 2017.

Enenkel, Karl A. E., and Koen Ottenheym, eds. *The Quest for an Appropriate Past in Literature, Art and Architecture.* Intersections 60. Leiden; Boston: Brill, 2018.

Enenkel, Karl A. E., and Claus Zittel, eds. *Die* Vita *als Vermittlerin von Wissenschaft und Werk: Form- und Funktionanalytische Untersuchungen zu frühneuzeitlichen Biographien von Gelehrten, Wissenschaftlern, Schriftstellern und Künstlern.* Scientia Universalis, Abteilung I, Studien zur Wissenschaftgeschichte der Vormoderne 1. Berlin: LIT Verlag, 2013.

Esser, Raingard. *The Politics of Memory: The Writing of Partition in the Seventeenth-Century Low Countries.* Leiden; Boston: Brill, 2012.

Exalto, John. *Gereformeerde heiligen: De religieuze exempeltraditie in vroegmodern Nederland.* Nijmegen: Uitgeverij Vantilt, 2005.

Eyffinger, Arthur. '*De fide et perfidia / On Trust and Treachery*, Latin text constituted from the Ms., with English Translation and Introduction by Arthur Eyffinger'. *Grotiana* 36, no. 1 (2015): 79–117.

Fantazzi, Charles. 'Vives versus Erasmus on the Art of Letter-Writing'. In Van Houdt et al., *Self-Presentation and Social Identification*, 39–56.

Fileti Mazza, Miriam, and Bruna Tomasello, eds. *Antonio Cocchi primo antiquario della Galleria Fiorentina, 1738–1758.* Modena: F. C. Panini, 1996.

Findlen, Paula. *Possessing Nature: Museums, Collecting, and Scientific Culture in Early Modern Italy.* Berkeley: University of California Press, 1998.

Fortunette Witstein, Sonja. 'Funeraire poëzie in de Nederlandse renaissance: enkele funeraire gedichten van Heinsius, Hooft, Huygens en Vondel, bezien tegen de achtergrond van de theorie betreffende het genre'. PhD diss., Utrecht University, 1969.

Frampton, Stephanie Ann. *Empire of Letters: Writing in Roman Literature and Thought From Lucretius to Ovid.* New York: Oxford University Press, 2009.

Frank-van Westrienen, Anna. *De Groote Tour: Tekening van de educatiereis der Nederlanders in de zeventiende eeuw.* Amsterdam: Noord-Hollandsche Uitgeversmaatschappij, 1983.

Frijhoff, W. T. M. *La societé neerlandaise et ses gradués, 1575–1814: Une recherche sérielle sur le status des intellectuels à partir des registres universitaires.* Amsterdam: APA, 1981.

Frish, Andrea. *Forgetting Differences: Tragedy, Historiography and the French Wars of Religion.* Edinburgh: Edinburgh University Press, 2015.

Fuchs, Ralf-Peter. 'Trust as a Concept of Religious Plurality During the Thirty Years' War'. In *Trust and Happiness in the History of European Political Thought*, edited by László Kontler and Mark Somos, 302–319. Leiden; Boston: Brill, 2018.

Fuchs, Thomas. 'Reformation, Tradition und Geschichte: Erinnerungsstrategien der reformatorischen Bewegung'. In *Protestantische Identität und Erinnerung: Von der Reformation bis zur Bürgerrechtsbewegung in der DDR*, edited by Joachim Eibach and Marcus Sandl, 71–89. Göttingen: Vandenhoeck & Ruprecht, 2003.

Fumaroli, Marc. 'The Republic of Letters'. *Diogenes* 36 (1988): 129–152.
Fumaroli, Marc. *The Republic of Letters*. Translated by Lara Vergnaud. New Haven, CT: Yale University Press, 2018.
Fury, Constance M. *Erasmus, Contarini, and the Religious Republic of Letters*. Cambridge: Cambridge University Press, 2005.
Füssel, Marian. '"On the Means of Becoming Famous in the Learned World": Practices in Scholarly Constitution of Status and the Emergence of a Moral Economy of Knowledge in the Eighteenth Century'. In *Scholars in Action: The Practice of Knowledge and the Figure of the Savant in the 18th Century*, edited by André Holenstein, Hubert Steinke, and Martin Stuber, 123–144. History of Science and Medicine Library 34 / Scientific and Learned Cultures and their Institutions 9, vol. 1. Leiden; Boston: Brill, 2013.
Füssel, Marian. 'A Struggle for Nobility: 'Nobilitas literaria' as Academic Self-Fashioning in Early Modern Germany'. In Kirwan, *Scholarly Self-Fashioning*, 103–120.
Gallo, Daniela, ed. *Filippo Buonarroti e la cultura antiquaria sotto gli ultimi Medici*. Florence: Cantini Edizioni d'Arte, 1986.
Galluzzi, Paolo. 'I sepolcri di Galileo: le spoglie "vive" di un eroe della scienza'. In Berti, *Il pantheon di Santa Croce*, 145–182.
Galluzzi, Paolo. 'The Sepulchers of Galileo: The "Living" Remains of a Hero of Science'. Translated by Michael John Gorman. In *The Cambridge Companion to Galileo*, edited by Peter Machamer, 417–447. Cambridge: Cambridge University Press, 1998.
Gama Oliveira, João, and Albert-László Barabási. 'Human Dynamics: Darwin and Einstein Correspondence Patterns'. *Nature* 437 (2005): 1251.
Garberson, Eric. 'Libraries, Memory and the Space of Knowledge'. *Journal of the History of Collections* 18, no. 2 (2006): 105–136.
Gat, Azar, with Alexander Yakobson. *Nations: The Long History and Deep Roots of Political Ethnicity and Nationalism*. Cambridge: Cambridge University Press, 2012.
Gaukroger, Stephen. 'Biography as a Route to Understanding Early Modern Natural Philosophy'. In Söderqvist, *The History and Poetics of Scientific Biography*, 37–49.
Gibson, Abraham, and Cindy Ermus. 'The History of Science and the Science of History: Computational Methods, Algorithms, and the Future for the Field'. *Isis* 110, no. 3 (2019): 555–566.
Gildenhard, Ingo. 'A Republic in Letters: Epistolary Communities in Cicero's Correspondence, 49–44 BCE'. In Ceccarelli et al., *Empire of Letters*, 43–80.
Ginggras, Yves. 'Mapping the Structure of the Intellectual Field Using Citation and Co-Citation Analysis of Correspondence'. *History of European Ideas* 36 (2010): 330–339.
Giometti, Christiano. 'Professori e cavalieri: Ceremonie funebri e monumenti del Seicento nel segno della continuità'. In Sicca, *Scultura a Pisa nell'età moderna*, 75–90.
Giurescu Heller, Ena. 'Access to Salvation: The Place (and Space) of Women Patrons in Fourteenth-Century Florence'. In *Women's Space: Patronage, Place, and Gender in the Medieval Church*, edited by Virginia Chieffo Raguin and Sarah Stanbury, 161–184. Albany: State University of New York Press, 2005.
Goggioli, Maria Mannelli. *La Biblioteca Magliabechiana: libri, uomini, idee per la prima biblioteca pubblica a Firenze*. Florence: Leo S. Olschki, 2000.
Goldgar, Anne. *Impolite Learning: Conduct and Community in the Republic of Letters, 1680–1750*. New Haven, CT; London: Yale University Press, 1995.
Goodman, Dena. 'Governing the Republic of Letters: The Politics of Culture in the French Enlightenment'. *History of European Ideas* 13, no. 3 (1991): 183–199.
Goodman, Dena. *The Republic of Letters: A Cultural History of the French Enlightenment*. Ithaca, NY; London: Cornell University Press, 1994.

Gordon, Bruce, ed. *Protestant History and Identity in Sixteenth-Century Europe*. 2 vols. St Andrews Studies in Reformation History. Aldershot; Brookfield; Singapore; Sydney: Ashgate, 1996.

Gordon, Bruce. '"This Worthy Witness of Christ": Protestant Uses of Savonarola in the Sixteenth Century'. In Bruce, *Protestant History and Identity in Sixteenth-Century Europe*, vol. 1: *The Medieval Inheritance*, 93–107.

Grafton, Anthony. *Athenae Batavae: The Research Imperative at Leiden, 1575–1650*. Scaliger Lectures 1. Leiden: Primavera Pers, 2003.

Grafton, Anthony. *Defenders of the Text: The Traditions of Scholarship in an Age of Science, 1450–1800*. Cambridge, MA; London: Harvard University Press, 1991.

Grafton, Anthony. *Joseph Scaliger: A Study in the History of Classical Scholarship*, vol 1: *Textual Criticism and Exegesis*. Oxford: Clarendon, 1983.

Grafton, Anthony. 'A Sketch Map of a Lost Continent: The Republic of Letters'. *Republic of Letters: A Journal for the Study of Knowledge, Politics, and the Arts* 1, no. 1 (2009): n. pag.

Grafton, Anthony. 'Texts, Antiquities, and Their Scholarly Transformations in the Early Modern West'. In *World Philology*, edited by Sheldon Pollock, Benjamin A. Elman, and Kuming Kevin Chang, 154–177. Cambridge, MA; London: Harvard University Press, 2015.

Graham, Shawn, Ian Milligan, and Scott Weingart. *Exploring Big Historical Data: The Historian's Macroscope*. London: Imperial College Press, 2016.

Grandi, Renzo. 'Dottori, scultori, pittori: ancora sui monumenti bolognesi'. In *Skulptur und Grabmal des Spätmittelalters in Rom und Italien: Akten des Kongresses 'Scultura e Monumento sepolcrale del tardo medioevo a Roma e in Italia' (Rom, 4.–6. Juli 1985)*, edited by Jörg Garms and Angiola Maria Romanini, 353–365. Vienna: Verlag der Österreichischen Akademie der Wissenschaft, 1990.

Grandi, Renzo. *I monumenti dei dottori e la scultura a Bologna (1267–1348)*. Colla studi e richerche, Nuova serie: Testi del Museo Civico Medievale di Bologna. Bologna: Commune di Bologna / Istituto per la Storia di Bologna, 1982.

Grasman, Edward. 'Het omstreden gedenkteken voor Coccejus'. In Den Hartog, *De Pieterskerk in Leiden*, 396–402.

Green, Anna. *Cultural History*. Theory and History. Basingstoke; New York: Palgrave Macmillan, 2008.

Greenblatt, Stephen. *Renaissance Self-Fashioning: From More to Shakespeare*. Chicago; London: University of Chicago Press, 1980.

Grell, Ole Peter. 'The Attraction of Leiden University for English Students of Medicine and Theology, 1590–1642'. In *The Great Emporium: The Low Countries as a Cultural Crossroads in the Renaissance and the Eighteenth Century*, edited by C. C. Barfoot and Richard Todd, 83–104. Studies in Literature 10. Amsterdam; Atlanta, GA: Rodopi, 1992.

Grosser, Th. 'Reisen und soziale Eliten: Kavalierstour—Patrizierreise—bürgerliche Bildungsreise'. In *Neue Impulse der Reiseforschung*, edited by Michael Maurer, 136–176. Berlin: Akademie Verlag, 1999.

Guerini, Luigi. *Antonio Cocchi naturalista e filosofo*. Biblioteca di medicina e storia 2. Florence: Polistampa, 2002.

Habermas, Jürgen. *The Structural Transformation of the Public Sphere: An Inquiry into a Category of Bourgeois Society*. Translated by Thomas Burger. Cambridge: Polity Press, 1989.

Halbwachs, Maurice. *On Collective Memory*. Edited, translated, and with an introduction by Lewis A. Coser. The Heritage of Sociology. Chicago; London: University of Chicago Press, 1992.

Hall, Marcia B. *Renovation and Counter-Reformation: Vasari and Duke Cosimo in Sta Maria Novella and Sta Croce, 1565–1577*. Oxford–Warburg Studies. Oxford; New York: Oxford University Press, 1979.

Hall, Marcia B. 'The *Tramezzo* in Santa Croce, Florence, Reconstructed'. *The Art Bulletin* 56, no. 3 (1974): 325–341.

Hampton, Timothy. *Writing From History: The Rhetoric of Exemplarity in Renaissance Literature*. Ithaca, NY: Cornell University Press, 1990.

Hankins, James. *Virtue Politics: Soulcraft and Statecraft in Renaissance Italy*. Cambridge, MA: Harvard University Press, 2019.

Harris, Jonathan. *Greek Émigres in the West, 1400–1520* (Camberley: Porphyrogenitus, 1995).

Hartog, Elizabeth den, ed. *De Pieterskerk in Leiden: Bouwgeschiedenis, inrichting en gedenktekens*. Zwolle: WBooks, 2011.

Hartog, Elizabeth, and John Veerman. 'Inleiding: Het interieur na de Reformatie'. In Den Hartog, *De Pieterskerk in Leiden*, 185–209.

Haynes, Roslynn D. *From Faust to Strangelove: Representations of the Scientist in Western Literature*. Baltimore; London: Johns Hopkins University Press, 1994.

Heering, Jan Paul. *Hugo Grotius as Apologist for the Christian Religion: A Study of His Work* De Veritate Religionis Christianae *(1640)*. Studies in the History of Christian Tradition 111. Leiden; Boston: Brill, 2004.

Heering, Jan Paul. 'Kunstbroeders en rivalen: Heinsius en Grotius'. In *Voorbeeldige vriendschap: Vrienden en vriendinnen in theologie en cultuur*, edited by H. J. Heering, 47–56. Groningen: Styx, 1993.

Heesakkers, Chris L. 'Driemaal Cicero: De Neolatijnse Cicero-cultus van Petrarca tot Lipsius'. *Lampas* 26, no. 2 (1993): 131–158.

Heesakkers, Chris L. 'Scaliger en de Dousa's'. In *Adelaar in de wolken: De Leidse jaren van Josephus Justus Scaliger, 1593–1609*, edited by P. G. Hoftijzer, 36–57. Kleine publicaties van de Leidse Universiteitsbibliotheek 69. Leiden: Scaliger Instituut; Universiteitbibliotheek Leiden, 2005.

Hellerstedt, Andreas, ed. *Virtue Ethics and Education from Late Antiquity to the Eighteenth Century*. Knowledge Communities. Amsterdam: Amsterdam University Press, 2018.

Helmrath, Johannes. 'Perception of the Middle Ages and Self-Perception in German Humanism: Johannes Trithemius and the *Cathalogus illustrium virorum Germaniam ... exornantium*'. In Baker, *Biography, Historiography, and Modes of Philosophizing*, 177–208.

Hendrix, Harald. 'The Early Modern Invention of Literary Tourism: Petrarch's Houses in France and Italy'. In Hendrix, *Writers' Houses and the Making of Memory*, 15–29.

Hendrix, Harald, ed. *Writers' Houses and the Making of Memory*. Routledge Research in Cultural and Media Studies 11. New York: Routledge, 2008.

Herklotz, Ingo. 'Ianus Nicius Erythraeus und Jean-Jacques Bouchard: Zur schweren Geburt einer neulateinischen Vitensammlung des 17. Jahrhunderts'. *Neulateinisches Jahrbuch: Journal of Neo-Latin Language and Literature* 10, edited by Marc Laureys and Karl August Neuhausen (2008): 145–176.

Hoekstra, Rik, and Marijn Koolen. 'Data Scopes for Digital History Research'. *Historical Methods* 52, no. 2 (2019): 79–94.

Holt, Mack P. *The French Wars of Religion, 1562–1629*. 2nd ed. New Approaches to European History. Cambridge; New York: Cambridge University Press, 2005.

Hoogewerff, G. J. 'Laurentius Theodorus Gronovius en zijn reizen naar Italië, 1680–82 en 1693–95: I. De reiziger, zijn reisgenooten en zijn wedervaren'. *Mededeelingen van het Nederlandsch Historisch Instituut te Rome*, 3rd series, 1 (1942): 35–56.

Hoogewerff, G. J. 'Laurentius Theodorus Gronovius en zijn reizen naar Italië: III. De briefwisseling met Antonio Magliabechi'. *Mededeelingen van het Nederlandsch Historisch Instituut te Rome*, 3rd series, 3 (1943): 160–167.

Hoogewerff, G. J. *De twee reizen van Cosimo de' Medici, prins van Toscane, door de Nederlanden (1667–1669): Journalen en documenten*. Amsterdam: Johannes Müller, 1919.

Horman, Henrik Theodoor. *Gezagsdragers Amersfoort 1405–1795: Gecompileerd alfabetisch register*, Historische toegangen van de Gelderse Vallei 18. Woudenberg: Bureau voor Familie-Historie, 1998.

Horst, D. J. H. ter. 'Daniel Heinsius (1580–1655)'. PhD diss., Utrecht University, 1934.

Hotson, Howard, and Thomas Wallnig, eds. *Reassembling the Republic of Letters in the Digital Age: Standards, Systems, Scholarship*. Göttingen: Göttingen University Press, 2019.

Hulsenboom, Paul, and Alan Moss. 'Tracing the Sites of Learned Men: Places and Objects of Knowledge on the Dutch and Polish Grand Tour'. In Scholten, Van Miert, and Enenkel, *Memory and Identity in the Learned World*, 257–306.

Hyatte, Reginal. *The Arts of Friendship: The Idealization of Friendship in Medieval and Early Renaissance Literature*. Brill's Studies in Intellectual History 50. Leiden: Brill, 1994.

Ianziti, Gary. 'Leonardo Bruni, the Medici, and the Florentine Histories'. *Journal of the History of Ideas* 69, no. 1 (2008): 1–22.

Israel, Jonathan. *The Dutch Republic: Its Rise, Greatness, and Fall, 1477–1806*. Oxford: Oxford University Press, 1995.

Jacobs, Richard Ivan. *Backpack Ambassadors: How Youth Travel Integrated Europe*. Chicago: University of Chicago Press, 2017.

James, Carolyn. 'In Praise of Women: Giovanni Sabadino degli Arienti's *Gynevera de le clare donne*'. In *The Intellectual Dynamism of the High Middle Ages*, Knowledge Communities, edited by Clare Monagle, 297–316. Amsterdam: Amsterdam University Press, 2021.

Jansen, W. L. 'Appropriating *Peri hypsous*: Interpretations and Creative Adaptations of Longinus' Treatise *On the Sublime* in Early Modern Dutch Scholarship'. PhD diss., Leiden University, 2019.

Jardine, Lisa. *Erasmus, Man of Letters: The Construction of Charisma in Print*. Princeton, NJ: Princeton University Press, 1993.

Jardine, Lisa. '"O decus Italiae virgo", *or* The Myth of the Learned Lady in the Renaissance'. *The Historical Journal* 28, no. 4 (1985): 799–819.

Jaumann, Herbert, ed. *Die europäische Gelehrtenrepublik im Zeitalter des Konfessionalismus / The European Republic of Letters in the Age of Confessionalism*. Wolfenbütteler Forschungen 96. Wiesbaden: Harrassowitz, 2001.

Jaumann, Herbert. '*Respublica litteraria / Republic of Letters*: Concept and Perspective of Research'. In Jaumann, *Die europäische Gelehrtenrepublik im Zeitalter des Konfessionalismus*, 11–19.

Jensen, Lotte, ed. *The Roots of Nationalism: National Identity Formation in Early Modern Europe, 1600–1815*. Heritage and Memory Studies. Amsterdam: Amsterdam University Press, 2016.

Jones, Henry. 'Searching for Statesmanship: A Corpus-Based Analysis of a Translated Political Discourse'. *Polis* 36, no. 2 (2019): 216–241.

Jones, Matthew L. *The Good Life in the Scientific Revolution: Descartes, Pascal, Leibniz, and the Cultivation of Virtue*. Chicago; London: University of Chicago Press, 2006.

Jonge, Henk Jan de. 'Grafsteen en graf van Scaliger'. *Jaarboekje voor geschiedenis en oudheidkunde van Leiden en omstreken* 70 (1978): 91–96.

Jonge, Henk Jan de. 'The Latin Testament of Joseph Scaliger, 1607'. *Lias* 2, no. 2 (1975): 249–263.

Jonge, Henk Jan de. 'Scaliger in Leiden'. *Jaarboekje voor geschiedenis en oudheidkunde van Leiden en omstreken* 71 (1979): 71–94.

Jong, Jan L. de. 'De Sepulcro Laurentii Vallae quid veri habeat: Tracing the Tomb Monument of Lorenzo Valla in St. John Lateran, Rome'. *Quellen und Forschungen aus italienischen Archiven und Bibliotheken* 94, no. 1 (2015): 94–128.

Jordan, Constance. *Renaissance Feminism: Literary Texts and Political Models*. Ithaca, NY; London: Cornell University Press, 1990.

Jordanova, Ludmilla. *Defining Features: Scientific and Medical Portraits, 1660–2000*. London: Reaktion Books, 2000.

Jordanova, Ludmilla. 'Science and Nationhood: Cultures of Imagined Communities'. In Cubitt, *Imagining Nations*, 192–211.

Kampinga, H. *De opvattingen over onze oudere vaderlandsche geschiedenis bij de Hollandsche historici der XVIe en XVIIe eeuw*. Utrecht: Hes, 1980.

Kan, A. H. 'Nicolaas Heinsius in Italië (1646–1648)'. *Onze Eeuw* 14, no. 3 (1914): 361–390.

Kan, A. H. 'Nicolaas Heinsius in Italië (1646–1648) (II slot)'. *Onze Eeuw* 14, no. 4 (1914): 57–74.

Kaplan, Benjamin J. *Divided by Faith: Religious Conflict and the Practice of Toleration in Early Modern Europe*. Cambridge, MA; London: Belknap Press of Harvard University, 2007.

Kenny, Neil. *The Uses of Curiosity in Early Modern France and Germany*. Oxford; New York: Oxford University Press, 2004.

King, Margaret L. 'Book-Lined Cells: Women and Humanism in the Early Italian Renaissance'. In *Beyond Their Sex: Learned Women of the European Past*, edited by Patricia H. Labalme, 66–90. New York; London: New York University Press, 1980.

Kingdon, Robert M. *Geneva and the Consolidation of the French Protestant Movement, 1564–1572*. Madison: University of Wisconsin Press, 1967.

Kinser, Samuel. *The Works of Jacques-Auguste de Thou*. International Archives of the History of Ideas 18. The Hague: Martinus Nijhoff, 1966.

Kircher, Timothy. 'At Play in the Republic of Letters: The Correspondence of Lapo da Castiglionchio the Younger'. *Renaissance Quarterly* 71, no. 3 (2018): 841–867.

Kirwan, Richard. 'Miscellanies of Memory: From Scholarly Biography to Institutional History in the Early German University'. In Scholten, Van Miert, and Enenkel, *Memory and Identity in the Learned World*, 235–254.

Kirwan, Richard, ed. *Scholarly Self-Fashioning and Community in the Early Modern University*. Farnham; Burlington, VT: Ashgate, 2013.

Kivistö, Sari. *The Vices of Learning: Morality and Knowledge at Early Modern Universities*. Education and Society in the Middle Ages and Renaissance 48. Leiden; Boston: Brill, 2014.

Kline Cohen, Samuel, Jnr. *The Cult of Remembrance and the Black Death: Six Renaissance Studies of Central Italy*. Baltimore; London: Johns Hopkins University Press, 1992.

Klinger, Linda S. 'The Portrait Collection of Paolo Giovio, 1545–1553'. PhD diss., Princeton University, 1991.

Kneppelhout van Sterkenburg, K. J. H. C. *De gedenkteekenen in de Pieters-kerk te Leyden*. Leiden: P. H. van den Heuvel, 1864.

Knöll, Stefanie A. 'Creating Academic Communities: Funeral Monuments to Professors at Oxford, Leiden and Tübingen, 1580–1700'. PhD diss., University of Sussex, 2003.

Kohl, Benjamin G. 'Petrarch's Prefaces to *De Viris Illustribus*'. *History and Theory* 13, no. 2 (1974): 132–144.

Kolb, Robert. *For All the Saints: Changing Perceptions of Martyrdom and Sainthood in the Lutheran Reformation*. Macon, GA: Mercer University Press, 1987.

Kolsky, Stephen. *The Ghost of Boccaccio: Writings on Famous Women in Renaissance Italy*. Late Medieval and Early Modern Studies 7. Turnhout: Brepols, 2005.

Koomen, Arjan de. 'L'età dei Lumi e il sublime: I monumenti a Niccolò Machiavelli e a Vittorio Alfieri'. In Berti, *Il pantheon di Santa Croce*, 183–220.

Koslofsky, Craig M. *The Reformation of the Dead: Death and Ritual in Early Modern Germany, 1450–1700*. Early Modern History: Society and Culture. Basingstoke and London: Macmillan; New York: St. Martin's Press, 2000.

Kristeller, Paul Oskar. 'Learned Women of Early Modern Italy: Humanists and University Scholars'. In *Beyond Their Sex: Learned Women of the European Past*, edited by Patricia H. Labalme, 91–106. New York; London: New York University Press, 1980.

Kuijpers, Erika, Judith Pollmann, Johannes Müller, and Jasper van der Steen, eds. *Memory Before Modernity: Practices of Memory in Early Modern Europe*. Studies in Medieval and Reformation Traditions 176. Leiden; Boston: Brill, 2013.

LaGuardia, David P., and Cathy Yandell, eds. *Memory and Community in Sixteenth-Century France*. Farnham; Burlington, VT: Ashgate, 2015.

Lalli, Roberto, Riaz Howey, and Dirk Wintergrün. 'The Dynamics of Collaboration Networks and the History of General Relativity, 1925–1970'. *Scientometrics* 122, no. 2 (2020): 1129–1170.

Lamping, A.J. *Johannes Polyander: Een dienaar van kerk en universiteit*. Kerkhistorische bijdragen IX. Leiden: E. J. Brill, 1980.

Langereis, Sandra. 'Geschiedenis als ambacht: oudheidkunde in de Gouden Eeuw: Arnoldus Buchelius en Petrus Scriverius'. PhD diss., University of Amsterdam, 2001. Published in Hollandse Studiën 37. Hilversum: Historische Vereniging Holland and Uitgeverij Verloren, 2001.

Laubichler, Manfred D., Jane Maienschein, and Jürgen Renn. 'Computation History of Knowledge: Challenges and Opportunities'. *Isis* 110, no. 3 (2019): 502–512.

Leeb, Isidore Leonard. *The Ideological Origins of the Batavian Revolution: History and Politics in the Dutch Republic, 1747–1800*. The Hague: Martinus Nijhoff, 1973.

Leerssen, Joep, and Ann Rigney, eds. *Commemorating Writers in Nineteenth-Century Europe: Nation-Building and Centenary Fever*. Basingstoke; New York: Palgrave Macmillan, 2014.

Lefèvre, Eckard. 'Daniel Heinsius' *Manes Lipsiani* nebst einer Erklärung von Joseph Scaligers *Epicedium* auf Lipsius'. In *Daniel Heinsius: Klassischer Philologe und Poet*, NeoLatina 3, edited by Eckard Lefèvre and Eckart Schäfer, 203–248. Tübingen: Gunter Narr Verlag, 2008.

Lemercier, Claire, and Claire Zalc. *Quantitative Methods in the Humanities: An Introduction*. Translated by Arthur Goldhammer. Charlottesville; London: University of Virginia Press, 2019.

Leopold, H. M. R. 'Laurentius Theodorus Gronovius en zijn reizen naar Italië: II. Oudheidkundige waarnemingen en bevindingen'. *Mededeelingen van het Nederlandsch Historisch Instituut te Rome*, 3rd series, 1 (1942): 58–87.

Lesaffer, Randall, and Janne E. Nijman, eds. *The Cambridge Companion to Hugo Grotius*. Cambridge: Cambridge University Press, 2021.

Levering, Matthew. *Predestination: Biblical and Theological Paths*. Oxford; New York: Oxford University Press, 2011.

Levitin, Dmitri. 'Introduction: Confessionalisation and Erudition in Early Modern Europe: A Comparative Overview of a Neglected Episode in the History of the Humanities'. In *Confessionalisation and Erudition in Early Modern Europe: An Episode in the History of the Humanities*, edited by Nicholas Hardy and Dmitri Levitin, 1–94. Proceedings of the British Academy 225. Oxford: Oxford University Press, 2019.

Lindeman, Ruud, Yvonne Scherf, and Rudolf Dekker. *Reisverslagen van Noord-Nederlanders van de zestiende tot begin negentiende eeuw: een chronologische lijst*. Haarlem: Stichting Egodocument, 1994.

Lochman, Daniel T., and Maritere López. 'Introduction: The Emergence of Discourses: Early Modern Friendship'. In Lochman, López, and Hutson, *Discourses and Representations of Friendship*, 1–26.

Lochman, Daniel T., Maritere López, and Lorna Hutson, eds. *Discourses and Representations of Friendship in Early Modern Europe, 1500–1700*. Farnham; Burlington, VT: Ashgate 2011.

Lotz-Heumann, Ute. 'The Concept of "Confessionalissation": A Historiographical Paradigm in Dispute'. *Memoria y Civilización* 4 (2001): 93–114.

Lotz-Heumann, Ute, and Matthias Pohlig. 'Confessionalization and Literature in the Empire, 1555–1700'. *Central European History* 40, no. 1 (2007): 35–61.

Lunardi, Roberto, and Oretta Sabbatini, eds. *Il remembrar delle passate cose. Una casa per memoria: Galileo e Vincenzo Viviani*. Testi e Studi 18, vol. 3. Florence: Polistampa, 2009.

Lux, David S., and Harold J. Cook. 'Closed Circles or Open Networks? Communicating at a Distance During the Scientific Revolution'. *History of Science* 36 (1998): 179–211.

Lynn, Michael. 'Enlightenment and the Public Sphere: The Musé de Monsieur and Scientific Culture in Late Eighteenth-Century Paris'. *Eighteenth Century Studies* 32 (1999): 463–476.

MacDonald, Katherine. *Biography in Early Modern France, 1540–1630: Forms and Functions*. London: Maney, 2007.

Maclean, Ian. *Scholarship, Commerce, Religion: The Learned Book in the Age of Confessions, 1560–1630*. Cambridge, MA; London: Harvard University Press, 2012.

Maffioli, Monica, ed. *Santa Croce nell'800*. Florence: Alinari, 1986.

Markham Schulz, Anne. *The Sculpture of Bernardo Rossellino and his Workshop*. Princeton, NJ: Princeton University Press, 1977.

Marr, Alexander, Raphaële Garrod, José Ramón Marcaida, and Richard J. Oosterhoff. *Logodaedalus: Word Histories of Ingenuity in Early Modern Europe*. Pittsburgh: University of Pittsburgh Press, 2019.

Martines, Lauro. *The Social World of the Florentine Humanists, 1390–1460*. Renaissance Society of America Reprint Texts 17. Toronto; Buffalo, NY; London: University of Toronto Press, 2011.

Mason, Cheryl L., Jane Butler Kahle, and April L. Gardner. 'Draw-A-Scientist Test: Future Implications'. *School Science and Mathematics* 91, no. 5 (1991): 193–198.

Mayer, Thomas F., and Daniel R. Woolf, eds. *The Rhetorics of Life-Writing in Early Modern Europe: Forms of Biography from Cassandra Fedele to Louis XIV*. Studies in Medieval and Early Modern Civilization. Ann Arbor: University of Michigan Press, 1995.

Mayhew, Robert. 'British Geography's Republic of Letters: Mapping an Imagined Community, 1600–1800'. *Journal of the History of Ideas* 65, no. 2 (2004): 251–276.

Mayhew, Robert. 'Mapping Science's Imagined Community: Geography as a Republic of Letters, 1600–1800'. *The British Journal for the History of Science* 38, no. 1 (2005): 73–92.

McCormick, Andrew P., and Henk Th. van Veen. *Tuscany and the Low Countries: An Introduction to the Sources and an Inventory of Four Florentine Libraries*. Italia e i Paesi Bassi 2. Florence: Centro Di, 1985.

McKitterick, David. *The Invention of Rare Books: Private Interest and Public Memory, 1600–1840*. Cambridge: Cambridge University Press, 2018.

McMahon, Darrin M. *Divine Fury: A History of Genius*. New York: Basic Books, 2013.

McManamon, John M. *Funeral Oratory and the Cultural Ideals of Italian Humanism*. Chapel Hill; London: University of North Carolina Press, 1989.

Miert, Dirk van, ed. *Communicating Observations in Early Modern Letters (1500–1675): Epistolography and Epistemology in the Age of the Scientific Revolution*. Warburg Institute Colloquia 23. London: Warburg Institute; Turin: Nino Aragno, 2013.

Miert, Dirk van. 'The Curious Case of Isaac Casaubon's Monstrous Bladder: The Networked Construction of Learned Memory within the Seventeenth-Century Reformed World of Learning', in Scholten, Van Miert, and Enenkel, *Memory and Identity in the Learned World*, 307–341.

Miert, Dirk van. 'Letters and Epistolography'. In *The Classical Tradition*, edited by Anthony Grafton, Glenn W. Most, and Salvatore Settis, 520–523. Cambridge, MA: Belknap Press, 2010.

Miert, Dirk van. 'The Limits of Transconfessional Contact in the Republic of Letters around 1600: Scaliger, Casaubon, and their Catholic Correspondents'. In De Landtsheer and Nellen, *Between Scylla and Charybdis*, 367–408.

Miert, Dirk van. 'Trommius's Travelogue: Learned Memories of Erasmus and Scaliger and Scholarly Identity in the Republic of Letters'. *Early Modern Low Countries* 1, no. 1 (2017): 51–70.

Miert, Dirk van. 'What was the Republic of Letters? A brief Introduction to a Long History (1417–2008)'. *Groniek* 204–205 (2010): 269–287.

Minonzio, Franco. *Studi Giovani: Scienza, Filosofia e Letteratura nell'opera di Paolo Giovio*. 2 vols. Como: Presso la Società a Valla Gallia, 2002.

Moisè, Francesco. *Santa Croce illustrazione storico-artistica: Con note e copiosi documenti inediti*. Florence: Tipografia Galileiana, 1845.

Molhuysen, P. C. *De komst van Scaliger in Leiden*. Leiden: A. W. Sijthoff, 1913.

Monfasani, John. *Byzantine Scholars in Renaissance Italy: Cardinal Bessarion and Other Émigres*. Variorum Collected Studies Series 485. Aldershot; Brookfield: Ashgate, 1995.

Moretti, Franco. *Distant Reading*. London: Verso, 2013.

Moss, Alan. 'Gemaakt op reis: Identiteitsvorming in verslagen van de Nederlandse educatiereis (1648–1713)'. PhD diss., Nijmegen: Radboud University, 2022.

Mulsow, Martin. *Die unanständige Gelehrtenrepublik: Wissen, Libertinage und Kommunikation in der Frühen Neuzeit*. Stuttgart; Weimar: J. B. Metzler, 2007.

Müntz, Eugène. 'Le musée de portraits de Paul Jove: Contributions pour servir à l'iconographie du Moyen Âge et de la Renaissance'. *Mémoires de l'Institut national de France, Académie des Inscriptions et Belles-Lettres* 36, no. 2 (1901): 249–343.

Najemy, John R. 'Civic Humanism and Florentine Politics'. In *Renaissance Civic Humanism: Reappraisals and Reflections*, edited by James Hankins, 75–104. Cambridge: Cambridge University Press, 2000.

Natali, Antonio. 'I sepolcri di Leonardi Bruni e Carlo Marsuppini'. In Berti, *Il pantheon di Santa Croce*, 17–55.

Nellen, Henk. *Hugo de Groot: Een leven in strijd om vrede, 1583–1645*. Amsterdam: Balans, 2007.

Nellen, Henk. *Hugo Grotius: A Lifelong Struggle for Peace in Church and State, 1583–1645*. Translated by J. C. Grayson. Leiden; Boston: Brill, 2015.

Nellen, Henk. *Geen vredestichter is zonder tegensprekers: Hugo de Groot, geleerde, staatsman, verguisd verzoener*. Amsterdam: Athanaeum, 2021.

Nelson, Jonathan. 'Dante Portraits in Sixteenth Century Florence'. *Gazette des Beaux-Arts* 119 (1992): 59–77.

Neumeister, Sebastian, and Conrad Wiedemann, eds. *Res Publica Litteraria: Die Institutionen der Gelehrsamkeit in der frühen Neuzeit*. 2 vols. Wolfenbütteler Arbeiten zur Barockforschung. Wiesbaden: Harrassowitz, 1987.

Neurdenburg, Elisabeth. *De zeventiende eeusche beeldhouwkunst in de noordelijke Nederlanden: Henrick de Keyser, Artus Quellinus, Rombout Verhulst*. Amsterdam: Meulenhoff, 1948.

Newstok, Scott L. *Quoting Death in Early Modern England: The Poetics of Epitaphs Beyond the Tomb*. Early Modern Literature in History. Basingstoke; New York: Palgrave Macmillan, 2009.

Niekus Moore, Cornelia. *Patterned Lives: The Lutheran Funeral Biography in Early Modern Germany*. Wiesbaden: Harrassowitz, 2006.

Nisard, Charles. *Le Triumvirat littérarie au XVIe siècle: Juste Lipse, Joseph Scaliger et Isaac Casaubon*. Paris: Amyot, [1852].

Nora, Pierre. 'Between Memory and History: Les Lieux de Mémoire'. *Representations* 26 (1989): 7–24.

Nora, Pierre, ed. *Les lieux de mémoire*. 7 vols. Paris: Gallimard, 1984–1992.

Notten, M. van. *Rombout Verhulst beelhouwer 1624–1698, een overzicht zijner werken*. The Hague: Nijhoff, 1907.

Nova, Alessandro. 'Paragone-Debatte und gemalte Theorie in der Zeit Cellinis'. In *Benvenuto Cellini: Kunst und Kunsttheorie im 16. Jahrhundert*, edited by Alessandro Nova and Anna Schreurs, 183–202. Cologne: Böhlau, 2003.

O'Connor, Anne. 'Dante Alighieri—From Absence to Stony Presence: Building Memories in Nineteenth-Century Florence'. *Italian Studies* 67, no. 3 (2012): 307–335.

O'Connor, Anne. *Florence: City and Memory in the Nineteenth Century*. Florence: Città di Vita, 2008.

Olszewski, Edward J. *Cardinal Pietro Ottoboni (1667–1740) and the Vatican Tomb of Pope Alexander VIII*. Philadelphia: American Philosophical Society, 2004.

Ommen, Kasper van. ' "Tous mes livres de langues estrangeres": Het oosterse legaat van Josephus Justus Scaliger in de Universiteitsbibliotheek Leiden'. PhD diss., Leiden University, 2020.

Oosterhoff, Richard J. 'From Pious to Polite: Pythagoras in the *Res publica litterarum* of French Renaissance Mathematics'. *Journal of the History of Ideas* 74, no. 4 (2013): 531–552.

Ortwin Rave, Paul. 'Paulo Giovio und die Bildnisvitenbücher des Humanismus'. *Jahrbuch der Berliner Museen* 1 (1959): 119–154.

Otterspeer, Willem. *Groepsportret met dame* I. *Het bolwerk van de vrijheid: de Leidse universiteit, 1575–1672*. Amsterdam: Uitgeverij Bert Bakker, 2000.

Oy-Mara, Elisabeth. *Florentiner Ehrengrabmäler der Frührenaissance*. Frankfurter Forschungen zur Kunst 18. Berlin: Gebr. Mann Verlag, 1994.

Paatz, Walter, and Elisabeth Paatz. *Die Kirchen von Florenz: Ein kunstgeschichtliches Handbuch*. 6 vols. Frankfurt a.M.: Vittorio Klostermann, 1940–1954.

Pal, Carol. *Republic of Women: Rethinking the Republic of Letters in the Seventeenth Century*. Ideas in Context 99. Cambridge: Cambridge University Press, 2012.

Paolini, Claudio. 'Francesco di Giovanni detto il Francione, *Dante, Petrarca e libri; Vasi di fiori e sportelli aperti*'. In *Inganni ad arte: Meraviglie del* trompe-l'œil *dall'antichità al contemporaneo* III.4, edited by Annamaria Giusti, 141–144. Florence: Mandragora, 2009.

Papy, Jan. 'Justus Lipsius, Rome en de Romereis: zoektocht naar een oude mythe?' *Kleio* 26, no. 3 (1997): 111–126.

Parish, Helen L. *Monks, Miracles and Magic: Reformation Representations of the Medieval Church*. London; New York: Routledge, 2016.

Parker, Charles H. 'Reformed Protestantism'. In *The Cambridge Companion to the Dutch Golden Age*, edited by Helmer Helmers and Geert H. Janssen, 189–207. Cambridge Companions to Culture. Cambridge; New York; Melbourne; New Delphi; Singapore: Cambridge University Press, 2018.

Parker, Deborah. 'Vasari's *Portrait of Six Tuscan Poets*: A Visible Literary History'. *Lectura Dantis* 22–23 (1998): 45–62.

Paul, Herman. *Historians' Virtues: From Antiquity to the Twenty-First Century*. Elements in Historical Theory and Practice. Cambridge: Cambridge University Press, 2022.

Paul, Herman. 'Performing History: How Historical Scholarship is Shaped by Epistemic Virtues'. *History and Theory* 50, no. 1 (2011): 1–19.

Paul, Herman. 'What is a Scholarly Persona? Ten Theses on Virtues, Skills, and Desires'. *History and Theory* 53, no. 3 (2014): 348–371.

Pavoni, R. 'Paolo Giovio et son musée de portraits à propos d'une exposition'. *Gazette des Beaux-Arts* 105, no. 1394 (1985): 109–116.

Pearl, J. L. 'The Role of Personal Correspondence in the Exchange of Scientific Information in Early Modern France'. *Renaissance and Reformation* 8, no. 2 (1984): 106–113.

Peters Bowron, Edgar. 'Giorgio Vasari's Portrait of Six Tuscan Poets'. *The Minneapolis Institute of Arts Bulletin* 60 (1971–1973): 42–53.

Plahte Tschudi, Victor. *Baroque Antiquity: Archaeological Imagination in Early Modern Europe*. New York: Cambridge University Press, 2016.

Pohlig, Matthias. *Zwischen Gelehrsamkeit und konfessioneller Identitätsstiftung: Lutherische Kirchen- und Universalgeschichtsschreibung, 1546–1617*. Spätmittelalter und Reformation, Neue Reihe 37. Tübingen: Mohr Siebeck, 2007.

Pohlig, Matthias, Ute Lotz-Heumann, Vera Isaiasz, Ruth Schilling, Heike Bock, and Stefan Ehrenpreis. *Säkularisierungen in der Frühen Neuzeit: Methodische Probleme und empirische Fallstudien*. Zeitschrift für Historische Forschung, Beihefte 41. Berlin: Duncker & Humblot, 2008.

Pollmann, Judith. 'Another Road to God: The Religious Development of Arnoldus Buchelius (1565–1641)'. PhD diss., University of Amsterdam, 1998.

Pollmann, Judith. *Memory in Early Modern Europe, 1500–1800*. Oxford: Oxford University Press, 2017.

Pollmann, Judith. 'No Man's Land: Reinventing Netherlandish Identities, 1585–1621'. In *Networks, Regions and Nations: Shaping Identities in the Low Countries, 1300–1650*, edited by Robert Stein and Judith Pollmann, 241–261. Studies in Medieval and Reformation Traditions 149. Leiden; Boston: Brill, 2010.

Pope-Hennessy, John. *Italian Renaissance Sculpture*. London: Phaidon, 1996.

Post, Anna-Luna. 'Claiming Fame for Galileo: Reputation and Scholarly Credibility in Early Modern Italy'. PhD diss., Utrecht University, 2020.

Posthumus Meyjes, G. H. M. 'Twee vrienden, Isaac Casaubon en Hugo de Groot'. In *Voorbeeldige vriendschap: Vrienden en vriendinnen in theologie en cultuur*, edited by H. J. Heering, 37–46. Groningen: Styx, 1993.

Price Zimmermann, T. C. *Paolo Giovio: The Historian and the Crisis of Sixteenth-Century Italy*. Princeton, NJ: Princeton University Press, 1995).

Price Zimmermann, T. C. 'Paolo Giovio and the Rhetoric of Individuality'. In Mayer and Woolf, *The Rhetorics of Life-Writing in Early Modern Europe*, 39–62.

Rademaker, C. S. M. *Life and Work of Gerardus Joannes Vossius (1577–1649)*. Translated by H. P. Doezema. Respublica literaria neerlandica 5. Assen: Van Gorcum, 1981.

Radstone, Susannah, and Bill Schwarz, eds. *Memory: Histories, Theories, Debates*. New York: Farnham University Press, 2010.

Raeburn, Gordon D. 'The Reformation of Burial in Protestant Churches'. In Booth and Tingle, *A Companion to Death, Burial, and Remembrance*, 156–174.

Rahm, Irène, and Paul Charbonneau. 'Probing Stereotypes Through Students' Drawings of Scientists'. *American Journal of Physics* 65, no. 8 (1997): 774–778.

Randall, Catharine. 'A Protestant Poetics of Process: Reformation Rhetorics of the Self in Sponde, de Bèze, and d'Aubigné'. In Mayer and Woolf, *Rhetorics of Life-Writing*, 243–282.

Rasmussen, Tarald, and Jon Øygarden Flæten, eds. *Preparing for Death, Remembering the Dead*. Gottingen; Bristol: Vandenhoeck & Ruprecht, 2015.

Ravenek, Walter, Charles van den Heuvel, and Guido Gerritsen. 'The ePistolarium: Origins and Techniques'. In *CLARIN in the Low Countries*, edited by Jan Odijk and Arjan van Hessen, 317–323. London: Ubiquity Press, 2017.

Reeve, M. D. 'Heinsius's Manuscripts of Ovid'. *Rheinisches Museum für Philologie*, Neue Folge 117, nos. 1–2 (1974): 133–166.

Reeve, M. D. 'Heinsius's Manuscripts of Ovid: A Supplement'. *Rheinisches Museum für Philologie*, Neue Folge, 119, no. 1 (1976): 65–78.

Reinhard, Wolfgang. 'Konfession und Konfessionalisierung in Europa'. In *Bekenntnis und Geschichte: Die Confessio Augustana im historischen Zusammenhang*, edited by Wolfgang Reinhard, 165–189. Munich: Vögel, 1981.

Ridder-Symoens, H. de. 'Brabanders aan de rechtsuniversiteit van Orléans, 1444–1556: Een socio-professionele studie'. *Bijdragen tot de Geschiedenis* 61 (1978): 195–347.

Ridderikhoff, C. M. 'Orléans and the Dutch Revolt'. In *The Great Emporium: The Low Countries as a Cultural Crossroads in the Renaissance and the Eighteenth Century*, edited by C. C. Barfoot and R. Todd, 59–82. Studies in Literature 10. Amsterdam; Atlanta, GA: Rodopi, 1992.

Risbjerg Eskildsen, Kasper. 'Exploring the Republic of Letters: German Travellers in the Dutch Underground, 1690–1720'. In *Scientists and Scholars in the Field: Studies in the History of Fieldwork and Expeditions*, edited by Kristian H. Nielsen, Michael Harbsmeier, and Christopher J. Ries, 101–122. Aarhus: Aarhus University Press, 2012.

Rocco, Alessandra. *Carlo Marsuppini traduttore d'Omero: la prima traduzione umanistica in versi dell'Iliade (primo e non libro)*. Richerche 13. Padua: Il poligrafo, 2000.

Rockwell, Geoffrey, and Stéfan Sinclair. *Hermeneutica: Computer-Assisted Interpretation in the Humanities*. Cambridge, MA; London: MIT Press, 2016.

Rogge, Y. H. 'De reis van Isaac Vossius (1641–1645)'. *Oud Holland* 18, no. 1 (1900): 3–20.

Roodenburg, Herman, ed. *Forging European Identities, 1400–1700*. Cultural Exchange in Early Modern Europe 4. Cambridge: Cambridge University Press, 2007.

Rößler, Hole. 'Character Masks of Scholarship: Self-Representation and Self-Experiment as Practices of Knowledge Around 1770'. In *Scholars in Action: The Practice of Knowledge and the Figure of the Savant in the 18th Century*, edited by André Holenstein, Hubert Steinke, and Martin Stuber, 459–480. History of Science and Medicine Library 34 / Scientific and Learned Cultures and Their Institutions 9, vol. 1. Leiden; Boston: Brill, 2013.

Roussel, Bernard. 'Le *Novum Testamentum* de Théodore de Bèze: L'édition, la traduction et l'annotation de l'*Epître de Jude*'. In Backus, *Théodore de Bèze*, 185–194.

Rummer, Erika. *The Confessionalization of Humanism in Reformation Germany*. Oxford Studies in Historical Theology. New York: Oxford University Press, 2000.

Rummel, Erika, ed. *Erasmus on Women*. Toronto; Buffalo, NY; London: University of Toronto Press, 1996.

Rupert Hall, Alfred. *The Abbey Scientists*. London: R. & R. Nicholson, 1966.

Rusnock, Andrea. 'Correspondence Networks and the Royal Society, 1700–1750'. *British Journal of the History of Science* 32 (1999): 155–169.

Sabelli, Rita, Laura Giovannini, and Carlo Biliotti, eds. *Le tombe terragne della Basilica di Santa Croce*. Florence: Città di Vita, 2001.

Sarnelli, Mauro. 'Spettacolarità, commercio e scienza nelle raccolte biografiche del medio Seicento'. *Proteo* 4, no. 2 (1998): 7–20.

Sas, N. C. F. Van, ed. *Vaderland: Een geschiedenis van de vijftiende eeuw tot 1940*. Reeks Nederlandse Begripsgeschiedenis 1. Amsterdam: Amsterdam University Press, 1999.

Schalk, Fritz. 'Von Erasmus' Res publica literaria zur Gelehrtenrepublik der Aufklärung'. In *Studien zur französischen Aufklärung*, 143–163. Das Abendland: Neue Folge 8. Frankfurt a.M.: Vittorio Klostermann, 1977.

Schilling, Heinz. 'Die Konfessionalisierung im Reich: Religiöser und gesellschaftlicher Wandel in Deutschland zwischen 1555 und 1620'. *Historische Zeitschrift* 246 (1988): 1–45.

Schilling, Heinz. 'Konfessionelle und politische Identität im frühneuzeitlichen Europa'. In *Nationale und ethnische Minderheiten und regionale Identitäten in Mittelalter und Neizeit*, edited by Antoni Czacharowski, 103–123. Torún: Wydawnictwo Uniwersytetu Mikołaja Kopernika, 1994.

Schilling, Heinz. 'Nationale Identität und Konfession in der europäischen Neuzeit'. In *Nationale und kulturelle Identität: Studien zur Entwicklung des kollektiven Bewußtseins in der Neuzeit*, edited by Bernard Giesen, 192–252. Frankfurt a.M.: Suhrkamp, 1991.

Schlüter, Lucy. *Standbeelden van Erasmus in Rotterdam: 1549–2008*. Rotterdam: Stichting Erasmushuis Rotterdam, 2008.

Schmid, Philippe. 'Catalogues in Catalogues: Imitation and Competition in Early Modern Book Collecting'. In *Book Trade Catalogues in Early Modern Europe*, edited by Arthur der Weduwen, Andrew Pettegree, and Graeme Kemp, 399–425. Library of the Written Word 93. Leiden; Boston: Brill, 2021.

Schmidt, Victor. 'A Humanist's Life Summarized—Leonardo Bruni's Epitaph'. *Humanistica Lovaniensia* 47 (1998): 1–14.

Schneider, Gary. *The Culture of Epistolarity: Vernacular Letters and Letter Writing in Early Modern England, 1500–1700*. Newark: University of Delaware Press, 2005.

Schneider, Ulrich Johannes, ed. *Kultur der Kommunikation: Die europäische Gelehrtenrepublik im Zeitalter von Leibniz und Lessing*. Wiesbaden: Harrassowitz, 2005.

Schneider, Ulrich Johannes, ed. *Kulturen des Wissens im 18. Jahrhundert*. Berlin; New York: Walter de Gruyter, 2008.

Schöch, Christof. 'Big? Smart? Clean? Messy? Data in the Humanities'. *Journal of Digital Humanities* 2, no. 3 (2013) n. pag.

Schöffer, I. 'The Batavian Myth During the Sixteenth and Seventeenth Centuries'. In *Some Political Mythologies: Papers Delivered to the Fifth Anglo-Dutch Historical Conference*, edited by J. S. Bromley and E. H. Kossmann, 78–101. Britain and the Netherlands 5. The Hague: Martinus Nijhoff, 1975.

Scholten, Frits. *Rombout Verhulst in Groningen: Zeventiende eeuwse praalgraven in Midwolde en Stedum*. Stad en Lande Historische Reeks 1–2. Utrecht: Stichting Matrijs, 1983.

Scholten, Frits. 'Sumptuous Memories: Studies in Seventeenth-Century Dutch Tomb Sculpture'. PhD diss., University of Amsterdam, 2003. Published in Studies in Netherlandish Art and Cultural History 5, Zwolle: Waanders, 2003.

Scholten, Frits, Joanna Woodall, and Dulcia Meijers, eds. *Art and Migration: Netherlandish Artists on the Move, 1400–1750*. Netherlands Yearbook for History of Art 63. Leiden; Boston: Brill, 2014.

Scholten, Koen. 'Introduction: Memory and Identity in Learned Communities'. In Scholten, Van Miert, and Enenkel, *Memory and Identity in the Learned World*, 1–27.

Scholten, Koen. 'De Republiek der Letteren als de totale intellectuele wereld?' *Tijdschrift voor Geschiedenis* 131, no. 4 (2018): 713–714.

Scholten, Koen, Dirk van Miert, and Karl A. E. Enenkel, eds. *Memory and Identity in the Learned World: Community Formation in the Early Modern World of Learning and Science*. Intersections 81. Leiden; Boston: Brill, 2022.

Scholten, Koen, and Asker Pelgrom. 'Scholarly Identity and Memory on a Grand Tour: The Travels of Joannes Kool and his Travel Journal (1698–1699) to Italy'. *Lias* 46, no. 1 (2019): 93–136.

Schröder, Peter. 'Trust (*Fides*)'. In Lesaffer and Nijman, *The Cambridge Companion to Hugo Grotius*, 118–137.

Schudt, Ludwig. *Italienreisen im 17. und 18. Jahrhundert*. Veröffentlichung der Biblioteca Hertziana (Max-Planck-Institut) in Rome. Vienna; Munich: Schroll-Verlag, 1959.

Schwandt, Silke. 'Digitale Methodes für die Historische Semantik: Auf den Spuren von Begriffen in digitalen Korpora'. *Geschichte und Gesellschaft* 44 (2018): 107–134.

Sebregondi Fiorentini, Ludovica. 'Francesco Buonarroti, cavaliere gerosolimitano e architetto dilettante'. *Rivista d'arte: studi documentari per la storia delle arti in Toscana* 38, no. 2 (1986): 49–86.

Segre, Michael. *In the Wake of Galileo*. New Brunswick, NJ: Rutgers University Press, 1991.

Segre, Michael. 'Viviani's Life of Galileo'. *Isis* 80, no. 2 (1989): 206–231.

Serrai, Alfredo. *La biblioteca di Lucas Holstenius*. Scienze bibliografiche 2. Udine: Forum Editrice Universitaria Udinese, 2000.

Shank, Michael H. 'How Shall We Practice History? The Case of Mario Biagioli's *Galileo, Courtier*'. *Early Science and Medicine* 1, no. 1 (1996): 106–150.

Shapin, Steven. *A Social History of Truth: Civility and Science in Seventeenth-Century England*. Chicago; London: University of Chicago Press, 1994.

Shaw, Paul. 'Poggio Bracciolini, an Inscription in Terranuova, and the Monument to Carlo Marsuppini: A Theory'. In *Poggio Bracciolini and the Re(dis)covery of Antiquity: Textual and Material Traditions*, edited by Roberta Ricci, 149–162. Florence: Florence University Press, 2020.

Shelford, April G. *Transforming the Republic of Letters: Pierre-Danuel Huet and European Intellectual Life, 1650–1720*. Rochester, NY: University of Rochester Press, 2007.

Sherlock, Peter. 'The Reformation of Memory in Early Modern Europe'. In Radstone and Schwarz, *Memory*, 30–40.

Sherman, William H. 'Stirrings and Searchings (1500–1720)'. In *The Cambridge Companion to Travel Writing*, edited by Peter Hulme and Tim Young, 17–36. Cambridge: Cambridge University Press, 2002.

Shortland, Michael, and Richard Yeo, eds. *Telling Lives in Science: Essays on Scientific Biography*. Cambridge: Cambridge University Press, 1996.

Sicca, Cinzia Maria, ed. *Scultura a Pisa nell'età moderna: Le sepolture dei docenti dello Studio*. Studi e fonti per la storia della scultura 1. Pisa: Edizioni Plus / Pisa University Press, 2007.

Siegel, Steffen. 'Bildnisordnungen: Visuelle Pragmatik in Paul Frehers Gelehrtenlexikon "Theatrum virorum eruditione clarorum" (Nürnberg 1688)'. *Archiv für Kulturgeschichte* 90, no. 1 (2013): 79–108.

Siegert, Reinhart. 'Scientific "Patriotism" between Self-Importance, Self-Recommendation and Camouflage: The Enlightenment Requirement of Public Benefit as Reflected in the Title Pages and Prefaces of Popular Works by Scholars'. In *Scholars in Action: The Practice of Knowledge and the Figure of the Savant in the 18th Century*, edited by André Holenstein, Hubert Steinke, and Martin Stuber, 853–876. History of Science and Medicine Library 34 / Scientific and Learned Cultures and Their Institutions 9, vol. 2. Leiden; Boston: Brill, 2013.

Simar, Théophile. *Etude sur Erycius Puteanus, 1574–1646, consideré spécialemente dans l'histoire de la philologie belge et dans sons enseignement à l'Université de Louvain*. Recueil de travaux publiés par les membres des conférences d'histoire et de philologie 23. Leuven: Bureau du recueil, 1909.

Sisi, Carlo. 'Il cimitero romantico di Santa Croce'. In Maffioli, *Santa Croce nell'800*, 125–133.

Smet, Ingrid A. R. de. *Thuanus: The Making of Jacques-Auguste de Thou (1553–1617)*. Geneva: Librairie Droz, 2006.

Smits-Veldt, Mieke B. ' "Het vaderland" bij Hollandse rederijkers, circa 1580–1625: Grondgebied en identiteit'. In Van Sas, *Vaderland*, 83–107.

Söderqvist, Thomas, ed. *The History and Poetics of Scientific Biography*. Aldershot; Burlington, VT: Ashgate, 2007.

Somos, Mark. 'Virtue'. In Lesaffer and Nijman, *The Cambridge Companion to Hugo Grotius*, 91–117.

Spalletti, Ettore. 'Gli esordi del pantheon romantico dal monumento a Vittorio Alfieri al cenotafio di Dante'. In Berti, *Il pantheon di Santa Croce*, 221–243.

Stagl, Justin. *Apodemiken: eine räsonnierte Bibliograhie der reisetheoretischen Literatur des 16., 17. und 18. Jahrhunderts*. Paderborn: Schöningh, 1983.

Stanglin, Keith D. *Arminius on the Assurance of Salvation: The Context, Roots, and Shape of the Leiden Debate, 1603–1609*. Brill's Series in Church History 27. Leiden; Boston: Brill, 2007.

Steen, Jasper van der. *Memory Wars in the Low Countries, 1566–1700*. Leiden; Boston: Brill, 2015.

Stewart, Frank Henderson. *Honor*. Chicago: University of Chicago Press, 1994.

Strocchia, Sharon T. *Death and Ritual in Renaissance Florence*. Baltimore; London: Johns Hopkins University Press, 1992.

Strong, Roy. *Art and Power: Renaissance Festivals, 1450–1650*. Woodbridge: Boydell, 1984.

Suter, Rufus. 'Four Galileian Inscriptions'. *Isis* 43, no. 3 (1952): 231–236.

Suter, Rufus. 'The Galileian Inscriptions on the Façade of Viviani's House in Florence'. *Osiris* 12 (1956): 225–243.

Sutton, Geoffrey V. *Science for a Polite Society: Gender, Culture and the Demonstration of Enlightenment*. Boulder, CO: Westview Press, 1995.

Sweet, Rosemary. 'The Changing View of Rome in the Long Eighteenth Century'. *Journal for Eighteenth-Century Studies* 33, no. 2 (2010): 145–164.

Sweet, Rosemary, Gerrit Verhoeven, and Sarah Goldsmith, eds. *Beyond the Grand Tour: Northern Metropolises and Early Modern Travel Behaviour*. London; New York: Routledge, 2017.

Swigchem, C. A. van, T. Brouwer, and W. van Os. *Een huis voor het Woord: Het protestantse kerkinterieur in Nederland tot 1900*. The Hague: Staatsuitgeverij; Zeist: Rijksdienst voor de Monumentenzorg, 1984.

Tervoort, Ad. *The Iter italicum and the Northern Netherlands: Dutch Students at Italian Universities and their Role in the Netherlands' Society (1426–1575)*. Education and Society in the Middle Ages and Renaissance 21. Leiden; Boston: Brill, 2005.

Tex, Jan den. 'Nederlandse studenten in de rechten te Padua 1545–1700'. *Mededelingen van het Nederlands Historische Instituut te Rome*, 3rd series, 10 (1959): 45–165.

Thomas, Keith. *Ends of Life: Roads to Fulfilment in Early Modern England*. Oxford: Oxford University Press, 2009.

Timpe, Kevin, and Craig A. Boyd. *Virtues and Their Vices*. New York: Oxford University Press, 2014.

Touber, Jetze. ' "I am happy that Italy fosters such exquisite minds": Gijsbert Cuper (1644–1716) and Intellectual Life on the Italian Peninsula'. *Incontri* 30, no. 2 (2015): 91–106.

Touber, Jetze. 'Religious Interests and Scholarly Exchange in the Early Enlightenment Republic of Letters: Italian and Dutch Scholars, 1675–1715'. *Rivista di Storia della Chiesa in Italia* 68, no. 2 (2014): 411–43.

Tournoy, Gilbert, Jan Roegiers, and Chris Coppens, eds. *Vives te Leuven: Catalogus van een tentoonstelling in de Centrale Bibliotheek te Leuven, 28 juni–20 augustus 1993* (Leuven: Leuven University Press, 1993).

Ureña-Carrion, Javier, Petri Leskinen, Juoni Tuominen, Charles van den Heuvel, Eero Hyvönen, and Mikko Kivelä. 'Communication Now and Then: Analyzing the Republic of Letters as a Communication Network'. *Applied Network Science* 7 (2022).

Van Houdt, Toon, Jan Papy, Gilbert Tournoy, and Constant Matheeussen, eds. *Self-Presentation and Social Identification: The Rhetoric and Pragmatics of Letter-Writing in Early Modern Times*. Supplementa Humanistica Lovaniensa 18. Leuven: Leuven University Press, 2002.

Vauchez, André. *Sainthood in the Later Middle Ages*. Translated by Jean Birrell. Cambridge: Cambridge University Press, 1997.

Veen, Henk Th. van. 'A Tuscan Plan of Action for Joan Blaeu's Book of Italian Cities'. *Lias* 18, no. 2 (1991): 221–227.

Veen, Henk Th. van, and Andrew P. McCormick. *Tuscany and the Low Countries: An Introduction to the Sources and an Inventory of Four Florentine Libraries*. Florence: Centro Di, 1984.

Verborgh, Ruben, and Max de Wilde. *Using OpenRefine*. Birmingham: Packt, 2013.

Verhoeven, Gerrit. *Anders reizen? Evoluties in vroegmoderne reiservaringen van Hollandse en Brabantse elites (1600–1750)*. Hilversum: Uitgeverij Verloren, 2009.

Verhoeven, Gerrit. *Europe Within Reach: Netherlandish Travellers on the Grand Tour and Beyond, 1585–1750*. Translated by Diana Webb. Egodocuments and History Series 9. Leiden; Boston: Brill, 2015.

Vermij, Rienk. 'On the Function of Dedications in Early Modern Scientific Books'. *Nuncius* 33, no. 2 (2018): 171–197.

Visser, Arnoud. *Reading Augustine in the Reformation: The Flexibility of Intellectual Authority in Europe, 1500–1620*. Oxford Studies in Historical Theology. New York: Oxford University Press, 2011.

Vugt, Ingeborg van. 'The Structure and Dynamics of Scholarly Networks Between the Dutch Republic and the Grand Duchy of Tuscany in the 17th Century'. PhD diss., University of Amsterdam; Scuola Normale Superiore di Pisa, 2019.

Vugt, Ingeborg van. 'Using Multi-Layered Networks to Disclose Books in the Republic of Letters'. *Journal of Historical Network Research* 1, no. 1 (2017): 25–51.

Wagenaar, Lodewijk, ed. *Een Toscaanse prins bezoekt Nederland: De twee reizen van Cosimo de' Medici, 1667–1669*. Translated by Berti Eringa. Amsterdam: Uitgeverij Bas Lubberhuizen, 2014.

Wallace, William E. 'Matters of Life and Death: Galileo in the Afterlife of Michelangelo'. *Sources: Notes in the History of Art* 17, no. 3 (1998): 20–24.

Waquet, Françoise. *Latin or the Empire of a Sign: From the Sixteenth to the Twentieth Centuries*. London; New York: Verso, 2001.

Waquet, Françoise. 'Qu'est-ce que la République des Lettres? Essai de sémantique historique'. *Bibliothèque de l'école des chartes* 147, no. 1 (1989): 473–502.

Warnke, Martin. 'Das Bild des Gelehrten im 17. Jahrhundert'. In Neumeister and Wiedemann, *Res Publica Litteraria*, 1:1–31.

Warren, Christopher N. 'Historiography's Two Voices: Data Infrastructure and History at Scale in the Oxford Dictionary of National Biography (ODNB)'. *Journal of Cultural Analytics* (22 November 2018).

Waszink, Jan. 'Lipsius and Grotius: Tacitism'. *History of European Ideas* 39, no. 2 (2014): 151–168.

Waszink, Jan. 'Historical Writings'. In Lesaffer and Nijman, *The Cambridge Companion to Hugo Grotius*, 315–338.

Waszink, Jan. 'Shifting Tacitisms: Style and Composition in Grotius's *Annales*'. *Grotiana* 29, no. 1 (2009): 85–132.

Watson, David. 'Jean Crespin and the Writing of History in the French Reformation'. In Gordon, *Protestant History and Identity in Sixteenth-Century Europe*, vol. 2: *The Later Reformation*, 39–58.

Watson, Roísín. 'Funeral Monuments, Ritual and Print: Strategies of Memorialization at the Württemberg Court'. *Past & Present* 234, Issue supplement 12, 'Cultures of Lutheranism: Reformation Repertoires in Early Modern Germany' (2017): 139–164.

Watson, Roísín. 'Remembering and Forgetting the Dead in the Churches of Reformation Germany'. In Cummings et al., *Remembering the Reformation*, 154–172.

Weeks, Christopher. 'The Restoration of Desiderio Da Settignano's Tomb of Carlo Marsuppini in S. Croce, Florence'. *The Burlington Magazine* 141, no. 1161 (1999): 732–738.

Weingart, Peter, Claudia Muhl, and Petra Pansegrau. 'Of Power Maniacs and Unethical Geniuses: Science and Scientists in Fiction Film. *Public Understanding of Science* 12, no. 3 (2003): 279–287.

Weingarten, Susan. *The Saint's Saints: Hagiography and Geography in Jerome*. Ancient Judaism & Early Christianity 58. Leiden; Boston: Brill, 2005.

Weiss, James M. *Humanist Biography in Renaissance Italy and Reformation Germany: Friendship and Rhetoric*. Variorum Collected Studies Series 947. Farnham: Ashgate, 2010.

Welch, Christina. 'Images of Death in Art and Literature in Late Medieval and Early Modern Europe (1300–1700)'. In Booth and Tingle, *A Companion to Death, Burial, and Remembrance*, 272–299.

Weller, Emil. *Die falschen und fingirten Druckorte: Repertorium der seit Erfindung der Buchdruckerkunst unter falscher Firma erschienenen deutschen, lateinischen und französischen Schriften*. 2 vols. Leipzig: Wilhelm Engelmann, 1864.

Weststeijn, Thijs. *The Visible World: Samuel van Hoogstraten's Art Theory and the Legitimation of Painting in the Dutch Golden Age*. Amsterdam: Amsterdam University Press, 2008.

Wetzel, James. 'Predestination, Pelagianism and Foreknowledge'. In *The Cambridge Companion to Augustine*, edited by Eleonore Stump and Norman Kretzmann, 49–58. Cambridge: Cambridge University Press, 2001.

Wiesner-Hanks, Merry E. *Women and Gender in Early Modern Europe*. 4th ed. Cambridge: Cambridge University Press, 2019.

Wilson, N. G. *From Byzantium to Italy: Greek Studies in the Italian Renaissance*. London: Duckworth, 1992.

Winsor Sage, Paul. 'Vatic Admonition in Horace *Odes* 4.9'. *The American Journal of Philology* 115, no. 4 (1994): 565–586.

Winter, Jay. *Sites of Memory, Sites of Mourning: The Great War in European Cultural History*. Cambridge: Cambridge University Press, 2014. Originally published in 1995.

Winter, Jay. 'Sites of Memory'. In Radstone and Schwarz, *Memory*, 312–324.

Witt, Ronald G. *'In the Footsteps of the Ancients': The Origins of Humanism from Lovato to Bruni*. Studies in Medieval and Reformation Thought 74. Leiden; Boston: Brill, 2000.

Wolff, Ruth. 'Le tombe dei dottori al santo: considerazioni sulla loro tipologia'. In *Cultura, arte e committenza nella Basilica di S. Antonio di Padova nel trecento: Atti del Convegno internazionale di studi Padova, 24–26 maggio 2001*, edited by Luca Baggio and Michela Benetazzo, 277–297. Centro Studi Antoniani 36. Padua: Centro Studi Antoniani, 2003.

Wrigley, Richard, and Matthew Craske, eds. *Pantheons: Transformations of a Monumental Idea*. Subject/Object: New Studies in Sculpture. Aldershot; Burlington, VT: Ashgate, 2004.

Yoran, Hanan. *Between Utopia and Dystopia: Erasmus, Thomas More, and the Humanist Republic of Letters*. Lanham: Lexington Books, 2010.

Zanker, Paul. *The Masks of Socrates: The Image of the Intellectual in Antiquity*. Translated by Alan Shapiro. Sather Classical Lectures 59. Berkeley; Los Angeles; London: University of California Press, 1995.

Zuidervaart, Huib J. 'Science for the Public: The Translation of Popular Texts on Experimental Philosophy into the Dutch Language in Mid-Eighteenth Century'. In *Cultural Transfer Through Translation: The Circulation of Enlightened Thought in Europe by Means of Translation*, edited by Stefanie Stockhorst, 231–262. Internationale Forschungen zur Allgemeinen und Vergleichenden Literaturwissenschaft 131. Amsterdam; New York: Rodopi, 2010.

# Appendix 1

## Corpus and Keyword Analysis

What follows is a bibliography of the main corpus that has been analysed in Chapters 1 and 2. The main corpus contains the collective scholarly *vitae*. For the sake of comparison, I have also compiled a so-called reference corpus to probe the distinctive use of language in the main corpus. The reference corpus consists of the main corpus as well as a variety of early modern texts in a variety of categories.

All texts have been obtained from the online *Thesaurus Eruditionis* compiled by the Deutsche Forschungsgemeinschaft.[1] A short-title bibliography has been provided here for the main corpus; full bibliographic details can be found in this book's bibliography. For the reference corpus, a short-title bibliography is given, since a full bibliography is available in the online *Thesaurus*. The reference corpus consists of the entire *Thesaurus* besides irregular volumes such as lexica. The reference corpus also contains the main corpus so it has not been unnecessarily repeated in the bibliography of the reference corpus. The Optical Character Recognition (OCR) accuracy rate is high, since all the digitised texts have been manually checked against the printed original, avoiding the problems often associated with a low accuracy rate in OCR.[2]

## Main Corpus


Adam, Melchior. *Vitae Germanorum Philosophorum* [...]. *Vitae*, vol I. Frankfurt a.M. and Heidelberg, 1615.

Adam, Melchior. *Vitae Germanorum Medicorum* [...]. *Vitae*, vol. II. Heidelberg: Jona Rosa, 1620.

Adam, Melchior. *Vitae Germanorum iuresconsultorum et politicorum* [...]. *Vitae*, vol. III. Heidelberg: Jona Rosa, 1620.

Adam, Melchior. *Vitae Germanorum Theologorum* [...]. *Vitae*, vol. IV. Heidelberg: Jona Rosa, 1620.

Adam, Melchior. *Decades duae continentes Vitas Theologorum exterorum principum* [...]. *Vitae*, vol. V. Frank-furt a.M.: widow of Jona Rosa, 1623.


1 *Thesaurus Eruditionis*, compiled by the Deutsche Forschungsgemeinschaft, for the CAMENA and TERMINI projects between 1999 and 2008; see http://mateo.uni-mannheim.de/camenahtdocs/ camenaref_e.html, last edited on 26 November 2013.

2 For the problem with an uncritical use of OCR, see Tim Hitchcock, 'Confronting the Digital: Or How Academic History Writing Lost the Plot', *Cultural and Social History* 10, no. 1 (2013): 9–23, esp. pp. 12–15.

Beza, Theodor. *Icones, id est Verae Imagines virorum doctrina simul et pietate illustrium* [...]. [Geneva]: apud Ioannem Laonium, 1580.
Giovio, Paolo. *Elogia Virorum literis illustrium* [...]. Basel: Petrus Perna, 1577.
Ravisius, Johannes. *De memorabilibus et claris mulieribus: aliquot diversorum scriptorum opera*. Paris: Simon de Colines, 1521.
Reusner, Nicolaus; Tobias Stimmer (engraver). *Icones, sive Imagines virorum literis illustrium* [...]. Frankfurt a.M.: Balthasar Diehl, 1719.
Rossi, Giovanni Vittorio. *Pinacotheca Imaginum Illustrium, doctrinae vel ingenii laude, Virorum* [...]. 3 vols. Cologne: apud Jodocum Kalcovium et Socios, 1645–1648.
Spitzel, Gottlieb. *Templum Honoris Reseratum* [...]. Augsburg: Typis Koppmayerianis, Prostat apud Gottlieb Goebelium, 1673.
Tomasini, Giacomo Filippo. *Elogia Virorum Literis & Sapientia Illustrium ad vivum expressis imaginibus exornata*. Patavia: ex Typographia Sebastiani Sardi, 1644.

## Reference Corpus

Beyerlinck, Laurentius. *Magnum Theatrum Vitae Humanae*. Leiden: Huguetan & Ravaud, 1656.
Calepinus, Ambrosius. *Dictionarium*. Lyon: Prost, Borde, Arnaud, 1647.
Comenius, Johann Amos. *Opera didactica omnia*. Amsterdam: De Geer & à Roy, 1657.
Du Cange, Charles du Fresne. *Glossarium Ad Scriptores Mediae et Infimae Latinitatis*. Frankfurt a.M.: Zunneriana, 1462.
Dupuy, Jacques. *Nominum propriorum Virorum, Mulierum, Populorum, &c.* Coloniae Allobrogum: Aubertum, 1634.
Estienne, Robert. *Dictionarium Latinogallicum*. Paris: Stephanus, 1546.
Estienne, Robert; Johann Matthias Gesner. *Novus Linguae Et Eruditionis Romanae Thesaurus*. Leipzig: Fritschii Viduae & Breitkopfii, 1749.
Hofmann, Johann Jacob. *Lexicon Universale, Historiam Sacram Et Profanam Omnis aevi* [...] *explanans*. Leiden: Hackius et al., 1698.
Junius, Hadrianus. *Nomenclator octilinguis*. Paris: Douceur, 1606.
Kirsch, Adam Friedrich. *Abundantissimum Cornu Copiae Linguae Latinae Et Germanicae Selectum*. Leipzig: Svikertii, 1774.
Nolte, Johann Friedrich. *Lexicon Latinae Linguae Antibarbarum Quadripartitum*. Leipzig; Helmstedt: Weygand, 1744.
Pexenfelder, Michael. *Apparatus eruditionis*. Nürnberg: Endter, 1670.
Schurzfleisch, Konrad Samuel. *Nomenclator Strategicus*. Giessen: Müller, 1720.
Winiarczyk, Marek. *Sigla latina in libris impressis occurentia*. Bratislava: University, 1995.

# Acknowledgements

Never could I have imagined the research for this book would entail leafing through age-old manuscripts in poorly lit Roman libraries, muttering obscurely in Latin if someone could pass me the salt at the dinner table, or looking for a manuscript in a handwritten library catalogue from the seventeenth century. But what has surprised me the most was the continuous encouragement, enthusiasm, and support I have received from friends, family, and colleagues. I want to thank you for listening, reading, sharing ideas, and giving feedback, but also for bringing me back to earth and diverting my thoughts.

I am grateful to Arnoud Visser and Dirk van Miert, who motivated me, listened to my ideas, and diligently read and commented on all my drafts. Moreover, they helped me tremendously with the enigmatic Neo-Latin texts—not to mention deciphering the funerary inscriptions—and inspired me and gave me the confidence to look at these texts afresh. I am highly appreciative of their support and engagement.

The SKILLNET project team was a pleasant foundation on which I could rely, despite its changes throughout the years I worked on my book. I want to thank everyone who participated in the project, in particular Manuel Llano, Karen Hollewand, Ingeborg van Vugt, Robin Buning, and Liliana Melgar Estrada for their motivation, inspiration, and support.

This research benefited greatly from research stays at the Royal Netherlands Institute in Rome (KNIR), the Dutch University Institute for Art History (NIKI) in Florence, and the Max-Planck-Institut für Wissenschaftsgeschichte in Berlin. The staff and archivists at the many libraries, archives, and churches I visited for this research were helpful, patient, and kind. I am particularly grateful to librarians and curators of the Biblioteca di Archeologia e Storia dell'Arte in Rome, the Basilica di Santa Croce in Florence, and the Pieterskerk in Leiden.

I want to thank Rosalie Versmissen for her trust, love, and support during the completion of this book.

Finally, I want to thank my parents, Theo and Hettie, for always supporting my endeavours, encouraging the pursuit of learning, and making it possible for me to follow my curiosity and intuition.

# Index Nominum

# General Index